Ion Surface Interaction,
Sputtering and Related Phenomena

Ion Surface Interaction, Sputtering and Related Phenomena

Edited by

R. BEHRISCH, W. HEILAND, W. POSCHENRIEDER, P. STAIB AND H. VERBEEK

Max-Planck-Institut für Plasmaphysik
Garching bei München, Germany

GORDON AND BREACH SCIENCE PUBLISHERS
London   New York   Paris

G1542908√

Physics

The papers in this book were first published in *Radiation Effects*, Volume 18, Nos. 1/2 (1973); Nos. 3/4 (1973); Volume 19, No. 1 (1973); No. 2 (1973); No. 3 (1973) and No. 4 (1973).

Preface

All of the original papers, which are collected in these proceedings, were first read at the International Conference on Ion Surface Interaction, Sputtering and Related Phenomena, held at Max-Planck-Institut für Plasmaphysik, Garching bei München, Germany, in September 1972.

Ion Surface Interaction is therefore one in a series of conference proceedings available both in hard-backed form and through the journal *Radiation Effects* in the open literature.

LEWIS T. CHADDERTON
Editor, *Radiation Effects*

Foreword

This volume contains the proceedings of the International Conference on Ion Surface Interaction, Sputtering and Related Phenomena, held at Max-Planck-Institut für Plasmaphysik at Garching, Munich, Germany, on September 24-27 1972.

Plasma physics and the study of ion bombardment phenomena, though different fields today, originally were closely connected. In fact sputtering was first observed in plasma discharges. Only after the ion source and bombarded target became separated in an accelerator did the two fields start to diverge. Still, particle bombardment represents an important phenomenon in all plasma experiments. Especially in experiments on controlled nuclear fusion reactors it will probably constitute a serious problem.

The main emphasis of the conference was placed on the physics of sputtering of single crystalline, polycrystalline and amorphous materials. This phenomenon, though the first known of all ion bombardment effects, is still the least understood. Included in various international meetings as one of several topics, it had not been the main theme of an international conference for more than eight years. Since thin film production is now well covered by several special meetings, it was decided not to include this topic. Among related phenomena, changes in the surface region (damage, blistering) had been a major subject.

Other topics were secondary ion electron and photon emission, and ion backscattering.

The Garching Conference was attended by 172 scientists. During 12 sessions 68 papers were presented and discussed. These proceedings comprise 50 papers, including some which were submitted but could not be presented orally. The decision to avoid parallel sessions imposed a very tight time schedule. Most papers had to be restricted to 10 minutes with 5 additional minutes for discussion. By way of compensation several invited talks served as background for the respective sessions.

The Committee would like to thank all chairmen and contributors for their excellent cooperation. As members of the Surface Physics Division, W. Eckstein, H. Liebl and H. Vernickel were responsible for a major share of the organization. Appreciation is due to H. Schmidl for his successful time-keeping facility. We should also like to thank Miss I. Wunderlich for her patience with the secretarial work for the conference.

The Editors would like to offer a special word of thanks to all the reviewers for their quick and thorough comments. Professor L. Chadderton, editor of *Radiation Effects*, is gratefully acknowledged for his advice and collaboration in arranging early publication.

THE EDITORS

The Conference was organized by the Surface Physics Division of Max-Planck-Institut für Plasmaphysik, Garching

Conference Chairman: A. Schlüter

Conference Secretary: B. M. U. Scherzer

Contents

† List of papers presented at the conference. Contributions marked with an asterisk are not published in
these proceedings.

H. SCATTERING OF IONS II

I. SECONDARY ION EMISSION I

CONTENTS

BINDING ENERGIES IN CUBIC METAL SURFACES

D. P. JACKSON

Chalk River Nuclear Laboratories, Atomic Energy of Canada Limited, Chalk River, Ontario, Canada

Surface binding energies are calculated using Morse potentials for the (100), (110) and (111) surfaces of seven f.c.c. and 9 b.c.c. metals. Estimates are made for a further seven metals. Surface relaxation effects are included and their effect discussed. The surface binding energies are found to be greater than the sublimation energy per atom of the solid. Qualitative arguments are proposed to explain the ordering of binding energies between crystal faces. Comparison with experimental data for the adsorption energies of tungsten atoms on tungsten crystal faces indicates the plausibility of the calculated values.

1 INTRODUCTION

Interest in single crystal sputtering is increasing both as a phenomenon involved in many interesting atomic collision effects and as an ideal case to identify mechanisms present in the sputtering of materials of technical interest. A key quantity which determines the energies of surface ejection processes is the surface binding energy. That is, the energy required to remove an atom from a crystal face. Thus far there has not been a comprehensive calculation of these energies. The present paper is an attempt to remedy this.

We shall use the well-known Morse potentials due to Girifalco and Weizer[1]:

$$\varphi_M = D(e^{-2\alpha(r-r_0)} - 2e^{-\alpha(r-r_0)}) \qquad (1)$$

These potentials are summed over ideally perfect static lattices. The arguments for and against using simple potentials such as these and the justification for the summation of pairwise potentials (SPP) approximation have been discussed frequently in the literature.[2] This approach seems at the moment to be the only readily tractable possibility. It has the advantage that it facilitates comparisons between various metals and between their crystal faces. Although precise quantitative agreement is not expected for SPP models, it is plausible to assume that the qualitative aspects indicated by the calculations will be reflected in experimental results. Thus far there are not enough experimental data for direct comparisons of binding energies however we anticipate data will be available in the future. Furthermore the binding energies are an important component of the present models for single crystal sputtering.[3]

2 CALCULATIONS

Since we are interested in evaluating surface binding energies, the natural way to formulate the summations

is in terms of atomic layers. For the low index directions treated, the atomic layers are true planes, that is none of the atoms protrude above or below the plane. Therefore we can define the simple layer potential sum:

$$l(z) = \sum_{i,j}^{\pm \infty} \varphi_M(i, j, z) \qquad (2)$$

This is the potential energy of interaction between an atom and a plane of atoms at a distance z consisting of atoms at points (i, j). The plane is considered to be infinite in extent. The contents of the point set $\{i, j\}$ depends, of course, on the crystallographic type of the plane and hence $l(z)$ is a distinct function for each type.

Suppose now we fix upon a particular plane type and retain it for the remainder of this section. Furthermore we shall use the simple notation $l(1), l(2), \ldots$ etc., to indicate the value of l at $1, 2, \ldots$ interplanar spacings. The surface binding energy, U_1 is defined as the potential energy between an atom in the surface and the rest of the crystal. Similarly U_2 refers to the layer below the surface and so on for the U_N's. To be absolutely clear U_1 is the minimum energy required to remove an atom from the surface and take it to infinity. In terms of the layer sums:

$$U_1 = l(0) + \sum_{z=1}^{\infty} l^*(z) \qquad (3)$$

U_1 consists of two contributions: $l(0)$ is the potential energy due to all the atoms in the surface plane and the second term represents the potential due to the rest of the underlying semi-infinite solid. The notation l^* indicates that surface relaxations must be taken into account, for example $l(1)$ would in this case be $l(1 + \delta_1)$ where δ_1 is a small correction reflecting the increased distance between the first and second planes. We have used these corrections in the calculations but we shall not continue to indicate them in the development of this section in order not to overelaborate the equations. However it must be understood that they are implied.

Similarly one defines the binding energies of the sub-surface atoms:

$$U_2 = I(1) + I(0) + \sum_{z=1}^{\infty} I(z) \qquad (4)$$

In this case the first term is due to the surface layer, the second indicates the interlayer binding and the third the rest of the layers below. In general

$$U_N = I(N-1) + I(N-2) + \cdots + I(1)$$
$$+ I(0) + \sum_{z=1}^{\infty} I(z) \qquad (5)$$

The cohesive energy of a crystal is physically defined as the energy involved in crystal binding. It is the energy difference between the state where all the atoms are infinitely spaced, and have no interaction with each other, and the state where they are bound to each other in a crystal. We may express the cohesive energy in terms of an energy per atom, where the total cohesive energy is divided by the number of atoms present. This cohesive energy is usually identified with the thermodynamic quantities: the heat of sublimation and the heat of vaporization. It is often called the "sublimation energy" in sputtering discussions. However, it should be understood that this term does not refer to any aspect of the actual process of sublimation but it is used only in the context of crystal energetics as we have described above. We shall denote this quantity U_0 (energy per atom). In terms of the I's:

$$U_0 = \tfrac{1}{2}(I(0) + 2 \sum_{z=1}^{\infty} I(z)) \qquad (6)$$

There are several features of this equation which should be noted. The $I(z)$'s in this case have no relaxations involved. The equation is basically the extension of Eq. (5) with $N \to \infty$, with the important modification of the factor $\tfrac{1}{2}$. It is worthwhile explaining the origin of this factor. Call the expression inside the parentheses P; it is the potential felt by an atom in the centre of the crystal due to all the other atoms of the crystal. Suppose there are N atoms in the crystal. The total cohesive energy may be evaluated by summing all the potential experienced by all the atoms; this would be NP. However in doing this each bond is counted exactly twice. Hence we must include this factor of $\tfrac{1}{2}$. The total cohesive energy of the crystal is therefore $NP/2$ or $P/2$ per atom. The tacit assumption is made that the interaction potential between two atoms is stored equally in each.

From Eqs. (3–6) we deduce that

$$U_N > \cdots > U_2 > U_1 > U_0 \qquad (7)$$

Hence we expect the binding energies to be greater as we go into the crystal from the surface, which is physically sensible. And in general all the binding energies should be greater than the sublimation energy. A corollary of this is

$$U_N \to 2U_0, N \to \infty \qquad (8)$$

which follows from the discussion after Eq. (6). This is a good check on the calculations. Comparing Eq. (3) and Eq. (6) we have the approximation (because relaxations are not taken into account in Eq. (6)):

$$U_1 \simeq \overline{U}_1 = U_0 + I(0)/2 \qquad (9)$$

Here $\overline{U}_1$ is used merely to denote the approximation for convenience of discussion.

Finally suffice it to say that the summations were performed on a CDC 6600 computer taking into account many thousands of atoms. The program used was similar to those described in Refs. 4 and 5 and the same checks on accuracy were employed.

3 RESULTS AND DISCUSSION

Table I gives values of U_1 for the (100), (110) and (111) surfaces of seven f.c.c. and nine b.c.c. metals. Before discussing these results in detail, two other issues should be mentioned.

3.1 Energy Relationships
Table II shows the values of the energy quantities, discussed in Section 2, for the example of tungsten. The values of U_1 to U_4 are greater than the sublimation energy (Eq. (7)) and $U_N > U_{N-1}$ (Eq. (7)). The U_4 values approach $2U_0$ (Eq. (8)). Note that U_4 for the (111) surface orientation is somewhat less than $2U_0$ since the distance into the crystal at layer four is smaller relative to that of the other two surfaces. The approximation $\overline{U}_1 \simeq U_1$ holds quite well. These relationships have been verified for the other metals and we conclude that calculated values are behaving as predicted in Section 2.

3.2 The Effect of Surface Relaxations
The surface relaxations for the metals listed in Table I have been studied previously.[5] Table III indicates the effect of these relaxations on the surface binding energy for the examples: lead (least relaxation) and rubidium (most relaxation). The δ_1's are the percentage increases in the interplanar spacings between the surface and

TABLE I
Surface layer binding energies U_1 (eV/atom)

	(100)	(110)	(111)
Pb	2.71	2.44	2.79
Ag	3.98	3.61	4.08
Ni	5.55	5.11	5.61
Cu	4.62	4.26	4.65
Al	3.80	3.53	3.80
Ca	2.39	2.25	2.38
Sr	2.23	2.09	2.21
Mo	8.78	9.18	7.38
Cr	5.05	5.22	4.27
W	11.52	11.86	9.75
Fe	5.47	5.48	4.72
Ba	2.32	2.21	2.08
K	1.21	1.06	1.20
Na	1.44	1.26	1.46
Cs	1.03	0.90	1.05
Rb	1.13	0.97	1.17

TABLE II
Energy functions for tungsten in eV/atom

	$l(0)$	$\bar{U}_1$	U_1	U_2	U_3	U_4	$2U_0$
(100)	4.99	11.34	11.52	15.81	17.00	17.41	17.68
(110)	5.97	11.83	11.86	16.91	17.57	17.67	17.68
(111)	1.50	9.59	9.75	12.28	15.05	16.69	17.68

TABLE III
Effect of surface relaxations. The $\hat{U}_1$'s are the surface binding energies calculated with no relaxations

		δ_1	U_1	$\hat{U}_1$
(a)	(100)	5.549	2.705	2.691
	(110)	8.278	2.438	2.420
	(111)	2.371	2.792	2.788
(b)	(100)	54.238	1.125	1.060
	(110)	22.618	0.966	0.931
	(111)	95.355	1.167	1.042

second layers. $\hat{U}_1$ is the value of U_1 calculated without including relaxations. The difference between U_1 and $\hat{U}_1$ for lead is at most 0.7% and the maximum difference for rubidium is about 11% but the latter is an extreme case. We note also that the relaxations cause lower values of U_1. This is a result of the fact that the repulsive energy which leads to the outward surface relaxations is included in $\hat{U}_1$ but not in U_1. Therefore it is easier to remove an atom in the unrelaxed state because of the availability of this repulsive energy which otherwise would go into surface relaxation. The general conclusion is that surface relaxations do not

have a very significant effect on the U_1 values: this is similar to the situation encountered in calculations of surface energy (γ).[6]

3.3 The f.c.c. Metals

Examining the first seven (f.c.c.) metals of Table I, the following trend for the surface binding energies is evident:

$$U_1(111) \gtrsim U_1(100) > U_1(110) \qquad (10)$$

The exception to the first inequality occurs in strontium. Equation 10 may be qualitatively explained on the basis of a simple-minded coordination argument. An atom in a (111) surface has 9 nearest-neighbour (n.n.) bonds and 3 next-nearest-neighbour (n.n.n.) bonds. For the (100) plane the corresponding numbers are 8 n.n. and 5 n.n.n. and for the (110) plane; 7 n.n. and 4 n.n.n. On the basis of the number of bonds one expects the ordering of Eq. (10) and the relatively small differences between the (111) and (100) planes. These considerations may also be applied to the surface relaxations which are ordered[5]

$$\delta_1(111) < \delta_1(100) < \delta_1(110) \qquad (11)$$

showing the reverse of the order of Eq. (10) since tighter bonding results in greater U_1 but smaller δ_1. The exception to Eq. (10) is strontium where the first inequality is slightly violated. This is due to the effect of relaxations; the inequality is satisfied before relaxation and the slight but unequal relaxation corrections are enough to reverse the order of (111) and (100) in Eq. (10). Therefore the f.c.c. entries in Table I present a coherent and satisfying qualitative picture.

3.4 The b.c.c. Metals

The same qualitative arguments may be applied to the last nine (b.c.c.) metals in Table I. Looking at the bonding we have 6 n.n. and 4 n.n.n. in the (110) plane, 4 n.n. and 5 n.n.n. for the (100) plane and 4 n.n. and 3 n.n.n. in the (111) plane. Hence from a coordination argument we would expect

$$U_1(110) > U_1(100) > U_1(111) \qquad (12)$$

which is in fact true for the first four b.c.c. metals. Complementing this the surface relaxations have the order[5]

$$\delta_1(110) < \delta_1(100) < \delta_1(111) \qquad (13)$$

However for the metals from barium on the situation is very confused and by the time one arrives at rubidium the ordering is

$$U_1(111) > U_1(100) > U_1(110) \qquad (14)$$

which is the exact reverse of Eq. (12). This results from a combination of two factors: without relaxation the order is (see Table III)

$$\hat{U}_1(100) > \hat{U}_1(111) > \hat{U}_1(110) \qquad (15)$$

The difference between Eq. (14) and Eq. (15) occurs because the (111) surface relaxes more than the (100) surface. However the main effect which causes the differences is due to the behaviour of $l(0)$ as a function of β, a dimensionless parameter of the Morse function variables ($\beta = e^{\alpha r_0}$). This β plays an important role in determining the magnitude of the surface relaxations,[5] in particular for low β values the relaxations are very large. For the f.c.c. metals, β ranges from 82.96 for lead to 39.6 for strontium, both parts of Table I are arranged in order of descending β. For the b.c.c. metals, β goes from 88.9 for molybdenum to 59.95 for iron, but β(barium) = 34.1 and the last four metals have $\beta \approx 23$. Small β implies that more atoms must be considered in any summation for an energy quantity and hence the properties of U_1 etc. are determined by more than the n.n. and n.n.n. bonds. For example $l(0)$ is strongly affected by β and this is the main influence, in conjunction with relaxations which reverses the ordering. The reason the f.c.c. metals were so "well-behaved" was that no very low β values were encountered. It is difficult to relate β to any one physical property but it is clear that low β implies "soft" low melting point metals which are not of much interest experimentally. For example the preparation of single crystal surfaces of the alkali metals is a formidable task. For this reason it does not seem worthwhile at this point to continue at greater length about the differences in low β metals.

3.5 Comparison with Experimental Values
There do not seem to be any experimental data which refer directly to the binding of an atom *in* an ideal single crystal surface. However there have been field-ion microscope experiments which have measured the binding of an atom *to* a metal surface. These results have been recently summarized by Ehrlich[7] for the case of tungsten adsorption on tungsten single crystal planes, for which the data seem most complete. Now qualitatively one would expect that the binding of an atom forming part of the surface would be much stronger than one which sits upon the surface. The latter is clearly the situation in the adsorption experiments since the adatoms are seen to migrate on the surface without very much difficulty. Hence one would expect U_a the adsorption energy, required to remove the atom from above the surface to be considerably lower than U_1. In fact qualitatively:

$$U_1 - l(0) \leqslant U_a \leqslant U_1 \qquad (16)$$

The first inequality implies that the adatom sits at a distance above the surface comparable to the interplanar spacing, which may or may not be true. However we would expect U_a to be closer to $U_1 - l(0)$ than to U_1. Table IV gives the comparison with the results quoted by Ehrlich.[7] The (111) case does not agree closely, however for the other two planes these data add some degree of plausibility to the results given the limitations of the model.

TABLE IV
Comparison with tungsten surface adsorption energies (eV/atom)

	$U_1 - l(0)$	U_a(expt)	U_1
(100)	6.57	7.0	11.52
(110)	5.88	5.3–6.3	11.86
(111)	8.25	6.0	9.75

4 ESTIMATES OF BINDING ENERGIES FOR OTHER CUBIC METALS

The Morse parameters given by Girifalco and Weizer[1] do not cover several cubic metals of interest in sputtering. Some of these metals are shown in Table V

TABLE V
Estimates of U_1 for other cubic metals (eV/atom)

	(100)	(110)	(111)
Au	4.95	4.54	5.03
Pt	7.67	7.02	7.78
Pd	5.16	4.72	5.24
Ir	9.08	8.32	9.22
Nb	9.71	10.01	8.29
Ta	10.52	10.84	8.98
V	6.89	7.10	5.88

together with estimates of U_1 their low index surfaces. If we avoid the low β region then the following approximations for the U_1's may be deduced from the data in Table I.

$$\text{f.c.c.:} \quad U_1(100) \approx 1.31\, U_0$$
$$U_1(110) \approx 1.20\, U_0 \qquad (17)$$
$$U_1(111) \approx 1.33\, U_0$$

$$\text{b.c.c.:} \quad U_1(100) \approx 1.30\, U_0$$
$$U_1(110) \approx 1.34\, U_0 \qquad (18)$$
$$U_1(111) \approx 1.11\, U_0$$

Using these approximations together with values of the cohesive energy from Kittel[8] we estimate the values given in Table V.

ACKNOWLEDGEMENTS

I am grateful to Peter Sigmund for encouraging me to undertake this project and for his kind advice and interest over the last few years. I am also indebted to K. B. Winterbon and H. R. Glyde for helpful discussions.

REFERENCES

1. L. A. Girifalco and V. G. Weizer, *Phys. Rev.* **114,** 687 (1959).
2. P. C. Gehlen, J. R. Beller, Jr. and R. I. Jaffee eds., *Interatomic Potentials and Simulation of Lattice Defects* (Plenum Press, 1972).
3. For example the contributions of M. T. Robinson and I. M. Torrens, D. E. Harrison *et al.* and A. van Veen and J. M. Fluit to this symposium.
4. D. P. Jackson, *J. Chem. Phys.* **56,** 3178 (1972).
5. D. P. Jackson, *Can. J. Phys.* **49,** 2093 (1971).
6. J. J. Burton and G. Jura, *J. Phys. Chem.* **71,** 1937 (1967).
7. G. Ehrlich, Ref. 2, pp. 573–619.
8. C. Kittel, *Introduction to Solid State Physics* 3rd ed. (John W. Wiley and Sons, 1968), p. 78.

ANISOTROPIC EMISSION IN SINGLE CRYSTAL SPUTTERING MEASUREMENTS ON hcp SINGLE CRYSTALS

WOLFGANG O. HOFER[†]

Gesellschaft für Strahlen- und Umweltforschung mbH., and
Sektion Physik der Universität München, München, Germany.

The ejection characteristics of single crystals of zinc, magnesium (hcp) and aluminum (fcc) were determined by bombarding various crystal surfaces with a beam of inert gas ions. The energy of the bombarding ions varied between 10 and 40 keV. The sputtered particles were collected on cooled, hemicylindrical metal collectors; with the help of an electron probe the thickness of the deposit was measured. Pronounced anisotropic emission was observed in all cases, although random emission comprises the major part of the total ejection. By comparing the $\langle 20\bar{2}3 \rangle$ ejection, which is produced by pure surface collisions, and the $\langle 11\bar{2}0 \rangle$ ejection, where focusing collisions are involved, we deduce that surface collisions are mainly responsible for the anisotropic emission. The width of both emission distributions increases linearly to the same extent in the temperature regime between 100 and 370 K. The agreement of the Mg $\langle 11\bar{2}0 \rangle$ ejection with the $\langle 110 \rangle$ ejection from an aluminum (111) single crystal is good and thus allows a comparison with the gold $\langle 110 \rangle$ ejection reported in the literature.

The assisted focusing ejection in $\langle 0001 \rangle$, which is very pronounced in zinc single crystals and strongly influences the $\langle 20\bar{2}3 \rangle$ ejection from the basal plane, does not show up in magnesium. No $\langle 111 \rangle$ ejection occurred in aluminum either. In both cases the weak repulsive potential of the low-Z atoms accounts for the lack of this preferential emission.

1 INTRODUCTION

When single crystal targets are bombarded with energetic ions the spatial distribution of sputter-ejected atoms is anisotropic (Wehner 1955). Preferential ejection has been observed near low-index crystallographic lattice rows in a variety of crystal structures. Two mechanisms have been proposed for ejection in the direction of close-packed lattice rows:

In the focuson model of sputtering[1,2] enhanced ejection in close-packed lattice rows is due to correlated collision sequences along the row.

In the surface collision model[3] enhanced emission is a consequence of the increased probability of a surface atom exceeding the surface binding energy after a nearly head-on collision.

Though there is no doubt that both effects must influence the ejection characteristic,[4,5] their relative contribution still remained unknown. Numerous investigations on fcc and bcc crystals did not allow any definite conclusion in favour of the one or the other mechanism.[4-8]

On the other hand, it has long been known that hcp crystals[‡] show preferential ejection from the (0001) surface.[9] This plane is not intersected by any close-packed row, Figure 1. The observed sixfold spot-pattern is due to enhanced emission in $\langle 20\bar{2}3 \rangle$, a lattice direction which comprises only pairs of atoms. The separation of the pairs is six times the atomic distance in the pair, and so focusing collision sequences can definitely be excluded. At first a "single-event focusing mechanism"[10] was proposed, but this model fails to explain the experimentally observed sharp ejection characteristic; defocusing collisions, which would broaden the characteristic, are not considered. In fact, it is the surface binding energy which excludes from the ejection process the majority of those atoms which have experienced non-central collisions. The pronounced ejection along $\langle 20\bar{2}3 \rangle$ must therefore be assigned to the surface collision mechanism (which, of course, also includes focused binary collisions)—Preferential ejection in the direction

---

† Present address: Max-Planck-Institut für Plasmaphysik D-8046 Garching bei München, Germany.

‡ Throughout the paper the four-index (Miller–Bravais) system of crystallographic notation is used. Crystal orientations are specified by their surface planes (hkil), whereas the ensemble of crystallographically equivalent lattice rows is denoted by the symbol $\langle hkil \rangle$. Only where necessary are specified lattice directions [hkil] mentioned. Note that, in general, the normal to a lattice plane (hkil) is not given by [hkil].

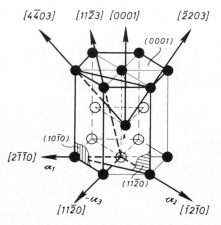

FIGURE 1   hcp lattice structure with some lattice elements relevant to anisotropic sputtering. In the text collision directions are, in general, given by the type of the lattice row, neglecting their actual crystallographic orientation. Thus $\langle 20\bar{2}3 \rangle$ comprises $[20\bar{2}3]$, $[2\bar{2}02]$, $[02\bar{2}3]$...

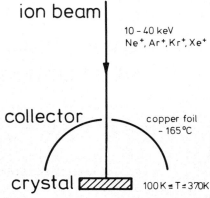

FIGURE 2   Schematic drawing of the target-collector setup. The application of hemicylindrical collectors avoids distortion of the ejection characteristic by geometrical effects and by an incidence-angle-dependent sticking factor. Using cooled metal collectors eliminates distortion by nonuniform condensation. The ratio of the collector radius and the beam diameter was about 25, while the beam divergence was less than $1°$.

of the "staggered chain" $\langle 11\bar{2}3 \rangle$ has been considered by Frère[11] but has not been experimentally verified.[5,12]

Ejection in $\langle 11\bar{2}0 \rangle$, on the other hand, may involve focusing collision sequences just as the $\langle 110 \rangle$ lattice row in fcc crystals. Thus, by comparing the emission characteristic in $\langle 11\bar{2}0 \rangle$ and $\langle 20\bar{2}3 \rangle$ the contribution of correlated collision sequences to anisotropic emission may be obtained. For this reason sputtering of hcp single crystals appears to be particularly suitable for affording further information on the prevailing ejection mechanism.

In addition to the above mentioned ejection mechanisms, preferential emission through symmetric rings of atoms is often observed (e.g. $\langle 100 \rangle$ ejection from fcc crystal surfaces). This emission became known as assisted focusing emission since momentum focusing is caused by rings of atoms surrounding the lattice row. As will be shown below, assisted focusing ejection in the $\langle 0001 \rangle$ direction is of particular importance in hcp crystals.

2   EXPERIMENTAL PROCEDURE

Single crystal targets of zinc, magnesium, and aluminum were bombarded with inert gas ions in the energy range between 10–40 keV. The sputtered material was collected on hemicylindrically bent copper foils (Figure 2), which were cooled to $-165°$ to guarantee a sticking factor independent of the flux density of the impinging particles.[13]

The sputtering experiments were performed under conventional vacuum conditions (2×10^{-6} torr). With current densities of 3 mA/cm² the projectile bombard-

ment rate exceeded the oxidation rate in the beam spot by about two orders of magnitude. In addition, during bombardment the collector mounting shielded the space above the target, which resulted in considerably lower oxygen partial pressure in this region because of oxygen gettering by the first sputtered atoms. To obtain spot profiles of reasonable thickness, fluences up to

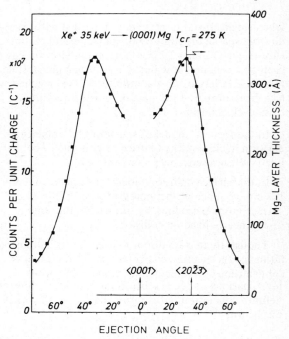

FIGURE 3   Ejection profile of a magnesium (0001) surface Azimuth: $(\bar{1}2\bar{1}0)$ Ion dose: 7.5×10^{16} Xe⁺.

10^{19} ions/cm^2 had to be used when Mg and Al crystals were sputtered.

The thickness of the sputtered deposit was measured with an electron probe. This technique allows the deposit profile to be determined, irrespective of whether or not oxidation has occurred.[13] Both the nonuniform condensation on insufficiently cooled collectors and the difficulty of evaluating oxidized spot patterns are probably the reasons why measurements of the ejection characteristics of hcp crystals have not yet been published. Figure 3 gives an example of a spot profile obtained after bombardment of a magnesium (0001) crystal surface with 35 keV Xe$^+$ ions.

3 EXPERIMENTAL RESULTS

Our first investigations on hcp crystals were performed on a zinc (0001) plane. It soon became clear that an assisted focusing ejection in ⟨0001⟩, which was unknown at that time, interferes with the ⟨20$\bar{2}$3⟩ ejection, thus preventing exact determination of the ⟨20$\bar{2}$3⟩ emission characteristic;[12] under certain experimental conditions (e.g. bombardment of a Zn crystal at 100 K with 20 keV Ne$^+$ ions) the ⟨0001⟩ ejection may even dominate the whole pattern. Furthermore, it can be theoretically shown[13] that a ⟨0001⟩-ejected Zn-atom destroys the surrounding ring of atoms in the surface while penetrating it. These atoms subsequently make soft collisions in the surface and deteriorate the pattern.

The final experiments were therefore performed on magnesium single crystals because Mg is the only hcp crystal where preferential ⟨0001⟩ emission is not observed.[5] In addition, Mg has an ideal hcp lattice structure, giving equal focusing energies in ⟨11$\bar{2}$0⟩ and ⟨20$\bar{2}$3⟩.

3.1 Bombardment of the Mg (0001) Plane

On the left-hand side of Figure 4 the normalized emission characteristics of the ⟨20$\bar{2}$3⟩ lattice direction is shown for Ar$^+$ and Xe$^+$ bombardment.

This reveals a typical feature of all Mg and (see Section 4) Al spots: the half width at half maximum of the ejection characteristics increases with decreasing ion mass (observed for Xe$^+$, Ar$^+$ and Ne$^+$ ions).

The position of the spot maxima is slightly shifted towards lower polar angles, but it does not depend on projectile mass or energy, contrary to the results for zinc.[12] No indication of preferential ⟨0001⟩ emission has been observed in the energy, mass and temperature ranges investigated, Figure 3.

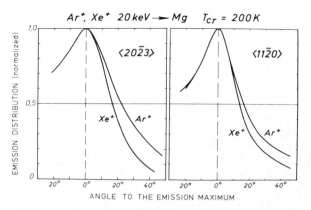

FIGURE 4   Normalized emission characteristic of the ⟨20$\bar{2}$3⟩ and ⟨11$\bar{2}$0⟩ ejection. Left-hand side: ⟨20$\bar{2}$3⟩ spot obtained on a (0001) crystal surface. Right-hand side: ⟨11$\bar{2}$0⟩ spot obtained on a (10$\bar{1}$0) crystal surface.

3.2 Bombardment of the Mg (10$\bar{1}$0) Plane

The ⟨11$\bar{2}$0⟩ emission distribution can best be studied when sputtering a (10$\bar{1}$0) prism plane. A (10$\bar{1}$0) crystal surface is penetrated by ⟨11$\bar{2}$0⟩ lattice rows at a polar angle of 30°. Thus, the difference in ejection angle between ⟨20$\bar{2}$3⟩ from the basal plane and ⟨11$\bar{2}$0⟩ from the prism plane is only 5°, which is of minor importance in this context.

The right-hand side of Figure 4 shows two normalized ⟨11$\bar{2}$0⟩ ejection profiles. It is again observed that spots from Xe$^+$ bombardment are sharper.

A comparison between ⟨20$\bar{2}$3⟩ and ⟨11$\bar{2}$0⟩ emission distributions is made in Figure 5. The spots observed

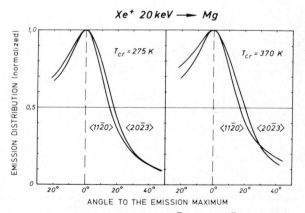

FIGURE 5 Comparison between ⟨20$\bar{2}$3⟩ and ⟨11$\bar{2}$0⟩ emission characteristic of magnesium obtained on (0001) and (10$\bar{1}$0) surfaces respectively.

in the direction which permits focusing collision sequences are sharper. However, the difference in spot width is surprisingly small, being approximately 20% for Xe⁺ bombardment.

The half-width at half-maximum, $\theta_{1/2}$, increases linearly with temperature in the range between $100 \text{ K} \leqslant \text{T} \leqslant 370 \text{ K}$, Figure 6. The slope is the same

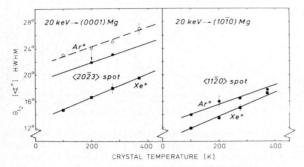

FIGURE 6   Temperature dependence of the half-width at half-maximum of the $\langle 20\bar{2}3 \rangle$ and $\langle 11\bar{2}0 \rangle$ emission distribution obtained on magnesium (0001) and (10$\bar{1}$0) surfaces respectively. The Debye temperature of magnesium is 400 K.

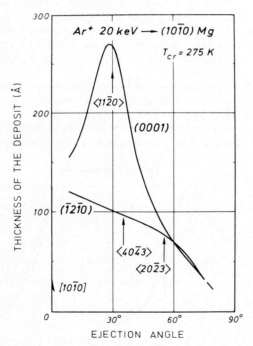

FIGURE 7   Ejection profiles of a Mg (10$\bar{1}$0) surface measured in two perpendicular planes. Ion dose: 9.4×10^{16} Ar⁺.

for the two spots; this fact is remarkable because, if long-range focusons substantially contribute to the $\langle 11\bar{2}0 \rangle$ emission, one would expect a slower increase for that spot than for the pure surface collision along

$\langle 20\bar{2}3 \rangle$; the normal component of the thermal mean square displacement on the surface increases with temperature up to three times faster than in the bulk.[14]

For Xe⁺ bombardment the slope is slightly larger than for Ar⁺ sputtering. The same effect has been observed on Au[7] and may be due to some dynamic randomization in dense cascades.[15]

In principle, it should be possible to compare the $\langle 11\bar{2}0 \rangle$ emission with the $\langle 20\bar{2}3 \rangle$ emission by rotating a (10$\bar{1}$0) crystal 90° around its surface normal: in the (0001) azimuth the emission characteristic of close-packed rows can be studied and in the (11$\bar{2}$0) azimuth the $\langle 20\bar{2}3 \rangle$ and/or $\langle 40\bar{4}3 \rangle$ ejection can be obtained. Two deposit profiles obtained in this way are shown in Figure 7. The dominance of the $\langle 11\bar{2}0 \rangle$ emission is, of course, only partly due to correlated collisions because emission in $\langle 20\bar{2}3 \rangle$ is reduced owing to both the atomic arrangement in the (11$\bar{2}$0) azimuthal plane and the larger polar angle.

3.3 Bombardment of the Mg (11$\bar{2}$0) Plane

The strong influence of the polar angle is illustrated in Figure 8, where the ejection characteristic of a (11$\bar{2}$0)

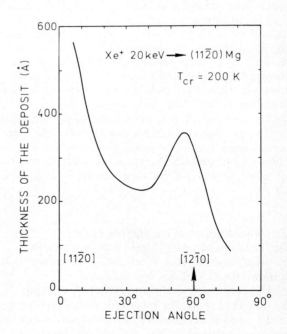

FIGURE 8   Ejection profile of a Mg (11$\bar{2}$0) surface. Azimuth: (0001). Ion dose: 9.4×10^{16} Xe⁺.

surface in the (0001) azimuth is plotted. Even crystallographically equivalent rows give different spot intensities: ejection perpendicular to the surface along [11$\bar{2}$0] is enhanced compared with oblique $[1\bar{2}\bar{1}0]$ ejection. This fact strongly corroborates the suggestion of Nelson and Thompson[16] that the polar angles should be equalized if the emission of two lattice directions is to be compared. This was done by bombarding the $(36\bar{9}8)$ plane of a Mg crystal.

3.4 Bombardment of the Mg (36$\bar{9}$8) Plane

A $(36\bar{9}8)$ crystal surface is intersected by the $[11\bar{2}0]$ and $[02\bar{2}3]$ lattice directions at a polar angle of 30°. The atomic arrangement in the plane defined by these two directions is sketched in Figure 9. It is seen that

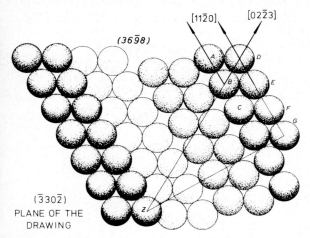

FIGURE 9 Atomic arrangement in the $(\bar{3}30\bar{2})$ lattice plane. This plane comprises $[11\bar{2}0]$ and $[02\bar{2}3]$ lattice rows at equal polar angles. The $(36\bar{9}8)$ lattice plane is perpendicular to the plane of the drawing. Brighter spheres should indicate positions below the plane of the drawing.

in this plane the Mg atoms are positioned along pairs of close-packed rows both of which contribute to the $[11\bar{2}0]$ emission. The $[02\bar{2}3]$ ejection, on the other hand, is only possible for atoms in the upper row DEFG. We thus expect a spot intensity ratio of:

$$S^{[11\bar{2}0]}:S^{[02\bar{2}3]} = 3:1$$

provided the momentum distribution of the subsurface atoms is isotropic.[†]

Figure 10 shows an experimental result. To obtain the ejection intensity of the two collision directions in question, one has to subtract the random part of the emission. This was done by assuming a cosine distribution and fitting it (dashed line) to the outer part of the

---

[†] Of course, equal ejection thresholds and no surface deterioration in the lattice-constant scale have thereby to be assumed.

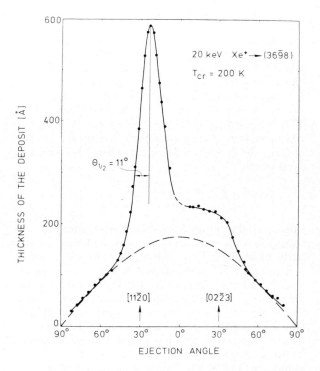

FIGURE 10 Ejection profile of a Mg $(36\bar{9}8)$ surface. Azimuth: $(\bar{3}30\bar{2})$. Ion dose: 5.5×10^{16} Xe⁺.

profile. In this way one gets a nearly rotational-symmetric spot in $[11\bar{2}0]$. The ejection intensity of a Gaussian like emission distribution is proportional to the area F below the experimental curve and the half-width $\theta_{1/2}$.

It was found that for Xe⁺ bombardment

$$F^{[11\bar{2}0]}:F^{[02\bar{2}3]} = 3.7 \pm 0.2.$$

Thus, with approximately equal half-widths of the two ejection profiles one can attribute at most 25% of the anisotropic emission in $[11\bar{2}0]$ to correlated collisions (i.e. deviation from the isotropic momentum distribution of the subsurface atom).[†]

A similar experiment on α-uranium has been performed by Nelson and von Jan.[17] The spot intensity ratio found for this high-Z metal is about twice as large as the authors expected from pure surface collisions. We feel that the difference is not so large since we expect the [021] ejection to be more suppressed than is reported. Anyway, the more complicated arrangement of the U-atoms in the relevant plane makes it difficult to calculate the impact probabilities.

4 DISCUSSION OF THE EXPERIMENTAL RESULTS

It has been pointed out in Section 3 that an unambiguous characterization of the $\langle 20\bar{2}3 \rangle$ ejection is prevented in most hcp crystals owing to preferential $\langle 0001 \rangle$ emission. Magnesium does not show this assisted focusing emission because of the weak repulsive potential of the three atoms forming a ring around $\langle 0001 \rangle$. Following the concept of assisted focusing collisions developed by Nelson and Thompson[1] a ring of three atoms behaves like a converging lens with the focal length[13]

$$f^{\langle 0001 \rangle} = \frac{1}{3} \frac{2a}{L-2a} \sqrt{\frac{2}{\pi} La} \frac{E}{A} \cdot e^{L/a}$$

where A and a are the Born–Mayer parameters, L ($=a_0/\sqrt{3}$) is the ring radius and a_0 is the lattice constant in the basal plane. From this it can be shown that the focal length $f^{\langle 0001 \rangle}$ in magnesium is about one order of magnitude larger than in zinc. Obviously, only trajectories of the very low energy recoils can be influenced by the Mg $\langle 0001 \rangle$ lens. The same applies to the $\langle 111 \rangle$ ejection from the fcc crystal aluminum if only the two topmost surface layers are regarded. Indeed, in experiments for measuring the emission characteristic of an aluminum (111) single crystal in pursuance of the magnesium investigations, no preferential $\langle 111 \rangle$ ejection was found, Figure 11.

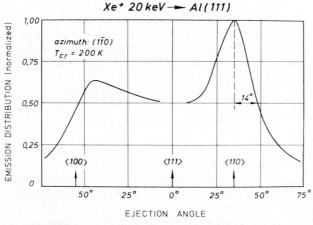

FIGURE 11   Normalized ejection profile of an aluminum (111) surface.

Thus, generally speaking, assisted focusing ejection is governed by two opposite effects: high transparency of the ring and strong focusing power.

The similarity between preferential ejection from Mg and Al single crystals is, in fact, more far-reaching: both of these light elements attain the highest energy

transfer in a two-body collision under Ne$^+$ bombardment and for both elements the energy transfer decreases with increasing ion mass. For gold targets, on the other hand, the situation is reversed. In the surface model it is now easy to interpret the opposite dependence of the spot width of light (Figure 6, 11, 12) and heavy[7] targets on the ion mass: the more the mass ratio of the target and projectile atoms approaches unity the more efficient is the energy transfer. This increases the number of recoils near the surface with somewhat higher energy and this, in turn, broadens the emission characteristic.

From Figure 12 it can be seen that Mg $\langle 11\bar{2}0 \rangle$ and Al $\langle 110 \rangle$ have the same ejection characteristics. (The

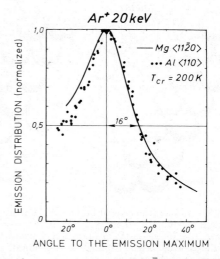

FIGURE 12   Comparison of a Mg $\langle 11\bar{2}0 \rangle$ and Al $\langle 110 \rangle$ emission characteristic. The side to the left of the maximum belongs to low polar angles. The $\langle 11\bar{2}0 \rangle$ emission was obtained from measurements on a (10$\bar{1}$0) surface.

left-hand side of the Mg $\langle 11\bar{2}0 \rangle$ ejection is slightly higher because of a region of low potential in the surface). This fact is expected since the ratio of the focusing and surface binding energy E_f/E_b, which primarily determines the spot width in both models of anisotropic ejection, is nearly the same in Mg and Al. Following the general theory of focusing collision chains developed by Lehmann and Leibfried it can be shown that a focusing energy of $E_f^{\langle 110 \rangle} = 8$ eV is reasonable in Al. This yields $E_f/E_b \approx 3$ and spot widths in the experimentally observed range.

A comparison between the $\langle 110 \rangle$ ejection characteristic of aluminum and gold has already been proposed by Silsbee. Since it was supposed that the small core size of the Al atoms reduces the focusing energy to an insignificant level, information on the contribution of focusons to sputtering could be expected. Up to now the difficulty of evaluating oxidized Al patterns has prevented the relevant experiments. Comparing the spot profiles of Mg $\langle 11\bar{2}0\rangle$ and Al $\langle 110\rangle$ presented in this work with results from previous investigations on Au,[7] it can be seen that

$$\Theta_{1/2}^{Al,Mg} \approx 1.8 \times \Theta_{1/2}^{Au}$$

provided bombarding conditions with equal transferred energy are considered (e.g. $Xe^+ \to Al$, $Ar^+ \to Au$). Neglecting this condition leads to inexplicably small differences in spot width for Xe^+ bombardment; for instance, the half-width $\Theta_{1/2}$ in Figure 10 is only 2° larger than for a Au $\langle 110\rangle$ spot obtained under similar conditions.

The reported experiments clearly show that we can obtain quite sharp spots with the surface collision mechanism. In cases where the ejection direction is a close-packed row surface collisions contribute predominantly to preferential ejection, but correlated collisions sharpen the ejection characteristic. Obviously, we cannot dispense with focusons in light metals. This is due to the very low energy loss to neighbouring rows ΔE, a fact which has been pointed out by Nelson and v. Jan.[18] According to these authors the relevant quantity determining the contribution of focusons to sputtering is $E_f/\Delta E$; in fcc metals ΔE can be calculated according to Leibfried[19] and Sanders and Fluit.[20] In hcp metals the energy loss to a static lattice is given by[13]

$$\Delta E^{\langle 11\bar{2}0\rangle} = 2E_f \cdot e^{-(a_0/2a)}\{e^{(a_0-x)/a} - 1\} \quad \text{with}$$

$$x = a_0\sqrt{\frac{3}{4} + \left(\frac{a}{a_0}\right)^2 \ln^2\left(\frac{E_f}{E}\right)}$$

In both Al and Mg we arrive at $E_f/\Delta E \approx 50$, a value at least as large as in Cu; this shows that focusing collision sequences may very well play an important role in the energy dissipation process.

5 SUMMARY AND CONCLUSIONS

Our conclusions from measurements of the spatial anisotropy of sputtered atoms on Zn, Mg, and Al single crystals can be summarized as follows:

Zinc single crystals show pronounced preferential ejection in the $\langle 20\bar{2}3\rangle$, $\langle 11\bar{2}0\rangle$ and $\langle 0001\rangle$ crystal directions; the latter ejection is much more enhanced than the $\langle 111\rangle$ ejection in Cu, thus indicating that this assisted focusing ejection cannot be interpreted merely by emission through a potential minimum in the surface layer.

Magnesium does not show preferential $\langle 0001\rangle$ ejection owing to the weak atomic potential. This is advantageous

for determining the $\langle 20\bar{2}3\rangle$ ejection characteristic.

Sharp $\langle 20\bar{2}3\rangle$ spots prove the effectiveness of the surface collision mechanism. Compared with the $\langle 11\bar{2}0\rangle$ ejection, where focusing collision sequences are involved, the $\langle 20\bar{2}3\rangle$ emission is broader, but reveals the same dependence on temperature and ion mass.

From experiments on the $(36\bar{9}8)$ surface it can be estimated that at most 25% of the anisotropic ejection in $\langle 11\bar{2}0\rangle$ can be attributed to correlated collisions along this close-packed row.

The anisotropic ejection of aluminum is largely equal to the magnesium ejection characteristic.

Random ejection contributed in all cases the major part of the total emission in agreement with Smith et al.[21] and recent measurements of Buchanov et al.[22]

Investigations of the anisotropic ejection in sputtering in general requires high fluences of energetic projectiles (10^{18} cm^{-2} and more) which cause damage in the crystal. Since the defect concentration in the surface layers is much smaller compared to the deeper region, the ejection characteristic should not be influenced significantly by the radiation induced damage. In order to prove this statement low-fluence sputtering experiments in an ion microprobe are in progress.

ACKNOWLEDGEMENTS

I should like to express particular appreciation to Professor R. L. Sizmann, who suggested this work and inspired it in many valuable discussions. I am also indebted to Dr. F. Schulz, whose findings and experimental set-up afforded the starting point for this investigation. On several occasions W. Weber, G. Dürr, and Th. Ligon provided help with experimental details and measurements.

REFERENCES

1. R. S. Nelson and M. W. Thompson, *Proc. Roy. Soc.* **A259**, 458 (1961).
2. R. S. Nelson and M. W. Thompson, *Phil. Mag.* **7**, 1385 (1962).
3. Chr. Lehmann and P. Sigmund, *Phys. Stat. Sol.* **16**, 507 (1966).
4. J. J. Ph. Elich and H. E. Roosendaal, *Phys. Letters* **33A**, 235 (1970).
5. M. T. Robinson and A. L. Southern, *J. Appl. Phys.* **39**, 3463 (1968).
6. M. T. Robinson and A. L. Southern, *J. Appl. Phys.* **38**, 2969 (1967).
7. F. Schulz and R. Sizmann, Proc. 8th Int. Conf. Ion. Phen. Gases, Vienna 1967, p. 35.
8. L. T. Chadderton, A. Johansen et al., *Rad. Effects* **13**, 75 (1972).
9. B. Perovic, Proc. 5th Int. Conf. Ion. Phen. Gases, München 1961, p. 1172.
10. R. R. Hasiguti, R. Hanada and S. Yamaguchi, *J. Phys. Soc. Japan* **18**, Suppl. III, 164 (1963).
11. R. Frère, *Phys. Stat. Sol.* **3**, 1252 (1963).

12. W. O. Hofer and R. Sizmann in *Atomic Collision Phenomena in Solids*. Eds. D. W. Palmer, M. W. Thompson and P. D. Townsend (North Holland Publ. Co., Amsterdam 1970) p. 298.
13. W. O. Hofer, Thesis, Universität München 1972.
14. P. Masri and L. Dobrzynski, *Surface Sci.* **32,** 623 (1972).
15. R. S. Nelson, *Rad. Effects* 7, 263 (1971).
16. R. S. Nelson and M. W. Thompson, *Phil. Mag.* **7,** 1425 (1962).
17. R. S. Nelson and R. v. Jan, *Can. J. Phys.* **46,** 747 (1968).
18. R. v. Jan and R. S. Nelson, *Phil. Mag.* **17,** 1017 (1968).
19. G. Leibfried, *J. Appl. Phys.* **30,** 1388 (1959).
20. J. B. Sanders and J. M. Fluit, *Physica* **30,** 129 (1964).
21. N. Th. Olson and H. P. Smith, *AIAA-Journal* **4,** 916 (1966); *Phys. Rev.* **157,** 241 (1967).
22. V. M. Buchanov, V. G. Morozov and V. E. Yurasova, Int. Conf. Ion-Surface Interaction, Garching, Sept. 1972.

DEPENDENCE OF THE SPACE DISTRIBUTION OF PARTICLES SPUTTERED FROM MONOCRYSTALLINE COPPER ON THE ION INCIDENCE ANGLE

V. M. BUKHANOV, V. G. MOROZOV and V. E. YURASOVA

Department of Physics, Moscow State University, Moscow 117234, U.S.S.R.

The dependence of the angular distribution of atoms sputtered from the (001) monocrystalline copper face on the incidence angle of 22 keV neon ions has been studied. Use is made of well known ideas defining the observed distribution as a result of superposition of the directed atomic yield along the close-packed axes and of the random yield along all directions. The values obtained for the relative contribution of the directed and random sputterings indicate that the dependence of these values on the ion incidence angle is distinctly anisotropic. The yield of the directed sputtering had a minimum value when the incidence directions of the bombarding ions coincided with the low-index crystallographic axes. In addition, the width of $\langle 011 \rangle$ and $[001]$ pattern spots was also found to be an anisotropic function of the ion incidence angle. The spots have the smallest width when the direction of ion incidence coincides with the low-index axes. The observed dependences are discussed on the basis of the focusing mechanism of the sputtered atom yield.

Increased interest has been taken in recent years in studies of some features of sputtering of single crystals with emphasis on the sputtering in certain crystallographic directions. Irrespective of the object of such studies (whether they pertain to the fine structure of the $\langle 011 \rangle$ spots found by L. Chadderton when bombarding Au single crystal with a well collimated ion beam in a random direction[1] or to the energy spectra of sputtered particles which M. W. Thompson and co-workers[2] are successfully studying or to the very interesting observations carried out by J. Kistemaker and coworkers[3] of essential changes in surface structure caused by very small changes in the angles of ion incidence on the single crystal and the influence of this effect on the sputtering ratio), they permit substantial progress to be made in understanding the sputtering mechanism and relative importance of focusing, channelling, dechannelling, etc. for this process.

The present paper continues the series of our studies bearing on the ion incidence angle dependence of sputtering of single crystals in individual close-packed directions (within individual spots) $\langle 011 \rangle$. It has been shown in the first report of this series delivered at the International Conference at Belgrade in 1965[4] and in subsequent papers[5,6] that when the (001) Cu face is sputtered by 22 keV Ne ions the intensity, P, of particle ejection in each of the preferential directions $\langle 011 \rangle$ is not monotonically changed by changing the ion beam incidence angle, φ, from 0° to 65° in the (110) and (100) planes (φ refers to the normal to the surface). In the case of coincidence of the beam incidence direction

with crystal open channels $\langle 110 \rangle$, $\langle 100 \rangle$, $\langle 111 \rangle$, $\langle 114 \rangle$, etc. the curves $P(\varphi)$ displayed a minimum for any of the $\langle 011 \rangle$ spots (recently a similar dependence was also observed by L. Chadderton[1]). Besides that, it has been shown in Refs. 4, 5, 6 that the $P(\varphi)$ curves are not the same for the $P_1[011]$ spot at the beam side and for the $P_2[101]$ spot opposite the beam. This results in a more complex angular dependence of the intensity ratio in the above mentioned spots, $P_1/P_2(\varphi)$, than $P_1(\varphi)$ or $P_2(\varphi)$.

The features of the $P_1/P_2(\varphi)$ curves permitted us to draw some conclusions[6] on the depth from which the energy transferred to target atoms from bombarding ions is carried (5–6 atomic layers for our case of Ne, 22 keV → (001) Cu) as well as on focusing energy (>33 eV, as in[7]). It will be noted that an interesting dependence $P_3/P_2(\varphi)$ (where P_3 is the intensity of a $[101]$ spot located sidewards from the beam) within narrow limits of variation of angles φ in the (100) plane has been obtained by J. Elich and H. Roosendaal.[8] The results of Ref. 8 agree with the data in[5] and the method for describing sputtering due to ion beam incidence near close-packed directions as adopted in[8] is, in general, similar to that adopted in[4,5,6], where the role of focused cascades is examined (particularly in the direction perpendicular to the ion beam).

It may also be of help in estimating the relative importance of the focused collisions to study the directed and random components of sputtering at different angles of ion incidence on single crystals. This is the problem dealt with in the present paper.

It has been shown in Refs. 9, 10, 11 that, to analytically describe the density distribution of deposits, the latter should be treated as a result of superposition of the directed atom yield from single crystals in the close-packed directions and the diffuse (random) yield in all directions (a so-called background). The deposit density distribution due to the directed yield can be described by the Gaussian law:

$$\mathcal{T}_i^{(g)}(\psi) = \mathcal{T}_i^{(g)}(0) \exp\{-\psi^2/2\sigma_i^2\} \qquad (1)$$

where $\mathcal{T}_i^{(g)}(0)$ is the deposit density in the centre of the i-th spot, σ_i is the mean square angular deviation from the close-packed direction, ψ is the angle of deviation from the close-packed direction. The deposit density distribution due to diffuse sputtering obeys a cosine law:

$$\mathcal{T}_d(\theta) = \mathcal{T}_d(0) \cos\theta \qquad (2)$$

where $\mathcal{T}_d(0)$ is the background density in the deposit centre, θ is the polar angle of deviation from the direction of a normal to the sputtering face. It has been shown that such an approach produces the best agreement with the experimentally observed distributions and allows certain conclusions to be made about the sputtering mechanism by comparing the portions of the directed and diffuse yields of the sputtered particles.

In the present paper we studied the dependence of a percentage content of the directed and diffuse sputtering from the (001) monocrystalline copper face, and of σ_i defining the angular width of Wehner spots, on the ion incidence angle. The bombardment was performed by 22 keV neon ions. The ions were produced by a source of the Ardenne duoplasmatron type. The ion current density on the target was not less than 5 mA/cm^2 the maximum ion dose was 10^{21} Ne$^+$/cm^2 and the beam divergence was 1°. The angle of incidence varied from 0° to 64° by means of a 2° rotation of the sample around the axis lying in the surface plane and coinciding with the [110] axis. The target temperature was about 80°C. The residual gas pressure in the target chamber was 3×10^{-6} mm Hg.

The deposits sputtered on flat mica collectors consisted of 4 spots in the ⟨011⟩ direction and a central spot in the [001] direction. Neither the additional sputtering spots in the high-index directions, as in Ref. 12, for 4 keV sliding ion beams nor the patterns of sputtered particle blocking found in Ref. 13 when bombarding a single crystal with fast ions were observed by us in the vicinity of ⟨011⟩ and [001] spots.

The density distribution in each spot was determined by means of an optical microphotometer. The distributions obtained were converted into the equivalent

distributions on a spherical collector.[14] The density of the diffuse background component, $\mathcal{T}_d(\theta)$, corresponding to a given polar angle, θ, was determined from the deposit density in the regions most distant from the spots. The background density in the centre of a deposit, $\mathcal{T}_d(0)$, was found by formula (2), and the density distribution in each spot (i.e. parameters $\mathcal{T}_i^{(g)}(0)$ and σ_i) was determined by subtracting the value of the background from the value of the total density at each point. By integrating (1) and (2) over all angles of ejection we obtained the total number of particles (in arbitrary units) produced in each of the close-packed directions, $P_i^{(g)}$, and by a diffuse mechanism P_d; the percentage ratio

$$A = \left\{\sum_{i=0}^{4} P_i^{(g)} \middle/ \left(P_d + \sum_{i=0}^{4} P_i^{(g)}\right)\right\} \cdot 100\% \qquad (3)$$

was called the percentage contribution of the directed sputtering, and the value

$$B = \left\{P_d \middle/ \left(P_d + \sum_{i=0}^{4} P_i^{(g)}\right)\right\} \cdot 100\% \qquad (4)$$

was called the background contribution.

It will be noted that the comparison of the measured deposit densities with those fitted (cosine + Gaussian spots) shows that the difference in these values did not exceed 2–3% for the high-density sections (θ varying from 0° to 55°). The actual density for the peripheral sections ($\theta > 55°$), however, decreases much more rapidly than that fitted. This fact is probably connected, in part, with a weak oxidation of deposits which is insignificant for the high-density sections but markedly reduces the transparency of thin layers and, in part, with a rapid decrease in the sticking probability with increasing θ. It should be noted, however, that the factors mentioned above did not markedly distort the values A, B, and σ_i for their importance in the sections making the main contribution to the studied values is negligible. The inclusion of these factors in combination with the errors inserted by approximate integration results in a total error that does not exceed 10%.

Figure 1(a) shows the dependence of the value A on the ion incidence angle. The data for angles from 0° to 18° are not available since at these angles a hole in the collector for transmitting the ion beam cuts off a significant part of the deposit, thus making it impossible to calculate the percentage ratios. A peculiar feature of the dependences given above is the fact that when the ion beam coincides with the low-index directions (angles of about 19° – ⟨114⟩, 35° – ⟨112⟩, and 55° – ⟨111⟩) the value A has a minimum, while the

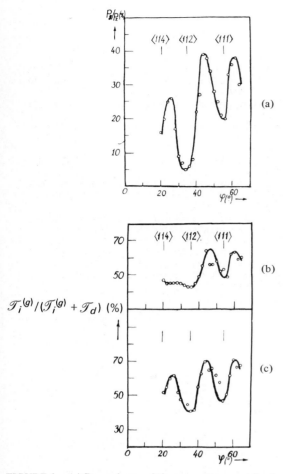

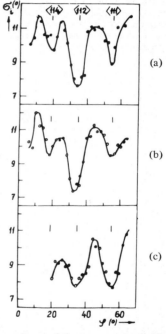

which these ratios are reduced to 40–50%. It is concluded in Ref. 15 from the data on the energy spectra of the sputtered particles that the contributions of random collisions and focusons to the sputtering along the ⟨011⟩ axes are approximately equal. Note that our data agree with the data of Ref. 15 if one assumes that mainly focusons are involved in the formation of Wehner spots.

Figure 2 presents the dependence of σ_i on the ion incidence angle. This dependence is also anisotropic since the spots have the smallest width when the ion beam coincides with the close-packed directions.

The angular width of Wehner spots is known to depend on the temperature,[16] energy, and type of bombarding ions.[17] It is the smaller the lower the target

FIGURE 1   (a) Dependence of the percentage contribution (value A) of the directed emission on the ion incidence angle. (b) Dependence of the percentage ratio of the central spot density at the [001] direction to the total deposit density at the same point on the ion incidence angle. (c) Dependence of the percentage ratio of the spot density in the [001] direction to the total deposit density at the same point on the ion incidence angle.

FIGURE 2   Dependence of the mean square halfwidth, σ_i, of the spots on the ion incidence angle for: (a) the [011] spot located at the beam side of the deposits; (b) the [101] spot opposite the beam; (c) the central [001] spot.

value B has a maximum. Note that in good agreement with data from Refs. 10, 11 the percentage contribution of the direct sputtering at all angles is noticeably less than the diffuse one, varying from 5% at 34° to 39% at 44°.

Figure 1(b, c) shows the dependence of the percentage ratios of the values $\mathcal{T}_i^{(g)}/(\mathcal{T}_i^{(g)} + \mathcal{T}_d)$ in the centre of the deposit spots on the ion incidence angle. Almost at all angles of incidence the density in the spot centre is seen to remain considerably greater than the background density at the same point. The only exception is for the angles corresponding to the ion incidence near low-index directions ⟨114⟩, ⟨112⟩, and ⟨111⟩, at

temperature is and the greater the energy and the smaller the ion mass are. The latter two dependences, as well as the one found in this paper, can be determined in the same way: Wehner spots are the narrower the deeper the ions penetrate a crystal.

The above result may be explained by assuming a focuson mechanism of sputtering in the close-packed directions. When moving along the low-index crystallographic axes the bombarding ions penetrate a lattice significantly deeply. This results in increasing the role of the long-range focusons in the formation of deposit spots, i.e. chains of collisions with the most perfect momentum focusing. For ion motion in the opaque directions the short-range and medium-range focusons are, apparently, of main importance; their focusing is less perfect, i.e. the momentum transmitted to the upper atoms of chains may have a larger angular dispersion relative to the chain axis.

REFERENCES

1. L. T. Chadderton, A. Johansen, L. Sarholt-Kristensen, S. Steenstrup, and T. Wohlenberg, *Rad. Effects* **13**, 75 (1972).
2. G. E. Chapman, B. W. Farmery, M. W. Thompson, and T. H. Wilson, *Rad. Effects* **13**, 121 (1972).
3. J. J. Ph. Elich, H. E. Roosendaal, H. H. Kersten, D. Onderde-linden and J. Kistemaker, *Rad. Effects* **8**, 1 (1971).
4. V. E. Yurasova, V. M. Bukhanov and M. V. Kuvakin, *Proc. 7th Int. Conf. on Phenom. in Ionized Gases*, Belgrade 1965, Amsterdam 1966, p. 155.
5. V. E. Yurasova and V. M. Bukhanov, *Sov. Radiotechnology and Electronics* **22**, 1853 (1967).
6. V. M. Bukhanov, D. D. Odintsov and V. E. Yurasova, *Sov. Sol. Phys. State* **12**, 2425 (1970).
7. C. Lehmann and G. Leibfried, *Z. Physik* **162**, 203 (1961).
8. J. J. Ph. Elich and H. E. Roosendaal, *Phys. Letters* **33A**, 235 (1970).
9. G. E. Chapman and J. C. Kelly, *Austr. J. Phys.* **20**, 283 (1967).
10. N. T. Olson and H. P. Smith, *Phys. Rev.* **157**, 241 (1967).
11. R. G. Musket and H. P. Smith, *J. Appl. Phys.* **39**, 3579 (1968).
12. V. E. Yurasova, I. G. Bunin, V. I. Shulga and V. M. Mamaev, *Rad. Effects* **12**, 175 (1972).
13. V. E. Dubinsky and S. Ya. Lebedev, *Phys. Letters* **32A**, 457 (1970); V. E. Dubinsky and V. G. Radionova (in press).
14. F. Schulz and R. Sizmann, *Phil. Mag.* **18**, 269 (1968).
15. M. W. Thompson, *Phil. Mag.* **18**, 377 (1968).
16. V. E. Yurasova and V. M. Bukhanov, *Sov. Crystallography* **7**, 257 (1962); R. S. Nelson, M. W. Thompson and H. Montgomery, *Phil. Mag.* **7, 8**, 1385 (1962).
17. M. Koedam, *Philips Res. Repts.* **16**, No. 2, 101; No. 3, 266 (1961).

A DEPTH ANALYSIS OF CLEAVED MICA SURFACES MONITORED BY AUGER SPECTROSCOPY

P. STAIB

Max-Planck-Institut für Plasmaphysik, EURATOM Association, 8046 Garching, Germany

Vacuum cleaved mica surfaces (muscovite) are etched by a 500 eV argon ion beam. The surface composition of the cleaved surface is known from the succession of atomic layers in the crystal. The surface structure and its chemical composition are analysed by LEED and AUGER electron spectroscopy. We show that the bombardment quickly destroys the crystal structure, making an accurate depth analysis of the sample impossible.

INTRODUCTION

Depth analyses are mainly realized by gradually sputtering a surface with an ion beam and by analysing the surface composition by backscattering of light ions,[1] by secondary ion mass spectroscopy,[2] or by AUGER electron spectroscopy.[3] The last method is already employed to determine the chemical composition and thickness of very thin films in the monolayer range.[4] However, the accuracy of the method in this range is doubtful. We therefore analyse mica samples that are cleaved in vacuum, and for which the absolute densities and depth distribution of the atoms are known from the crystal structure. The etching is realized by an argon ion beam and structure and surface composition of the uppermost atomic layers are controlled by low energy electron diffraction (LEED) and AUGER electron spectroscopy (AES). Great difficulties arise to interpret the changes in intensities of the AUGER lines, owing to the quick destruction of the crystal structure.

EXPERIMENTAL CONDITIONS

Natural mica samples (Muscovite) are cleaved in ultra-high vacuum at a background pressure of $5 \cdot 10^{-11}$ torr. Etching of the sample surface is performed by a 500 eV argon ion beam with a low current density of typically $0.2 \, \mu A \cdot cm^{-2}$ and a diameter of about 1.5 cm. The incidence angle is 43 deg. The ion current density, measured by a movable faraday cup, shows a flat distribution over the centre of the beam. The actual ion current reaching the mica surface—a good insulator —is not exactly known. The doses indicated refer to currents measured with the faraday cup. The ion gun works after backfilling the vacuum chamber with argon gas up to a pressure of about 10^{-5} torr. The argon is

changed after each etch period, i.e. after each AUGER measurement. A titanium sublimation pump traps the residual active gases.

The AUGER analyses are conducted by bombarding the surface at normal incidence with a defocused electron beam of 2372 eV energy, 6 μA intensity and a diameter of about 1.5 mm. A low current density is needed to reduce charging effects of the mica surface and to avoid electron impact desorption of potassium ions as reported by Poppa and Elliot.[5] The energy analysis of the secondary electron spectrum is performed by a conventional 4-grid retarding field analyser. The sample has to be rotated by 140 deg from its position for AUGER analysis to the position for etching. The mechanical reproducibility of the relative beam-sample positions after rotation is better than 0.2 mm. The modulation applied to the retarding grid is 5 V rms and the total spectrum is acquired from 40 eV to 750 eV without changing the range of the lock-in amplifier. The analog signal provided by the amplifier is converted by a high precision analog-to-digital converter (14 bits) and stored into a 1024 multichannel analyser (ND812). The spectra presented are the sum of 5 scans, each corresponding to a time constant of 0.1 s for the amplifier and a scan time of 102.4 s. The pressure during the AES measurements is in the 10^{-9} torr range.

EXPERIMENTAL RESULTS

The mica structure is depicted in Figure 1 (reprinted from Ref. 6). The cleavage takes place along a potassium layer which corresponds to the weakest bonds in the crystal between potassium and oxygen atoms. The cleavage plane is a monolayer of K ions arranged in a

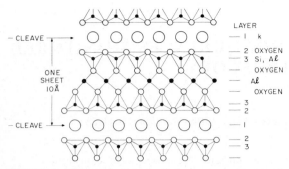

FIGURE 1 Schematic diagram of the muscovite structure (from Ref. 6).

hexagonal array and shared alike by the two cleavage faces. The distance between the K atoms is 5.2 Å. The next layer consists of oxygen atoms, the third of silicon and aluminium in the proportion $\frac{3}{4}$ to $\frac{1}{4}$. The fourth layer is oxygen, the fifth aluminium and the structure repeats itself symmetrically. The space between the potassium layers, about 10 Å thick, is filled by 7 other atomic layers. Structure and LEED analyses of cleaved mica surfaces are reported by Mueller and Chang in Ref. 6.

The experimental AUGER spectra are presented in Figure 2 for argon ion dose up to $3 \cdot 10^{-2}$ C/cm^2 corresponding to about 140 argon ions per oxygen atom of layer 2 (Figure 1). Although their amplitudes appear relatively small, the AUGER lines have a fine structure

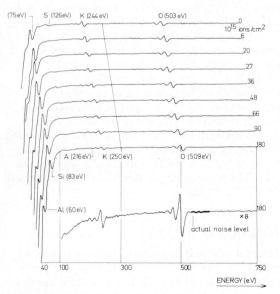

FIGURE 2 AUGER spectra of mica for increasing ion bombardment dose.

as the magnified curve shows. The spectra of Figure 3 offer a clear survey of the intensity evolutions of the lines. They correspond to the integration of the experimental spectrum followed by a suitable background

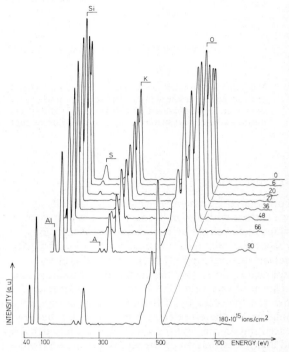

FIGURE 3 AUGER spectra obtained by integration of the measured distributions of Figure 2, followed by subtraction of the background.

substraction. The numerical program is written for the ND812 and the computation can be run immediately after the measurements. The resulting lines are not presently quantitatively exact: the bases of the lines situated at low energies (Al, Si) are truncated and their intensity is reduced by about 40 to 50%. The LEED patterns reproduced in Figure 4 correspond to a primary beam energy of 97.2 eV. At the beginning of the bombardment, we note only slight changes of the AUGER intensities, whereas the LEED patterns are quickly destroyed as shown in Figure 4b and c. The last picture corresponds to a dose of about $4.5 \cdot 10^{-4}$ C/cm^2, i.e. 7 argon ions per oxygen atom of layer 2, and shows no diffraction pattern at any primary energy. This indicates that the structure of at least the topmost layers of K and O is destroyed. The sample becomes much more insulating as a consequence of the destruction of the K layer. The specimen charges up much more during the AUGER measurements and the

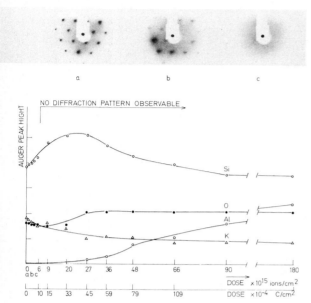

FIGURE 4   Dose dependence of the LEED diffraction patterns and of the intensity of the AUGER lines.

left the surface either sputtered or injected into deeper atomic layers. If we define the sputtering yield as the ratio between the number of K atoms having left the surface and the total number of incident ions, its value in the first stage of the etching is 0.8%. The value of the yield for a bulk potassium sample is expected to be about 60%.[7] Since there is only half a monolayer of potassium initially on the surface, about 60 argon ions must impinge on a unit cell of potassium in order to sputter the corresponding K atom. This value is very high and not very realistic. We can justify it by saying that the collision cascades in the mica lead quickly to an uniform distribution of constituent atoms so that all the different atoms are present in the uppermost layers. Experimentally, the AUGER analyses show only very slight changes in the intensities of the O, K and Si lines for doses greater than about $5.0 \cdot 10^{16}$ ions/cm^2 and steady state concentrations seem to be achieved for these elements.

AUGER lines are shifted by 6 eV to higher energies. This happens already at a dose of $1 \cdot 10^{-4}$ C/cm^2.

DISCUSSION

As a consequence of the heavy damages caused by the ion bombardment, we cannot attribute the intensity variations of the AUGER lines to sputtering effects alone. The increase of the silicon line at the beginning of the bombardment may be due not only to the sputtering of the K and O layers, but also to the increasing disorder carrying the silicon atoms to the top of the surface. The same explanation can be put forth to account for the increase of the low energy aluminium line at doses above $2 \cdot 10^{-4}$ C/cm^2 provided by the fifth atomic Al layer.

The measured decrease of the AUGER line of potassium gives us the possibility to evaluate the number of remaining potassium atoms as a function of the dose. This decrease corresponds to atoms having

CONCLUSIONS

We have shown the difficulties encountered in interpreting of the etching of mica surfaces by argon ions. LEED analyses show a rapid destruction of the topmost atomic layers. The exact knowledge of the surface composition vanishes and an ambiguity arises in explaining the intensity variations of the AUGER lines: these may be due to both the surface erosion and the disorder induced into the uppermost atomic layers. Depth analyses of very thin layers (about one monolayer) of light elements by etching the sample with a low energetic argon beam are therefore not reliable.

REFERENCES

1. J. W. Mayer, L. Eriksson and J. A. Davies, *Ion implantation in semiconductors* (Academic Press, New York, 1970).
2. A. Benninghoven, *Surf. Science* 28, 541–562 (1971).
3. M. L. Tarng and G. K. Wehner, *J. Appl. Phys.* 43, 2268-72 (1972).
4. R. N. Yasco and L. J. Fried, *Rev. Scient. Inst.* 43, 335-37 (1972).
5. H. Poppa and A. G. Elliot, *Surf. Science* 24, 149-63 (1971).
6. K. Mueller and C. C. Chang, *Surf. Science* 141, 39-51 (1969).
7. R. C. Bradley, *Phys. Rev.* 93, 719–728 (1954).

IMPLICATIONS IN THE USE OF SECONDARY ION MASS SPECTROMETRY TO INVESTIGATE IMPURITY CONCENTRATION PROFILES IN SOLIDS

F. SCHULZ, K. WITTMAACK and J. MAUL

Gesellschaft für Strahlen- und Umweltforschung mbH München,
Physikalisch-Technische Abteilung,
D-8042 Neuherberg, Germany

Using a recently developed high sensitivity secondary ion mass spectrometer the influence of the sputtering ion energy (5 to 50 keV argon) on the original range distribution has been investigated with a standard implantation profile as a reference (20 keV boron in amorphous silicon). It was found that recoil processes change the distribution only in the long range tail so that projected range and standard deviation are not influenced by this effect within the limits of experimental accuracy. Because of the low sputtering yield of silicon ($S \lesssim 2$ atoms/ion), however, there is a change in sputtering rate at the beginning of layer removal until equilibrium between implantation and release of argon is obtained. This effect results in an increase in the measured value of the projected range which can be corrected by extrapolation to zero sputtering energy ($R_p = 690$ Å, $\Delta R_p = 280$ Å).

Neither scanning electron microscopy nor talysurf measurements showed any indication for the development of a surface topography after sputtering silicon layers up to 5000 Å in thickness.

1 INTRODUCTION

Secondary ion mass spectrometry (SIMS) has not yet been demonstrated to allow a quantitative investigation of impurity concentration distributions in solids. This is somewhat astonishing because there is a great need for a universal method to determine implantation or diffusion profiles.

The main objections that have been put forward against SIMS as a reliable means for such measurements concerned both the influence of energy dissipation in the bulk of the specimen on the original concentration distribution and the changes in surface topography upon sputtering. More strictly speaking, it has been argued[1] that recoil processes might shift and broaden the impurity profile and that faceting[2] would prevent measurements with a high depth resolution.

Nevertheless sputtering with low energy ions (2 to 8 keV) has been used quite frequently as a means for layer removal in studies on the range distribution of radioactive ions in solids.[3,4] In some cases the reliability of this method has been checked and under certain conditions surface flattening instead of faceting could be achieved.[5] Thus many authors using sputtering energies between 10 and 20 keV relied on SIMS as a quantitative tool to investigate implantation[6-9] or diffusion[8,10-12] profiles. A detailed study of the influence of the sputtering energy on the original impurity distribution and on the possible development

of a certain surface topography does not seem to have been reported so far. It is very much likely that a limit for the sputtering energy exists, above which a noticeable influence of recoil processes on the original impurity distribution will be observed. There is, on the other hand, much interest in using sputtering energies as high as possible since increasing the energy results in a strong increase of the sputtering rate for two reasons: (i) because of the energy dependence of the sputtering yield[13] and (ii) because of the voltage dependence of the maximum current density of the ion beam, j, which may be estimated from the space charge law ($j \propto V^{3/2}$).[14]

The aim of this work, therefore, was to study in some detail the influence of the probe energy on the shape of impurity profiles. For that purpose we measured the concentration distribution of a standard implantation profile as a function of the sputtering energy. Somewhat arbitrarily we used the range distribution of 20 keV boron ions in silicon as a reference. Silicon as a target material was supposed to have the advantage of reducing problems concerning faceting because it renders amorphous upon prolonged ion bombardment.[15]

2 EXPERIMENTAL

Our experimental arrangement has been described in detail elsewhere.[16] A telefocus ion gun was designed to produce 5 to 50 keV ion beams with constant current densities across about 60% of the total beam diameter (up to 400 μA/cm^2).[17] Argon ions were used throughout as sputtering projectiles. Mass analysis of the secondary ions was obtained by means of a quadrupole filter. The necessary low background in the mass spectra could only be achieved after suppression of sputtered neutrals and high energy secondary or backscattered primary ions by use of a plate capacitor in front of the mass filter.[16,18] Sputtering was done at constant beam current and at an angle of incidence of 20° with respect to the target normal. In our set-up this is the optimum target orientation with respect to a high ratio of peak to background intensity (better than 10^8).[16,18] Edge effects[12,18] were avoided by removing an area sufficiently larger in diameter ($>$ 4 mm) than the implanted region (2 mm). To get a high degree of ionization of the secondary ions[19] the partial pressure of oxygen was increased to the 10^{-5} torr level (total background pressure 2 x 10^{-6} torr).

Talysurf measurements were carried out for each specimen after sputtering to calibrate the depth scale, to study the surface roughness and to check the rectangular shape of the beam profile. Figure 1 shows

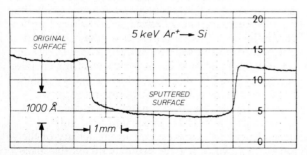

FIGURE 1   Typical sputtering relief of a silicon specimen after low energy bombardment, as measured with a talysurf.

that in the case of a proper beam profile depth calibration is quite easy although the surface flatness of the polished specimens (Wacker-Chemitronic) is sometimes not satisfactory for this purpose. In accordance with scanning electron microscope observations the original surface roughness is found not to change during sputtering.

3 RESULTS

Figure 2 gives an example of the range distribution of 20 keV boron ions in amorphous (predamaged) silicon as measured by SIMS. In this case sputtering was done with 10 keV argon ions. Each point corresponds to the

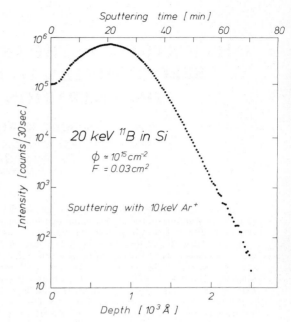

FIGURE 2   Range distribution of boron in amorphous silicon as measured by SIMS.

removal of a 20 Å layer. The concentration distribution could be measured through nearly five orders of magnitude. There is an indication for a beginning effect during the first minutes of sputtering which results in an enhanced ionization yield. This is obviously due to the fact that the surface layers contain a high amount of oxygen (silicon oxide). This effect, however, does not seem to prevent a determination of the most probable projected range R_p and the range straggling of the implanted ions (standard deviation ΔR_p). Thus R_p and ΔR_p were taken as a reference to determine the influence of the sputtering energy on the depth distribution.

Figure 3 shows that the measured projected range increases with increasing sputtering energy whereas the range straggling is scarcely influenced throughout the energy region investigated. Extrapolation to zero sputtering energy yields R_p = 690 Å. Moreover we find ΔR_p = 280 Å, assuming a Gaussian range distribution. These results agree very well with theoretical data (R_p = 650 Å and ΔR_p = 270 Å) obtained by Schiøtt[20] using the experimental value for the electronic stopping cross section of boron in silicon.[21]

The errors in R_p-determination found at non zero sputtering energies are believed to result from a change in sputtering yield at the beginning of layer removal until a saturation concentration of implanted

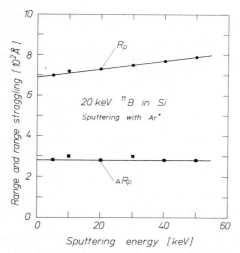

FIGURE 3   Measured range data versus sputtering energy.

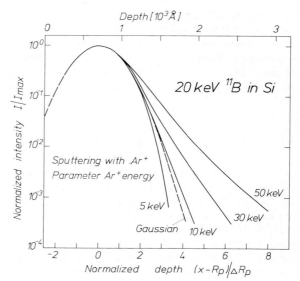

FIGURE 4   Effect of sputtering energy on the boron concentration in the long range tail.

argon is obtained. In the case of silicon as a target the equilibrium argon concentration C_∞ (atoms/atom) may become very large since C_∞ is simply given by the reciprocal of the equilibrium sputtering yield S_∞ of silicon,

$$C_\infty = 1/S_\infty \qquad (1)$$

and the maximum sputtering yield found between 15 and 20 keV is as small as 2.[22] A measurable increase of the initial sputtering yield S_0 during the removal of a layer corresponding to the projected range of the argon ions is thus not astonishing.

From the results shown in Figure 3 one can see that ΔR_p may be taken as a proper depth unit to compare the measured range distributions in the long range tail. To get rid of the effect due to changing sputtering yield we shifted the measured range distributions by such an amount (100 Å at the most, see Figure 4) that the most probable ranges of all profiles coincided. This is demonstrated in Figure 4. One can see that there is indeed an indication for a knock-on effect which produces a considerable shift in the tail of the distribution, especially for sputtering energies above about 10 keV. The relative magnitude of this effect will of course depend upon the absolute values of projected range and standard deviation and upon details of the mechanism of energy dissipation in the particular target material.

4   DISCUSSION

4.1   Effects Caused by Varying Sputtering Yield

One can estimate the possible errors occurring in the measured range data due to this effect. The true fluence dependence of layer removal is shown schematically in Figure 5 (thick curve). To calibrate the depth scale after sputtering a layer of thickness t by the fluence ϕ_t, however, one assumes a constant removal rate with a mean sputtering yield S_m (thin line) which gives an apparent depth x^* instead of the true depth x. The following relations are obvious for $x \geqslant R_{Ar}$,

$$x = x^* (1 + \delta R/t) - \delta R \qquad (2)$$

and

$$\Delta x = \Delta x^* (1 + \delta R/t) \qquad (3)$$

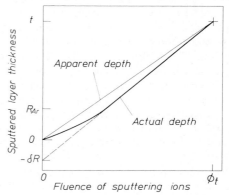

FIGURE 5   Actual and apparent fluence dependence of layer removal.

To calculate δR one has to measure the apparent and the actual depth of a characteristic range value. In this case one obviously will take the most probable projected ranges of boron, R_p^* and R_p, as a reference value because these data can be determined easily. For $R_p \geqslant R_{Ar}$ we have

$$\delta R = (R_p^* - R_p)/(1 - R_p^*/t) \qquad (4)$$

R_p^* follows directly from range distribution measurements at different sputtering energies (Figure 4) whereas R_p is found from an extrapolation of R_p^* to the limit of zero sputtering energy. From this we obtained R_p = 690 Å. The maximum value of R_p^* is found at a sputtering energy of 50 keV (R_p^* = 790 Å). Since t was 5000 Å in that case we have δR = 120 Å and $\delta R/t$ = 2.4%. At lower sputtering energies δR and $\delta R/t$ are even smaller. Thus we can conclude that the depth scale as calibrated by determining the total depth of the sputtered region must be corrected in case of a low sputtering yield by using Eq. (2). Differences in depth (Eq. (3)) usually do not need a correction since $\delta R/t$ is typically of the order of 1% whereas the inaccuracy in depth measurement is estimated to be 2 to 5% depending on the original flatness of the target surface. This conclusion is in agreement with the fact that the measured standard deviation ΔR_p is independent of the sputtering energy within the limits of experimental accuracy.

4.2 Profile Broadening Due to Recoil Processes

It was evident from Figure 4 that by increasing the sputtering energy an increasing amount of implanted boron atoms is knocked deeper into the target. A theory of this recoil phenomenon is not yet available although the present status of collision cascade theory is believed[23] to allow a calculation of this effect. Nelson[24] has carried out some calculations concerning recoil implantation from a thin evaporated layer. His results cannot be applied directly to our measurements. From Figures 2 and 3 in Ref. 24 one can conclude, however, that the main part of the "recoil tail" in Figure 4 of this work is due to repeated low energy recoil events, i.e. at recoil energies below about 0.1 to 0.2 of the bombarding energy. Direct knock-on of a boron atom by the sputtering projectile is much less likely. Since the range of 5 to 10 keV boron ions in silicon is 170 to 330 Å[20] one can see that several knock-on events are necessary to account for the large tail measured at a sputtering energy of 50 keV (depth difference about 1100 Å at $I/I_{max} = 10^{-3}$, see Figure 4).

4.3 Comparison With Other Range Measurements

The most thorough study of boron range in silicon has been carried out by Seidel[25] who measured the electrical profiles in the 30 to 300 keV energy region in random and channelling direction after annealing at 510 to 850 C. It has been pointed out earlier by the present authors[9] that the broadening observed in the electrical profiles[25] is likely to be caused by diffusion. In this study a noticeable influence due to either intrinsic[26] or radiation enhanced[27] diffusion is unlikely since the temperature of the specimens increased only slightly above RT during sputtering. These processes which occur during annealing after ion implantation may of course be studied easily with our high sensitivity method and will be investigated in the near future.

Very few SIMS measurements of boron range in silicon have been reported so far. They mainly concerned range distributions in the micron region[6–8] and were obtained at sputtering energies of 10 to 14 keV, so that effects like those reported here could not be detected. A comparison of the experimental data[7,8] with range calculations[20] shows satisfactory agreement. Our own measurements previously reported[9] yielded range values slightly too high due to a sputtering energy of 20 keV.

4.4 Optimum Sputtering Conditions

As mentioned at the beginning the sputtering conditions used in this work were optimized only with respect to a high ratio of peak to background intensity in the mass spectra. To reduce the effects reported above one should change the angle of incidence from $20°$ to the optimum value of about 70 to $80°$ and use high mass sputtering projectiles like xenon. Each alteration will increase the sputtering yield[13,22] roughly by a factor of three to four so that an order of magnitude reduction in saturation concentration of implanted sputtering ions and less change in sputtering yield at the beginning of bombardment will result. Moreover the range of the bombarding ions normal to the target surface will decrease so that saturation is obtained at lower fluences which also reduces δR (see Figure 5). Investigations to check these considerations are under way.

5 SUMMARY

It has been shown that SIMS can be used as a reliable means to investigate impurity concentration distributions in solids. Errors in profile measurement may be introduced by either changing removal rate due to target loading with sputtering ions or repeated recoil processes resulting from sputtering at elevated energy. These effects can be reduced to the usual limits of accuracy in measurement of the other experimental parameters by suitably adjusting the sputtering energy. Optimum conditions to minimize the above effects may be estimated from the well-known angular and mass dependence of the sputtering yield.

ACKNOWLEDGEMENTS

We thank Mr. Rothaug and Mr. Letzel of Rank Precision Industries GmbH, Nürnberg, for repeated talysurf measurements and Dr. H. Klingele, München, for carrying out the scanning electron microscope investigations.

This work was supported by grant of the German Federal Ministry of Education and Science within the frame of the Technology Program.

REFERENCES

1. J. A. Cairns, D. F. Holloway and R. S. Nelson, *Rad. Effects* **7**, 167 (1971).
2. J. J. Ph. Elich, H. E. Roosendaal, H. H. Kersten, D. Onderdelinden, J. Kistemaker and J. D. Elen, *Rad. Effects* **8**, 1 (1971).
3. H. Lutz and R. Sizmann, *Z. Naturforschung* **19a**, 1079 (1964).
4. R. Kelly, *Phys. stat. sol.* **30**, 37 (1968).
5. H. Heinen, H. Lutz and R. Sizmann, *Z. Naturforschung* **19a**, 1131 (1964).
6. V. M. Pistryak, A. K. Gnap, V. F. Kozlov, R. I. Garber, A. I. Fedorenko and Ya. M. Fogel, *Sov. Phys.-Sol. State* **12**, 1005 (1970).
7. D. P. Lecrosnier and G. P. Pelous, *European Conf. Ion Implantation* (P. Peregrinus Ltd., Stevenage, 1970), p. 102.
8. B. Blanchard, N. Hilleret and J. B. Quoirin, *Int. Meeting Chemical Analysis by Charged Particle Bombardment* (L.A.R.N. Namur, 1971), Preprints paper 29.
9. J. Maul, F. Schulz and K. Wittmaack, *Phys. Lett.* **41A**, 177 (1972).
10. A. E. Barrington, R. F. K. Kerzog and W. P. Poschenrieder, *J. Vac. Sci. Techn.* **3**, 239 (1966).
11. H. W. Werner, *Developm. Appl. Spectroscopy* **7A**, 239 (1969).
12. M. Croset, *Revue Techn. Thompson-CSF* **3**, 19 (1971).
13. P. Sigmund, *Phys. Rev.* **184**, 383 (1969).
14. W. L. Rautenbach, *Nucl. Instr. Meth.* **12**, 169 (1961).
15. D. J. Mazey, R. S. Nelson and R. S. Barnes, *Phil. Mag.* **17**, 1145 (1968).
16. K. Wittmaack, J. Maul and F. Schulz, *Int. J. Mass Spectrom. Ion Phys.* **11**, 23 (1973).
17. K. Wittmaack and F. Schulz, in *Proc. V. Int. Conf. Electron and Ion Beam Science and Technology,* edited by R. Bakish (The Electro-chemical Society, Inc., Princeton, N.Y., 1972), p. 181.
18. J. Maul, F. Schulz and K. Wittmaack, *Frühjahrstagung der Deutschen Physikalischen Gesellschaft, Regensburg 1972,* Verhandl. DPG (VI) **7**, 444 (1972), GSF-Bericht P 37.
19. J.-F. Hennequin, *J. de Physique* **29**, 655 (1968).
20. H. Schiøtt, private communication.
21. F. H. Eisen, unpublished work, see F. H. Eisen, B. Welch, J. E. Westmoreland and J. W. Mayer in *Atomic Collision Phenomena in Solids,* edited by D. W. Palmer, M. W. Thompson and P. D. Townsend (North Holland Publ. Co., Amsterdam, 1970), p. 111.
22. H. Sommerfeldt, E. S. Mashkova and V. A. Molchanov, *Phys. Letters* **38A**, 237 (1972).
23. P. Sigmund, private communication.
24. R. S. Nelson, *Rad. Effects* **2**, 47 (1969).
25. T. E. Seidel in *Proc. II. Int. Conf. Ion Implantation in Semiconductors,* edited by I. Ruge and J. Graul (Springer, Berlin, 1971), p. 47.
26. R. N. Ghoshtagore, *Phys. Rev.* **B3**, 389 (1971).
27. D. G. Nelson, J. F. Gibbons and W. S. Johnson, *Appl. Phys. Letters* **15**, 246 (1969).

NOTE ADDED IN PROOF

The technique studied in this work has recently been applied to an investigation of the energy dependence (10 to 250 keV) and the annealing behaviour (650 to 1000°C) of boron range distributions in amorphous silicon.[28] It was found that deviations from the Gaussian form occur at boron energies above about 40 keV, the relative deviations reaching a maximum around 100 keV.[29] Details of the measured profile shape are in good agreement with recent profile calculations.[30] A comparison of measured with calculated range data allowed a semiempirical determination of the electronic stopping cross section of boron in silicon.[28,29]

REFERENCES

28. K. Wittmaack, J. Maul and F. Schulz, in *Proc. Int. Conf. Ion Implantation in Semiconductors and Other Materials,* in press.
29. K. Wittmaack, F. Schulz and J. Maul, submitted to *Phys. Lett.*
30. K. B. Winterbon, private communication.

DEPTH DISTRIBUTIONS OF IMPLANTED ATOMS UNDER THE INFLUENCE OF SURFACE EROSION AND DIFFUSION

J. P. BIERSACK

*New York University, Physics Dept., New York, N.Y. 10003 and
Hahn-Meitner Institute, Nuclear Chemistry Division, Berlin-39, Germany.* [†]

Shallow range distributions are altered significantly by diffusion and surface erosion. It will be shown here that the simultaneous processes of implantation, diffusion, and surface erosion can be described by a second order differential equation which is mathematically tractable and yields general results in terms of analytical functions. The resulting depth distributions depend on *one* parameter only which is a combination of the mean particle range, the diffusion coefficient, and the velocity of surface erosion (e.g. sputtering). The theory is applied to practical cases, e.g. implantation of plasma particles in the wall of fusion reactors, or solar wind ions in the surface of meteorite materials.

1 INTRODUCTION

Recent improvements in the technique of measuring depth profiles have led to a resolution of a few atomic layers near the surface of a solid. With regard to future measurements of very shallow implantation profiles and their interpretation, the influence of diffusion and surface shift (due to erosion, corrosion or sputtering) shall be considered. There are also practical cases, where —within typical times—the implanted atoms migrate (by diffusion) further than the initial range, and/or the surface of the solid shifts (by erosion) further than the particle range. In these cases stationary distributions are of interest. Examples are:

i) solar wind ions (H, D, He) implanted into moon or meteorite materials,

ii) plasma particles (D, T, He) incident into the inner wall of fusion reactors, and

iii) atoms of the coolant (He, Li) knocked into fusion reactor materials by fast neutrons.[8]

2 GENERAL TREATMENT

The deposition of atoms according to a given range distribution $Q(x)$, the migration of these atoms with a diffusion coefficient D, and finally the shift of the momentary profile relative to the surface with the velocity v (speed of surface erosion), lead to the differential equation

$$\frac{\partial c}{\partial t} = D \frac{\partial^2 c}{\partial x^2} + v \frac{\partial c}{\partial x} + Q(x), \qquad (1)$$

where $c(x, t)$ denotes the concentration of implanted atoms as a function of time t and depth x.

For a general mathematical treatment the source term shall be written in the form

$$Q = q \cdot \frac{1}{r} f\left(\frac{x}{r}\right),$$

where the first factor, q, describes the total number of particles implanted per cm^2 and sec, i.e. particle current density times a factor $\alpha \leqslant 1$ accounting for backscattering which occurs especially at low energies and non-normal incidence. The second factor $(1/r)f(x/r)$ is the probability distribution of ranges with the mean range r; $f(\xi)$ is supposed to be any arbitrary function normalized to its zero and first order moments,

$$\int_0^\infty f(\xi)\, d\xi = \int_0^\infty \xi f(\xi)\, d\xi = 1,$$

for example $e^{-\xi}$ or $\delta(\xi - 1)$, describing the shape of the implantation profile. The existence of four independent parameters D, v, q, r in the differential equation, does not cause difficulties in a general treatment. Introducing the universal (dimensionless) variables

$$C = \frac{v}{q} c, \qquad \xi = \frac{x}{r}, \qquad \tau = \frac{v}{r} t \qquad (2)$$

and connecting the remaining three parameters to a single one,

$$p = \frac{D}{rv}, \qquad (3)$$

one obtains the universal equation

$$\dot{C} = pC'' + C' + f(\xi), \qquad (4)$$

---

† Present address.

29

where dot and prime denote the derivates $\partial/\partial\tau$ and $\partial/\partial\xi$ respectively.

The aim of the present treatment is the calculation of stationary profiles which in the above examples are most likely observed, and which yield the maximum obtainable concentration. Stationary profiles may be expected, if at the time of observation vt or/and $\sqrt{Dt}$ is large compared to the mean range.

In the stationary case the number of particles deposited per unit time is balanced by an equal number of particles leaving the surface by diffusion and/or erosion. The solution of Eq. (4) for stationary profiles, $\dot{C} = 0$, are obtained under the boundary conditions of zero concentration at the surface and at infinite depth,

$$C(0) = C(\infty) = 0. \tag{5}$$

These particular solutions are

$$C(\xi) = -\frac{e^{-\xi/p}}{p} \cdot \int_0^\xi e^{x/p} F(x)\,dx \tag{6}$$

with

$$F(x) = \int_\infty^x f(z)\,dz. \tag{7}$$

Any other boundary conditions may be complied with by adding the general homogeneous solution $C_H = a \cdot \exp(-\xi/p) + b$ to the above particular solution (6), where a and b are adjustable constants. From these solutions some general properties of the depth profiles can be derived:

i) The initial slope of the distribution (6) is

$$C'(0) = \frac{1}{p} \quad \text{or} \quad \left.\frac{dc}{dx}\right|_0 = \frac{q}{D} \tag{8a}$$

and the diffusional release of implanted particles through the surface equals q; there is no release due to erosion directly, as $c(0) = 0$ (diffusional release keeps $c(0)$ adjusted to zero at all times—according to boundary condition (5)).

For implanted material which is not gaseous or volatile at the temperature of the target, the appropriate boundary conditions would be $C(0) = 1$ and $C(\infty) = 0$, and the solution consists of Eq. (6) plus the term $\exp(-\xi/p)$ from the homogeneous solution ($a = 1$, $b = 0$). In this case one obtains

$$C'(0) = 0 \quad \text{or} \quad \left.\frac{dc}{dx}\right|_0 = 0 \tag{8b}$$

which means that no diffusional release occurs, and the total particle loss q (per sec and unit surface area) is due to slicing off surface layers at speed v.

ii) The maximum concentration which can locally occur, is

$$C_{\max} \leqslant 1 \quad \text{or} \quad c_{\max} \leqslant \frac{q}{v} \tag{9}$$

(under any boundary condition). A more restricting limit is obtained if the distribution is mainly controlled by diffusion, i.e. for $p > 1$; then

$$C_{\max} < \frac{1}{p} \quad \text{or} \quad c_{\max} < \frac{qr}{D} \tag{10}$$

is valid (under boundary condition(5)).

iii) The total number of particles implanted per unit surface area,

$$N = \int_0^\infty c(x)\,dx = \frac{qr}{v}\int_0^\infty C(\xi)\,d\xi,$$

can be calculated using C of Eq. (6). The form of the expression (6) suggests partial integration which yields

$$\int_0^\infty C\,d\xi = 0 - \int_0^\infty F(\xi)\,d\xi.$$

Applying partial integration a second time, one obtains

$$\int_0^\infty C\,d\xi = 0 - 0 + \int_0^\infty \xi \cdot f(\xi)\,d\xi \equiv 1$$

by virtue of the normalization requirement (on the first moment) of $f(\xi)$. The result

$$N = \frac{qr}{v} \tag{11a}$$

is surprising insofar as it does not depend on the diffusion coefficient D nor on the shape of the range distribution $f(\xi)$. Loosely speaking, the influence of D or f is such that any broadening of the profile is exactly compensated for by an appropriate reduction in height.

Dealing with the other important set of boundary conditions $C(0) = 1$, $C(\infty) = 0$ (non-volatile implant), the solution $C(\xi)$ contains the additional term $+\exp(-\xi/p)$ which upon integration yields the value p. Hence,

$$\int_0^\infty C\,d\xi = 1 + p,$$

and the number of implanted particles (per unit surface area) becomes

$$N = \frac{qr}{v}(1 + p). \tag{11b}$$

This yields larger values than Eq. (11a) due to the lack of diffusional particle losses through the surface, and correspondingly higher concentrations near the surface.[†]

3 APPLICATIONS OF THE THEORY

For illustration, the theoretical result (6) shall be applied to some special cases.

3.1 Normal Incidence of Light Particles in the MeV Range

The particles are mainly slowed down by electronic stopping power, range straggling is small, and hence the range distribution may be approximated by a delta function:

$$f(\xi) = \delta(\xi - 1). \tag{12}$$

By virtue of Eqs. (7) and (6) one obtains

$$F(\xi) = \begin{cases} -1 & \text{for} \quad \xi < 1 \\ 0 & \text{for} \quad \xi > 1 \end{cases} \tag{13}$$

and

$$C(\xi) = \begin{cases} 1 - e^{-\xi/p} & \\ (e^{1/p} - 1)\, e^{-\xi/p} & \end{cases} \text{for} \quad \begin{matrix} \xi < 1 \\ \xi > 1 \end{matrix} \tag{14}$$

which is depicted in Fig. 1. For values of the parameter $p = (D/rv) \ll 1$ the profiles are strongly influenced by surface erosion, for $p \gg 1$ they are governed mainly by diffusion.

The present case would apply to irradiation experiments with high target temperatures, where light atoms undergo rapid diffusion. Examples are the blistering experiments of Kaminsky and Das,[2] where D^+, He^+ are implanted into Niobium at 700 and 900°C. The observed precipitation of gas bubbles at depth $x = r$ (= projected range) corresponds to the position of the maximum gas concentration in Figure 1. This is a peculiarity of the delta function source term; any other range distribution would cause a shift in the position of the maximum, cf. Fig. 2, 3, 4. However, the theory predicts—according to Eq. (10)—that no blisters may

---

† During the preparation of the manuscript the author received from M. T. Robinson a reprint entitled *Deduction of Ion Ranges in Solids from Collection Experiments,*[1] and an extract from the M.S. thesis of F. R. O'Donnell (University of Tennessee, Knoxville, 1965). These papers contained a similar Ansatz as Eq. (1), and a derivation of Eq. (11b). Depth distributions, however, were not obtained for the general case $D \neq 0$ and $v \neq 0$. To the author's opinion, the application of Eq. (11b) to rare gases implanted into solids[1] was not justified. However, for $p \ll 1$ the boundary conditions become irrelevant, i.e. the results N or $c(x > 0)$ for different boundary conditions become indistinguishable (cf. Appendix).

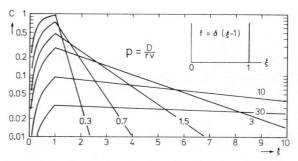

FIGURE 1   Equilibrium depth profiles in universal coordinates. Range distribution (insert) corresponds to normal incidence of particles at high energies. The curves are labeled with the value of the parameter p which is composed of mean projected range r, diffusion coefficient D, and velocity of surface erosion v.

occur, if qr/D is smaller than a gas concentration c_0 critical for precipitation. Hydrogen isotopes in Niobium at elevated temperatures ($\gtrsim 700°C$) reach their stationary concentrations within a fraction of a second, due to very fast diffusion (except for diffusional tails at $x \gg r$ which build up more slowly and require times of the order x^2/D).

3.2 Isotropic Incidence of Light Particles in the MeV Range

Assuming that the incident particles proceed a given distance along their (random) initial directions, one obtains a boxlike depth distribution as source term, as shown in the insert in Figure 2,

$$f(\xi) = \begin{cases} 1/2 & \text{for} \quad \xi \leq 2 \\ 0 & \quad\;\; \xi > 2 \end{cases} \tag{15}$$

Application of Eqs. (7) and (6) yields

$$F(\xi) = \begin{cases} -1 + \xi/2 & \text{for} \quad \xi \leq 2 \\ 0 & \quad\;\; \xi > 2 \end{cases} \tag{16}$$

and

$$C(\xi) = \begin{cases} 1 - \xi/2 - e^{-\xi/p} + \dfrac{p}{2}(1 - e^{-\xi/p}) \\[2ex] \left(\dfrac{p}{2} e^{2/p} - \dfrac{p}{2} - 1 \right) e^{-\xi/p} \end{cases}$$

$$\text{for} \quad \begin{matrix} \xi \leq 2 \\ \xi > 2 \end{matrix} \tag{17}$$

These final distributions are shown in Figure 2, where they can be compared with Figure 1. The maximum has been lowered, and shifted either to the left ($p < 1$) or to the right ($p > 1$).

This is a more hypothetical case; it may give a first

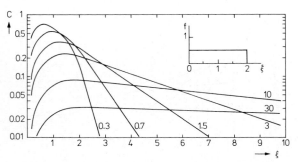

FIGURE 2   Universal stationary depth profiles of implanted particles for isotropic incidence at high energies (range distribution in the insert). The curves are labeled with the parameter $p = D/rv$.

rough estimate on the implantation of highly energetic He neutrals into the inner wall of a fusion reactor. According to an estimate of Kaminsky,[3] there might be a considerable fraction of He which temporarily picks up electrons from the plasma, and escapes from the magnetic confinement at energies ranging from about 100 keV to 1 MeV.

The time necessary to reach the maximum possible concentration, i.e. for complete build-up of a stationary profile, will be in the order of a day for 1 MeV He in the fusion reactor wall according to surface sputtering of about 1 mm/year; diffusion is very slow.[4]

When applying the generalized results, Eq. (17) or Figure 2, to specific cases, one should keep in mind that in the conversion formulae (2) and (3) the quantity r denotes the mean projected range along the x axis, i.e. half of the individual particle range in the case of isotropic incidence.

3.3 Normal Incidence of Light Particles in the keV Range

Trajectories of low energetic light particles slowed down in a heavy stopping material, exhibit large side straggling, a high ratio of total path length s to the mean projected range $\langle x \rangle = r$ (e.g. $s/r \approx 50$ for 1 keV T in Nb[5]), and a range straggling $\sqrt{\langle \Delta x^2 \rangle}$ about as large as the projected range itself.[5] In this case, the distribution of projected ranges may be approximated by an exponential function $(1/r) \cdot \exp(-x/r)$ which indeed yields $\sqrt{\langle \Delta x^2 \rangle} = \langle x \rangle = r$, and which has been used by other authors earlier.

In our universal notation this yields

$$f(\xi) = e^{-\xi}, \quad F(\xi) = -e^{-\xi}, \tag{18}$$

and

$$C(\xi) = \frac{e^{-\xi} - e^{-\xi/p}}{1 - p} \tag{19}$$

which is plotted in Figure 3.

This case applies for example to solar wind implantation, or to simulation experiments for the first wall of fusion reactors.

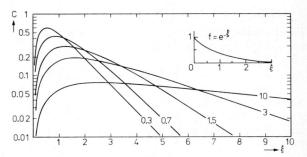

FIGURE 3   Universal stationary depth profiles for normal incidence of keV light particles (range distribution in the insert). The curves are labeled with the parameter $p = D/rv$.

3.4 Isotropic Incidence of Light Particles in the keV Range

Considering individual range distributions to be exponential as in Section 3.3, but along random directions instead of normal to the surface, one obtains the depth distribution for particle deposition,

$$f(\xi) = \tfrac{1}{2} E_1(\xi/2), \tag{20}$$

by applying the same principle as in Section 3.2. (Note that again for isotropic incidence the mean projected range is reduced by a factor $1/2$). E_n denotes the exponential integral functions of nth grade; they are interrelated by the useful formula $E'_n = -E_{n-1}$ ($n \geqslant 1$). In particular E_1, i.e. our source term, is a function steeply peaked at the surface, and hence extremely sensitive to slicing of surface layers, or to diffusional particle losses through the surface. These effects indeed remove the pole at $\xi = 0$ completely, and yield a stationary depth profile which does not resemble the source term at all:

$$F(\xi) = -E_2(\xi/2), \tag{21}$$

and hence

$$C(\xi) = E_2(\xi/2) + \frac{p}{2} E_1(\xi/2)$$

$$- \frac{p}{2} e^{-\xi/p} E_1(\xi/2 - \xi/p). \tag{22}$$

For negative arguments, i.e. for $p < 2$ in the last term, the E_1 function becomes undefined; then $E_1(-x) = -Ei(x)$ shall be used in Eq. (22). Special cases are $p = 2$ with the solution

$$C = E_2(\xi/2) + E_1(\xi/2) + e^{-\xi/2} \cdot \log (\xi/2), \qquad (23)$$

and $p = 0$ with the solution $C = E_2(\xi/2)$.

The family of solutions is depicted in Figure 4, in this case giving an example for finite material thickness. The boundary conditions $c(0) = c(20r) = 0$ are complied with by adding the appropriate homogeneous solution $a \cdot \exp (-\xi/p) + b$ to Eq. (22).

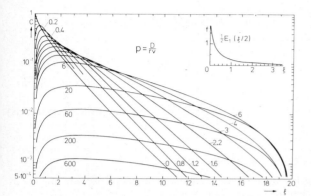

FIGURE 4   Universal stationary depth profiles for isotropic incidence of keV light particles (range distribution in the insert). Here also the influence of a different boundary condition is shown: finite thickness, $d = 20 \, r$. For infinite material thickness —as in Figures 1, 2, 3—the curves would continue as straight lines beyond $\xi \approx 5$.

From Figure 4 it is seen that with increasing values of $p = (D/rv)$ the distribution washes out, and the position of the maximum as well as the mean depth moves from the surface to a certain depth, as in Figures 2 and 3. For $p > 20$ the shape of profiles does not change any more, and actually cannot be distinguished from the pure diffusional profile (cf. Section 4, Eq. (33) with $A = 0$).

Applications may be found in the bombardment of the inner wall of fusion reactors by neutral particles escaping from the plasma. Typical data would be: Thermal energy of D, T, He about 10 keV, mean ranges in Nb of several 100 Å, surface erosion due to sputtering some 3.10^{-9} cm/s, diffusivities of 10^{-4} cm²/s for Deuterium,[6] 10^{-15} cm²/s for Helium[4] in Nb at 800°C. Thus we obtain $p \approx 10^{10}$ for Deuterium which gives rise to a very flat concentration profile ($C_{max} < 1/p$, $c_{max} < qr/D$, Eq. (10), at $x = 4r$) sitting on a background of dissolved Deuterium. The situation for Tritium would be similar. In view of these results, the formation of bubbles or blisters due to critical D or T concentrations seems very unlikely. For He, however, we obtain $p \approx 0.3$ which yields a distribution with a high peak directly under the surface. Figure 4 indicates a concentration maximum $C_{max} \approx 0.5$, $c_{max} \approx 0.5 \, q/v$ at $\xi \approx 0.4$,

i.e. at a depth of about 40 Å ($\xi = 1$ corresponds to the mean range of He, $r \approx 100$ Å). If there occurs a reduction of D due to trapping of He into radiation induced defects, the correspondingly smaller value of p will even increase the concentration and move it still closer to surface, according to Figure 4. This may be a serious condition for surface erosion (peeling, blistering). In addition, there are He$^+$ ions leaking out of the plasma fringes, with mean energies of only 1 keV, and mean ranges of about 15 Å in Niobium.

The case of this subsection, as described by the source term (21) and the solution (22), may also be applied—for other reasons—to the implantation of fast neutron knock-ons from the coolant (He, Li) into structural material of fusion reactors,[8,9] or to solar wind implantation during long term exposures of meteorites or lunar surface particles. This will be described in a forthcoming paper.

4 INFLUENCE OF TRAPPING

The diffusion coefficient in the region of implantation may be lowered considerably by the influence of radiation damage. Especially at lower irradiation temperatures, vacancies and their clusters have little chance to anneal and may become stabilized by He atoms which in turn are rendered immobile. Above certain temperatures, however, thermal release from such traps should also be considered.

As a first estimate we may assume a model with only one kind of trap homogeneously distributed in the region of interest. Then the chance per unit time that a mobile atom will be absorbed into a trap becomes[7]

$$A \approx 4\pi RDn \qquad (24)$$

where R denotes the effective radius of the traps, and n their concentration (traps per cm³). The probability per unit time for thermal release of a trapped atom is

$$B \approx \nu \, e^{-\Delta H/kT}, \qquad (25)$$

where ν is the frequency of the atom in the trap, and ΔH the change of enthalpy upon release.

No replacement collisions can occur when light particles are stopped in a heavy material. The particles come to rest in an interstitial position, where they are mobile. Under this assumption the basic differential equation—replacing Eq. (1)—will be in the stationary case

$$\dot{c}_f = 0 = Dc''_f + vc'_f - Ac_f + Bc_t + Q$$
$$\dot{c}_t = 0 = vc'_t + Ac_f - Bc_t \qquad (26)$$

where the subscripts f and t denote "free" (mobile)

atoms and "trapped" (immobile) atoms resp. ($\dot{} = d/dt$, $' = d/dx$). Together with the relation

$$c_f + c_t = c, \qquad (27)$$

c_f and c_t can be eliminated, and the last equations yield

$$vDc''' + (v^2 - BD)c'' + (A + B)vc'$$
$$- (A + B)Q + vQ' = 0, \qquad (28)$$

or upon integrating Eq. (28) once,

$$vDc'' + (v^2 - BD)c' + (A + B)vc$$
$$- (A + B) \int_{\infty}^{x} Q \, dx + vQ = 0. \qquad (29)$$

(The constant of integration has been chosen such that c, c', c'', and Q may vanish at infinity.) The universal treatment of Eq. (28) or (29) is possible along the same lines as before, but requires more parameters, and hence is not very illustrative. In the present paper, only limiting cases shall briefly be demonstrated:

i) Little influence of trapping, $A, B \to 0$; Eq. (29) immediately yields $Dc'' + vc' + Q = 0$. Here, of course, the original Eq. (1) has been retrieved.

ii) Little influence of erosion, $v \to 0$; Eq. (28) yields

$$\frac{D}{1 + A/B} c'' + Q = 0. \qquad (30)$$

In this case the problem can be treated as if the diffusion coefficient were reduced, i.e. by simply replacing

$$D_{\text{eff}} = \frac{D}{1 + A/B} \, † \qquad (31)$$

for D in all expressions (e.g. in the above solutions for diffusion controlled profiles, $p \gg 1$).

iii) Little influence of diffusion, $D \to 0$; trivially no influence of trapping should occur either. Indeed, summing the two Eqs. (26) with $D = 0$, yields $vc' + Q = 0$, and hence the former results with $p \to 0$ remain valid (erosion controlled profiles).

These considerations suggest the following approximate treatment: The parameter p is evaluated using D_{eff} of Eq. (31). Then, for $p \ll 1$, one obtains

$$C \approx -F(\xi) \quad \text{or} \quad c \approx -\frac{q}{v} F\left(\frac{x}{r}\right), \qquad (32)$$

i.e. the limiting case $p \to 0$ of Section 2 can be used as a first approximation.

For $p \gg 1$ erosion may entirely be neglected, and the solution of Eq. (30)

$$c \approx -\frac{qr}{D}\left(1 + \frac{A}{B}\right)\left[\int_0^{x/r} F(\xi) \, d\xi + ax + b\right] \qquad (33)$$

will be a useful approximation (a, b adjustable constants).

5 CONCLUSION

It has been shown that the problem of particle implantation with simultaneous diffusion and surface erosion is mathematically tractable, and that the number of independent parameters can be reduced to one (p, a combination of r, v, D). With trapping, however, only limiting cases find simple solutions (profiles predominantly controlled either by diffusion and trapping, or by surface erosion).

Applications to practical problems indicated for example that the danger of blistering and peeling in the vacuum chamber of a fusion reactor is serious due to He implantation ($p < 1$), whereas D or T profiles are extremely flat ($p \gg 1$) and add very little to the background of dissolved hydrogen isotopes.

ACKNOWLEDGEMENT

I am grateful to Dr. S. Roth and Dr. G. Basbas for many fruitful discussions. Mrs. S. Franke and especially Dr. S. Roth have been of invaluable help in preparing the graphs. I am very much obliged to Dr. M. T. Robinson and Dr. K. Schroeder for discussions and suggestions.

Appendices

A PRACTICAL SIGNIFICANCE OF LIMITING CASES

Figures 5 and 6 depict the transition of the general solution to the limiting cases $p \to 0$ and $p \to \infty$, where finally the influence of either diffusion or erosion becomes negligible.

The practical examples mentioned earlier (in Section 3) show that often profiles are strongly controlled either by diffusion ($p \gg 1$) or by erosion ($p \ll 1$). Occasions where both influences are of equal importance ($0.1 \lesssim p \lesssim 10$) may occur seldom. Thus, if one finds p to be very small or very large, the analysis may be simplified by neglecting the less important influence

---

† A similar expression is known to approximately describe the long time behaviour of gas release from an initially homogeneously filled sample.[10]

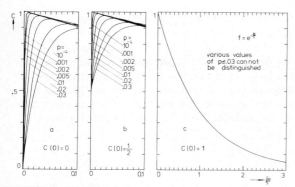

FIGURE 5   Vanishing influence of diffusion. For a study of the limiting case $p \to 0$, solutions are shown for small values of p under different boundary conditions. For $p \to 0$, always the same curve is approached independent of the left boundary condition; $C(0) = 0$ (a), $C(0) = 0.5$ (b), and $C(0) = 1$ (c). Scale expanded in (a) and (b) to emphasize difference.

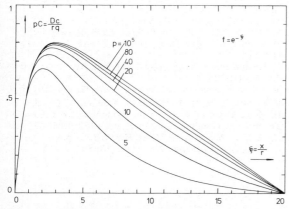

FIGURE 6   Vanishing influence of erosion. For a study of the limiting case $p \to \infty$, solutions are shown for large values of p (boundary conditions $c(0) = c(20) = 0$).

altogether. Proceeding along the same lines as in Section 4 one arrives at the results Eq. (32) for $D = 0$, and Eq. (33) for $v = 0$ (with $A = 0$ for no trapping conditions). These results are indeed simpler and more easily applied than the general solution, Eq. (6).

B   TIME TO REACH STEADY STATE

For erosion controlled profiles (diffusion neglected) the time to reach steady state will be of the order r/v (slightly dependent on the range distribution). In the case of pure diffusion (erosion neglected), equilibrium concentrations are approached after times of the order r^2/D in the region of implantation (regions more distant from the surface take correspondingly longer times). This consideration applies to volatile implants only

($c(0) = 0$); for other implants which are not released from the surface, the concentration builds up steadily, and no equilibrium state exists for $v = 0$ (for $v \to 0$ the general solution with $C(0) = 1$ yields $c(0) = q/v \to \infty$).

The time necessary to approach steady state concentration may be obtained for the example of plasma particles implanted in the first wall of fusion reactors:

i) $D \to Nb; r \approx 10^{-6}$ cm, $v \approx 3 \cdot 10^{-9}$ cm/sec, $D(800°C) \approx 10^{-4}$ cm^2/sec yields $p \approx 3 \cdot 10^{10}$ which indicates a purely diffusional profile. Hence, for a region of depth $10 \, r$, steady state may be expected after about $(10 \, r)^2/D \approx 10^{-6}$ sec.

ii) He $\to Nb; r \approx 10^{-6}$ cm, $v \approx 3 \cdot 10^{-9}$ cm/sec, $D(800°C) \approx 10^{-15}$ cm^2/sec or smaller due to radiation induced trapping. These data yield $p \lesssim 0.3$. Influence of erosion brings profile close to steady state after some $r/v \approx 300$ sec.

C   SOME REMARKS ON THE CHOICE OF BOUNDARY CONDITIONS

i) In Section 2 the general solution of Eq. (4) is given as one particular integral (6) plus the general homogeneous solution $a \exp(-\xi/p) + b$, a and b are constants which are adjustable to the requirements of two boundary conditions. For the rear part of "thick" targets the condition $C(\infty) = 0$ is generally assumed to be appropriate. For the front surface, then $C(0)$ can be chosen to equal any value s with $0 \leqslant s \leqslant 1$ (other values would be inconsistent with the steady state requirement). The physical meaning of s turns out to be the fraction of implanted material which is released directly by erosion ($1 - s$ is the fraction lost by diffusion through the surface, cf. Eqs. (8) and their discussions). $C(0) = 1$, for example, means 100% particle loss is due to erosion, none due to diffusion directly.

In the absence of diffusion, i.e. for $p = 0$, the differential equation is of first order. Therefore only one integration constant remains which reasonably will be chosen to comply with $C(\infty) = 0$, as in solution (32). It is quite interesting to note that in the limit $p \to 0$ the general solution yields a concentration profile which merges into the pure erosion profile, Eq. (32), *independent* of the choice of the boundary condition at $x = 0$. In fact, the limiting case $p \to 0$ is contained in the general solution, and the left boundary condition becomes irrelevant for $p \to 0$. Already for $p < 0.05$ the influence of diffusion becomes so small that only the immediate surface layers, but not the bulk gets affected, even in the most unfavourable case of $C(0) = 0$, cf. Figure 5.

ii) A matter of much controversy has been the influence of surface motion on the boundary condition at $x = 0$. Is the condition $c(0) = 0$ for gaseous implants still adequate when the surface moves? Beyond what speed v can we expect the surface to overrun particles which had no time for a diffusional jump, and what will be the final effect on the concentration profile? At first, one has to investigate whether a particle in the last lattice layer, i.e. at distance d from the surface, performs its random jump (to the outside or into the bulk) faster than the layer is removed by erosion. The mean dwell time τ within one lattice layer ($y - z$ plane) is connected to the diffusion constant by

$$D = \frac{d^2}{2\tau} \quad (d \text{ lattice constant}).^{\dagger}$$

The time to remove a lattice layer is

$$T = d/v.$$

In order to adjust to the free surface, $T > \tau$ must be required, or equivalently

$$p > \frac{d}{2r}^{\ddagger}$$

As $d/2r$ is a very small number (10^{-2} or less), our assumption of adjustment to the free surface is justified in all cases depicted in the graphs, Figures 1 through 4, and in our practical examples.

The adjustment to the free surface becomes incomplete ($p \approx d/2r$) or impossible ($p \ll d/2r$) only for extremely small values of p, where—according to paragraph (i)—the boundary condition $C(0)$ becomes irrelevant anyhow, and where profiles become practically indistinguishable from the case of pure erosion, cf. Figure 5.

---

† Assuming the diffusional jump distance equals the lattice constant.

‡ The same result can be obtained by means of the more rigorous transport theory.

REFERENCES

1. M. T. Robinson, *Appl. Phys. Letters* **1**, 49 (1962).
2. M. Kaminsky and S. K. Das, paper presented at International Conference on Ion-Surface Interaction, Garching, Germany, September 25–28, 1972, to be published in *Rad. Effects*, also: *Appl. Phys. Letters* **21**, 443 (1972) and references therein.
3. M. Kaminsky, private communication.
4. S. Blow, UKAEA, Report AERE-R 6845, Harwell 1971.
5. H. Schiøtt, *Mat. Fys. Medd. Dan. Vid. Selsk.* **35**, No. 9 (1966).
6. G. Schaumann, J. Volkl and G. Ahlefeld, *Phys. Stat. Sol.* **42**, 401 (1970).
7. M. v. Schmoluchowski, *Z. Phys. Chemie* **92**, 129 (1917).
8. F. K. Altenheim, H. Andresen, Ch. Donner, W. Lutze, H. Migge, Paper presented at the Symposium on Fusion Technology, Grenoble, 24-29 October 1972 (EUR 4938e).
9. H. Gaus, H. Migge, K. D. Mirus, *Rad. Eff.*, in press.
10. H. Gaus, *Z. Naturforsch.* **20a**, 1298 (1965).

THE SPUTTERING OF OXIDES
Part I: A Survey of the Experimental Results[†][‡]

ROGER KELLY and NGHI Q. LAM§

Institute for Materials Research, McMaster University, Hamilton, Ontario, Canada

The sputtering behavior of oxides is now fairly well understood in 12 cases. Three categories can be recognized, with some oxides belonging to more than one category. The eight oxides Al_2O_3, MgO, Nb_2O_5, SiO_2, Ta_2O_5, TiO_2, UO_2, and ZrO_2 have sputtering coefficients as expected from Sigmund's theory of collisional sputtering, at least provided the surface binding energy can be identified with the heat of atomization. The four oxides MoO_3, SnO_2, V_2O_5, and WO_3 show high and (except possibly with SnO_2) temperature-dependent sputtering coefficients such that the inferred surface binding energies are significantly less than the heats of atomization. For example, V_2O_5 has $S = 12.7$ atoms/ion for 10-keV Kr, so that E_b is apparently 1.3 eV/atom as compared with 5.7 eV/atom for atomization. This is taken as evidence for thermal sputtering. The five oxides MoO_3, Nb_2O_5, TiO_2, V_2O_5, and WO_3 show preferential oxygen sputtering, such that they first amorphize, then lose oxygen, and finally crystallize as the phases MoO_2, NbO, Ti_2O_3, V_2O_3, and $W_{18}O_{49}$. A comparison between corresponding oxides and metals is also possible with the information available. The main result is that the ratio S_{oxide}/S_{metal} is normally greater than unity (Al_2O_3, MgO and TiO_2 are exceptions), whereas the ratio (S_{oxide}/S_{metal}) (mole fraction of metal in the oxide) is normally less than unity (V_2O_5 is an exception).

1 INTRODUCTION

The sputtering of oxides has been traditionally either neglected or misunderstood. Part of the problem has probably been due to the difficulty of getting specimens, and, having obtained them, of bombarding them without charge build-up. Most work has therefore been carried out with only two oxides (Al_2O_3 and SiO_2 [1-9]), though MgO,[3,8] Ta_2O_5,[4] TiO_2,[8] and UO_2[10] have also been considered. Information on a further group of oxides is in principle available[11] though offers various difficulties (Section 3.1). More recently we have in our own work had considerable success using anodic oxide films, with the result that new sputtering information is available for the following ten oxides: Al_2O_3, TiO_2,[12] Nb_2O_5, Ta_2O_5, WO_3,[13] MoO_3, SiO_2, V_2O_5, ZrO_2,[14] and SnO_2.[14,15]

Less easy to explain is why the misunderstanding remains that oxides should resist sputtering in a significant manner and therefore act as protective barriers to metals. Such a protective action is certainly not predicted theoretically. According to Sigmund's[16] theory the most material-dependent quantity controlling S,

the sputtering coefficient, is E_b, the surface binding energy:

$$S \propto 1/E_b,$$

with the values of E_b being such that nearly all metals and oxides should have comparable values of S (Table I). Only with Al_2O_3 and MgO would a protective action be expected.

TABLE I
Comparison of surface binding energies
for oxides and metals

Oxide	$E_b(\text{oxide})/E_b(\text{metal})$[a]
Al_2O_3	1.9
MgO	3.4
MoO_3	0.8
Nb_2O_5	0.9
SiO_2	1.4
SnO_2	1.5
Ta_2O_5	0.9
TiO_2	1.4
UO_2	1.4
V_2O_5	1.1
WO_3	0.7
ZrO_2	1.2

[a] Deduced by identifying E_b(metal) with the heat of vaporization and E_b(oxide) with the heat of atomization (discussed in Section 3.3). Numerical values taken from Kubaschewski *et al.*[17]

† Supported by grants from the Defence Research Board, Ottawa, and the Geological Survey of Canada, Ottawa.

‡ Part II, which is in preparation, considers the problem of explaining thermal sputtering.

§ Present address: Materials Science Division, Argonne National Laboratory, Argonne, Illinois.

It will be the object of the present work to review the information on oxide sputtering as contained in Refs. 1–15. We will show that the behavior of oxides is in some cases "normal" in the sense of following Sigmund's theory of *collisional sputtering*. In other cases the behavior is "abnormal", in that the collisional sputtering is supplemented by *thermal sputtering* or by *oxygen sputtering*.

2 RESULTS WITH METALS

The only aspect of metal sputtering that need concern us here is the extent to which the sputtering is collisional

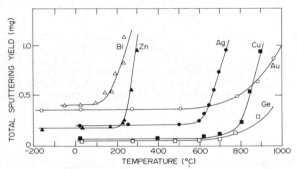

FIGURE 1   The sputtering yield (expressed as loss in weight for a dose of 2.9×10^{16} ions) for 45-keV Xe impact on various metals as a function of temperature. These experiments suggest that thermal sputtering is normally *unimportant* with a metal. Due to Nelson.[21]

or thermal. Thus, we seek information on how the overall sputtering coefficient can be divided as follows:

$$S = S_{\text{collisional}} + S_{\text{thermal}}$$

There are at least three instances in which S_{thermal} was determined or inferred. Thompson and Nelson[18] and Farmery and Thompson[19] found that a portion of the atoms sputtered from unheated Cu and Au targets had energies in the region 0 to 1.5 eV and took this as evidence for thermal sputtering. The conclusion was supported with a model in which thermal spikes were shown to be capable of evaporating atoms from the target. The results are summarized in Table II and suggest that thermal sputtering is normally *unimportant* with a metal.

In related but more direct work, Nelson[21] determined S for Ag, Au, Bi, Cu, Ge, and Zn as a function of temperature and found that within $\sim 250°C$ of the melting point there was a remarkable increase in S, such as would occur if thermal sputtering were setting in (Figure 1). The behavior of Ag was, in this respect, confirmed also by Almén and Bruce.[22] We again

conclude, however, that thermal sputtering is unimportant with a metal provided the metal is unheated.

Apparently only collisional sputtering need be considered with metals and we therefore expect, according to Sigmund's[16] theory,

$$S = S_{\text{collisional}} = \frac{\alpha M_2 k_\rho s_n(\epsilon)}{1.42 k_e E_b} \qquad (1)$$

Here M_1 and M_2 are the mass numbers of the ion and target, α is a quantity which depends only on M_2/M_1 (Figure 13 of Ref. 16), k_ρ is defined by $\rho = k_\rho R$ where R is in $\mu g/cm^2$ (tabulated by Winterbon[23]), $s_n(\epsilon)$ is the nuclear stopping cross-section in ϵ-ρ units (Table I of Ref. 16), and k_ϵ is defined by $\epsilon = k_\epsilon E$ where E is in keV (tabulated in Ref. 23). Finally, E_b is the surface binding energy in units of eV/atom, which Jackson[24] has shown to be similar to (though in fact slightly less than) the heat of vaporization.

The correlation between S and E_b will be one of the basic arguments used when oxide data are considered, so it is worthwhile considering to what extent it is borne out with metals. Support comes first of all from the periodicity in S observed by Almén and Bruce[22] for 45-keV Kr ions and by Rosenberg and Wehner[25] for various low-energy ions, such periodicity being mirrored also in $\Delta H_{\text{vaporization}}$. Secondly, it comes from the examples discussed by Sigmund[16] in which theory and experiment are shown to agree fairly well when E_b is assigned a value equal to $\Delta H_{\text{vaporization}}$. See, for example, Figure 2, in which it is shown that calculated and measured S's are numerically similar for 0.4-keV Xe incident on 30 different metals. Thirdly, there are the measurements of Thompson et al.[18,19] of the energy spectra of sputtered atoms together with the accompanying analysis which yields E_b directly.

3 RESULTS WITH OXIDES

3.1 Summary of Results

The most satisfactory way of working with oxides is probably that in which an anodized metal is used as the target, for such specimens are conductive by virtue of the extremely small oxide thicknesses and can be made as large as desired. Anodic films are formed on a metal by immersing the metal along with an inert counter-electrode in a non-solvent, oxygen-containing electrolyte and then applying a positive voltage (1 to 100 volts) to the metal. See, for example, Table III, which gives a summary of electrolytes and thickness calibrations. The resulting films allow sputtering coefficients to be determined using either the conven-

TABLE II
Experimental values of $S_{thermal}$ for unheated metal targets

Energy, ion, target	Total S (atoms/ion)	$S_{thermal}/S_{total}$	$S_{thermal}$ (atoms/ion)
43-keV Ar–Au	11[20]	0.04[18]	0.4
43-keV Xe–Au	38[20]	0.12[18]	5
47-keV Ar–Cu, ⟨121⟩	7[20]	0.5[19]	4
40-keV Ar–Cu, ⟨110⟩ or ⟨100⟩	7[20]	0.00 to 0.14[19]	0.0 to 1.0

TABLE III
Summary of anodizing procedures

Metal	Electrolyte	Thickness of oxide film ($\mu g/cm^2$)	Ref.
Al	5% ammonium pentaborate	0.451xV	27
Mo	glacial acetic acid; 0.02M $Na_2B_4O_7 \cdot 10H_2O$; 1M additional water	1.4 + 1.58xV	28
Nb	0.25% KF	1.4 + 1.41xV	13
Si	ethylene glycol; 4% water; 0.4% KNO_3	~0.1xV	29
Sn	1 liter ethylene glycol; 300 ml water; 330 g ammonium pentaborate	—	15
Ta	0.2% KF	1.32xV	30
Ti	1% KOH	~0.85xV	31
V	see Mo	2.0 + 1.14xV	32
W	water; 0.4M KNO_3; 0.04M HNO_3	3.0 + 1.28xV	13
Zr	3% ammonium pentaborate	~1.6xV	33, 69

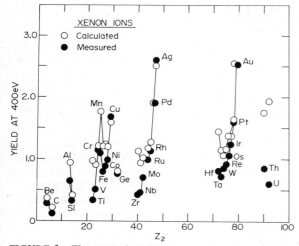

FIGURE 2 The sputtering yield (expressed as atoms/ion without a correction for secondary electrons) for 0.4-keV Xe impact on various metals as a function of the atomic number of the metal. The agreement between theory and experiment is here taken as suggesting that the relation $S \propto 1/E_b$ is correct. Due to Sigmund[16] and to Rosenberg and Wehner.[25]

tional weight-loss procedure or else one of the following approaches:

a) The anodic film is bombarded *in situ* with a homogeneous ion beam and the interference-color changes are noted (possible with MoO_3, Nb_2O_5, Ta_2O_5, TiO_2, V_2O_5, WO_3, and ZrO_2). The color changes are re-expressed as thickness changes by means of a three-way correlation between the color, the formation voltage, and the thickness-voltage relation. (This approach, which is one of the most powerful that exists for determining sputtering coefficients, was anticipated in work by Wehner et al.[4] and by Nielsen and Shepherd.[26])

b) The anodic film is stripped, layed on any anodized surface showing strong interference colors, and then bombarded until the colours show through and begin to alter (possible with Al_2O_3,[12] Nb_2O_5, and Ta_2O_5[13]).

Figure 3 gives sputtering coefficients for anodic Nb_2O_5 and WO_3 as obtained by Lam[13] using Kr ions with energies up to 30 keV. Likewise, Table IV gives, in the first three columns, values of S for 10-keV Kr incident on 12 different oxides. In certain instances the reported data were for ions and energies other than 10-keV Kr and these data have therefore been extrapolated to simulate 10-keV Kr using Sigmund's theory as in Eq. (3) (to be discussed). This involves evaluating ratios such as

(S for Kr ions at 10 keV)/(S for Ar ions at 3 keV)

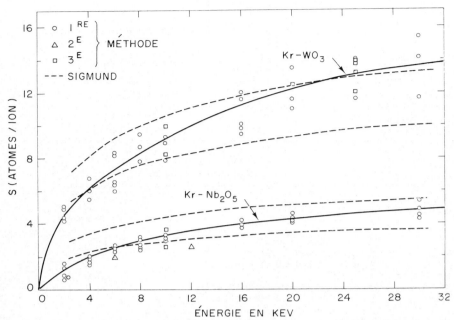

FIGURE 3   The sputtering yield for Kr impact on Nb_2O_5 and WO_3 as a function of the krypton energy. The three methods used in the experiments are, respectively, (a) of Section 3.1, (b) of Section 3.1, and weight loss. The dashed lines show the extent to which Eq. (3), thence the collisional theory of Sigmund,[16] fits the results for two arbitrary choices of E_b. Due to Lam and Kelly.[13]

and can be accomplished without knowledge of E_b. Excluded from Table IV are the Xe–SiO_2 results of Hines and Wallor,[1] which can be seen from Figure 4 of Davidse and Maissel[6] to be untypical, as well as the results of Panin and Tel'kovskii,[11] for which the units of S are unusual (volume units) and the bombardment currents were probably sufficient to induce vaporization (1000–2000 $\mu A/cm^2$). We also exclude the data of Wehner et al.[4] and of Jorgenson and Wehner[5] in view of the low energies used ($\leqslant$400 eV). The data of Akishin et al.[2] are used in spite of confusion in the labeling of the Figures.

3.2 Comparison of Oxides and Metals
Perhaps the first point to consider in connection with the results summarized in Table IV is a comparison of oxide and metal behavior. The reason is that oxide skins are sometimes assumed to have a significant protective action against sputtering. Columns 4 to 7 of Table IV give the necessary information and the following trends emerge:

a) In 9 cases S_{oxide}/S_{metal} is greater than unity. In effect, most oxides sputter more rapidly than the corresponding metals provided the *total yield* (atoms/ion) is considered.

b) Nevertheless, in 11 cases the quantity (S_{oxide}/S_{metal})x_{metal}, where x_{metal} is the mole fraction of metal in the oxide, is less than unity. A protective action does in fact exist, though (except for Al_2O_3, MgO, and TiO_2) is rather slight. The protective aspect of an Al_2O_3 film on Al also follows from the work of Andrews et al.,[35] where the ratio (S_{oxide}/S_{metal})x_{metal} for 2.5-keV Cs ions was determined directly to be 0.24.[†] Similarly, the value 0.20 has been reported for 1-keV Ar–Al_2O_3.[70]

3.3 Evidence for Thermal Sputtering
It will be apparent from Table IV that some oxides (MoO_3, SnO_2, V_2O_5, and WO_3) show very high sputtering coefficients, as if the collisional sputtering were being supplemented by another process. Let us therefore see to what extent the overall sputtering coefficient has components as follows:

$$S = S_{collisional} + S_{thermal}$$

The most direct information would be of the type shown in Figure 1, i.e. measurement of S as a function of temperature. Very little such work has been done, however, with results to date being confined to those

† Work which is approximately equivalent to a direct evaluation of (S_{oxide}/S_{metal})x_{metal} has been reported also for Ta[22] and Cu.[22,47] The ratio is in both cases less than unity.

TABLE IV
Comparison of sputtering behavior of oxides and metals for 10-keV Kr impact

Oxide	S for oxide (atoms/ion)	Ref.	S for metal (atoms/ion)	Ref.	$\dfrac{S_{\text{oxide}}}{S_{\text{metal}}}$	$\dfrac{S_{\text{oxide}}}{S_{\text{metal}}} \cdot x_{\text{metal}}$
Al_2O_3	1.6; 1.4 ± 0.2[a]	12; 6, 7, 8	3.2 ± 0.6[a]	34–37	0.5	0.2
MgO	1.8 ± 0.5[a]	3, 8	8.1[a]	3	0.2	0.1
MoO_3	9.6 ± 0.4	14	2.8 ± 1.0[a]	34, 38, 39, 40	3.4	0.9
Nb_2O_5	3.4 ± 0.5	13	1.6[a]; 2.0[b]	39; 41	1.9	0.5
SiO_2	4.2; 3.0 ± 1.5[a]	14; 2, 6, 9	2.1[a]	42	1.7	0.6
SnO_2	15.3 ± 1.8	14, 15	6.7; 6.4 ± 0.6[a]	22	2.3	0.8
Ta_2O_5	2.5 ± 0.5	13	1.6 ± 0.3[a]	34, 39, 43	1.6	0.4
TiO_2	1.9; 1.4[a]	12; 8	2.1 ± 0.8[a]	36, 39, 44	0.8	0.3
UO_2	3.8 ± 0.5[a]	10	2.4[a]	45	1.6	0.5
V_2O_5	12.7 ± 1.7	14	2.3 ± 0.4[c]	46	5.5	1.6
WO_3	9.2 ± 1.2	13	2.6 ± 1.0[a]	34, 38, 39, 40	3.6	0.9
ZrO_2	2.8 ± 0.1	14	2.3[a]	40	1.2	0.4

[a] Data other than for 10-keV Kr; extrapolated as if for 10-keV Kr using Eqs. (1) or (3).
[b] Data for 10-keV Nb.
[c] Taken as the mean of S for Mo, Ta, Ti, and W on the basis of the Hg-bombardment work of Wehner and Rosenberg.[46]

shown in Table V. We would interpret these results as showing that Al_2O_3 and Nb_2O_5 have, at least up to 200°C, a low, temperature-independent S due to undergoing mainly collisional sputtering. But the other three oxides have large, temperature-dependent S due to a mixture of collisional and thermal sputtering. (See note a at end.)

FIGURE 4   Transmission electron-diffraction patterns for oxygen-bombarded Nb_2O_5. (a): Anodic film of Nb_2O_5 which has been stripped and crystallized. (b): As before but after bombardment with 35-keV oxygen to a dose of 5×10^{16} ions/cm^2 (note the amorphous halos). (c): As before but with the bombardment extended to 1×10^{17} ions/cm^2 (note the pattern of NbO). These experiments suggest that *oxygen sputtering* occurs with Nb_2O_5. Due to Murti and Kelly.[53]

TABLE V
Effect of temperature on S for oxides (10-keV Kr impact)

Oxide	S at room temperature (atoms/ion)	S at 200°C (atoms/ion)	Ref.
Al_2O_3[b]	1.25[a]	1.1	7
Nb_2O_5	3.4 ± 0.5	3.9 ± 0.4	14
WO_3	9.2 ± 1.2	13.2 ± 1.0	14
MoO_3	9.6 ± 0.4	15.1 ± 1.0	14
V_2O_5	12.7 ± 1.7	17.2 ± 1.1	14

[a] Liquid-nitrogen temperature.
[b] 10-keV Cr impact.

A second type of information to consider involves comparing the apparent E_b and the thermodynamic E_b. Here the "apparent E_b" is the quantity obtained by substituting experimental S values into the analytical expression for S deduced by Sigmund[16] and retrieving E_b. A severe problem enters in that the theory applies only to unary targets, whereas oxides are binary, and we have therefore had to consider two approximations. The one is to use Eq. (1) with the *mean atomic weight* of the oxide substituted:

$$S = S(\overline{M}) \qquad (2)$$

while the other is to use Eq. (1) in a form which is weighted for the mole fractions of metal and oxygen,

$$S = x_{\text{metal}}S(M_{\text{metal}}) + x_{\text{oxygen}}S(M_{\text{oxygen}}) \qquad (3)$$

(Note that the approximation used by Lam[13] is numerically equivalent to Eq. (3).) The "thermodynamic E_b" will be taken as the heat of atomization, as in the following example, this being equivalent to the use of the heat of vaporization with metals:

$$Al_2O_3(s) = 2Al(g) + 3O(g)$$

$$\Delta H = 5 \times \Delta H_{\text{atomization}}$$

The necessary heats are tabulated by Kubaschewski *et al.*[17] The results are shown in Table VI and we note first of all that the eight oxides showing low S have apparent E_b's in rough agreement with the thermodynamic values; in effect, the sputtering must be mainly collisional. At the other extreme, the four oxides showing high S (MoO_3, SnO_2, V_2O_5, WO_3) have apparent E_b's which are much lower than the thermodynamic values. A further process, such as thermal sputtering is evidently occurring.

Table VI is also of interest in that the Ta_2O_5 and UO_2 entries suggest Eq. (3) to be preferable to Eq. (2).

3.4 Evidence for Oxygen Sputtering

We will now show that still a further type of sputtering occurs with oxides. This is sputtering in which there is a stoichiometry change due to the preferential loss of oxygen, and as a result of which the overall sputtering coefficient can be written:

$$S = S_{\text{collisional}} + S_{\text{oxygen}} + (S_{\text{thermal}})$$

Here S_{oxygen} should be regarded as the amount by which the oxygen removal per ion exceeds $x_{\text{oxygen}}S$.

The basic experiment for demonstrating oxygen sputtering consists of studying bombarded oxide surfaces by reflection electron diffraction. The most usual result in such studies has been that the oxide remained or became crystalline, there being 18 such examples: BeO, CaO, Cu_2O, Fe_3O_4, HfO_2, MgO, MoO_2, NbO, NiO, SnO_2, ThO_2, TiO, Ti_2O_3, UO_2, VO, V_2O_3, $W_{18}O_{49}$, and ZrO_2.[48][†] Nearly as common have been the instances in which the surface region of the oxide was amorphized, there being 10 examples: Al_2O_3, Bi_2O_3, Cr_2O_3, GeO_2, $9Mg_2SiO_4 \cdot Fe_2SiO_4$, SiO_2, Ta_2O_5, TeO_2, $ThSiO_4$, and $ZrSiO_4$.[48][‡] Quite unexpected have been recent results in which oxides were found first to amorphize in the usual way but then to lose oxygen. For example, Murti[53] has demonstrated that the impact of 35-keV oxygen ions on Nb_2O_5 leads first to an amorphous phase of undeterminable stoichiometry (complete by 1×10^{16} ions/cm^2) but then to crystalline NbO (complete by 1×10^{17} ions/cm^2) (Figure 4). Likewise, Bach[8] has described changes with bombarded single-crystalline TiO_2 (polycrystalline diffraction pattern; darkened color) which imply that he had a conversion to a lower oxide.

The known instances in which oxygen loss occurs now number five, though additional systems (CuO, Fe_2O_3, U_3O_8) are being studied and the list may grow:

$$MoO_3 \rightarrow MoO_x(\text{amorphous}) \rightarrow MoO_2 \text{ [54]}$$
$$Nb_2O_5 \rightarrow Nb_2O_x (\text{amorphous}) \rightarrow NbO \text{ [53]}$$
$$TiO_2 \rightarrow TiO_x (\text{amorphous}) \rightarrow Ti_2O_3 \text{ [52]}$$
$$V_2O_5 \rightarrow V_2O_x (\text{amorphous}) \rightarrow V_2O_3 \text{ [54]}$$
$$WO_3 \rightarrow WO_x (\text{amorphous}) \rightarrow W_{18}O_{49} \text{ [52]}$$

As far as the magnitude of S_{oxygen} is concerned there is the problem that the most direct method of evaluation is not yet feasible. Thus, estimating it from a knowledge of the thickness of the altered layer at a given dose fails due to the continuing lack of information on thickness.

---

† The examples of Cu_2O,[49,50] Fe_3O_4,[51] TiO,[52] and VO[53] are additional to those discussed in Ref. 48.

‡ The example of Ta_2O_5[53] is additional to those discussed in Ref. 48.

TABLE VI
Comparison of surface binding energies

Oxide	Apparent E_b (from Eq. (2)) (eV/atom)	Apparent E_b (from Eq. (3)) (eV/atom)	Thermodynamic E_b (eV/atom)
Al_2O_3	9.4	10.0	6.4
MgO	7.8	7.9	5.2
TiO_2	11.3	10.3	6.6
Ta_2O_5	13.3	10.0	7.2
ZrO_2	8.3	7.8	7.6
UO_2	10.6	8.1	7.2
Nb_2O_5	6.8	5.9	6.8
SiO_2	3.9	4.1	6.4
WO_3	3.3	2.6	6.3
MoO_3	2.4	2.0	5.6
V_2O_5	1.5	1.3	5.7
SnO_2	1.8	1.6	4.8

An alternative approach is the following: considering bombardments just with 30–40 keV ions, oxygen loss occurs with MoO_3 for both Kr and O ions[54] but with TiO_2 only for Kr ions.[52] This result is understandable provided S_{oxygen} for 30–40 keV O ions has the values

$$S_{oxygen} > 1 \quad \text{(with } MoO_3\text{)}$$
$$< 1 \quad \text{(with } TiO_2\text{)}$$

Generalizing somewhat, it would appear that the value of S_{oxygen} lies in the vicinity of unity for various oxides.

3.5 Other Effects

Various other categories of experiments involving the sputtering of oxides have been described. How these experiments relate to the theme presented here, namely that one can distinguish collisional, thermal, and oxygen sputtering, is not always clear but they will probably prove to be of future importance.

Firstly there are experiments in which S is determined as a function of dose in the region *below* 1×10^{16} ions/cm^2, much as was done by MacDonald and Haneman[55] for Ge. This is a difficult though not impossible experiment with oxides when proper use is made either of dissolution curves[32] or of interference-color changes.[53] It is anticipated that S may be unusually high at the doses where oxygen loss is occurring (as is implied by Figure 4 of Ref. 32), and, if so, that one could perhaps conclude the loss to be thermal rather than collisional. This work is, however, still incomplete (Giani[15]; Murti[53]; Arora[56]), so will not be discussed further.

Wehner *et al.*[57] have studied the effect of Hg bombardment of CuO and Fe_2O_3 at a temperature of about 300°C. The CuO was found to evolve to Cu_2O and Cu while the Fe_2O_3 evolved to Fe_3O_4, FeO, and Fe. This is a more extreme effect than the formation of intermediate oxides, as discussed in Section 3.4 for room-temperature experiments, and it will be important to decide the extent to which it is chemical (thermodynamic) in origin.

Information will undoubtedly also come from work involving oxygen bombardment of metals to yield *intermediate* oxides. For example, Murti[53] has shown that with Nb the electron-diffraction pattern evolves as in Figure 5, confirming the following stoichiometry change:

$$Nb + 35\text{-keV oxygen ions} \rightarrow NbO$$

Likewise, Parker[52] has demonstrated the following change for Ti:

$$Ti + 30\text{-keV oxygen ions} \rightarrow TiO$$

FIGURE 5   Transmission electron-diffraction patterns for oxygen-bombarded Nb. (a): Sputtered film of Nb (in part a normal b.c.c. pattern). (b): As before but after bombardment with 35-keV oxygen to a dose of 5×10^{16} ions/cm^2 (pattern of NbO). These experiments, when combined with those of Figure 4, suggest that oxygen sputtering may be analogous to congruent vaporization and therefore possibly thermal in origin. Due to Murti and Kelly.[53]

while other results include the conversion of Cu to Cu_2O,[49,50,58] FeS_2 to Fe_3O_4[51] (here assumed to be equivalent to the conversion of Fe to Fe_3O_4), Ni to NiO,[71] Ta to "TaO_z",[53] and V to VO.[53] These experiments, when combined with those in which higher oxides become lower oxides, could in principle be taken to suggest that oxygen sputtering is analogous to congruent vaporization. For example, the conversion of both Ti and TiO_2 to stoichiometries in the region of TiO to $TiO_{1.5}$ may parallel the result of Gilles *et al.*[59] that both Ti_2O and TiO_2 approach Ti_3O_5 when heated in vacuum. If this parallelism should prove to be correct, one would have an important insight into the origin of oxygen sputtering.

Less clearly relevant is work in which Cu was converted to Cu_2O at 350°C[60,72] and Fe to FeO at 600°C,[61] since the oxidation may have been in part thermal, thence thermodynamically controlled. Likewise, work which led to the conversion of Al to what was probably Al_2O_3,[62] Si to SiO_2,[63,64] and Ti to a high-resistivity oxide such as TiO_2[62] was done with rather high oxygen energies (50–100 keV). Under these conditions sputtering effects would be less important than doping effects and a high oxide would occur naturally.

The observation of spot patterns in sputter deposits is frequently taken as evidence for collision sequences, though with some uncertainty as to whether the sequence length involves a few atoms or a large number of atoms. The study of spot patterns with oxides would thus be of particular importance as it might serve to distinguish collisional from thermal effects in a most direct manner. In fact, most such experiments are bound to fail: of the 12 oxides considered here, three amorphize under impact (Al_2O_3, SiO_2, Ta_2O_5), five lose oxygen and take up a fine-grained polycrystalline habit (MoO_3, Nb_2O_5, TiO_2, V_2O_5, WO_3), and one tends to a polycrystalline habit (MgO[65]). This leaves only SnO_2, UO_2, and ZrO_2 for a spot-pattern experiment and it is significant that Nelson[66] has indeed confirmed that UO_2 single crystals yield a spot pattern. We would take this result as substantiating the view that UO_2 shows dominantly collisional sputtering.

A further effect has been described by Jones et al.[67] and by Hasseltine et al.[7] on the basis of experiments in which SiO_2 and Al_2O_3 were sputtered in the presence of a background pressure of oxygen. With SiO_2 S decreased even for very small amounts of oxygen and this was explained in terms of oxygen adsorbing on the surface and thus covering the Si ions. With Al_2O_3 S again decreased though only slightly, a result which was attributed to an increase in the binding energy for the Al ions. Whether or not the effects (as well as the explanations) are equivalent need not concern us here. More important is the fact that in both cases we are probably dealing more with a chemical effect than with an impact effect, and it is therefore likely that the experiments are not relevant in the present context.

4 CONCLUSIONS

The sputtering behavior of oxides turns out to be rather more complex than that of metals and a number of categories can be proposed. Firstly, there are those oxides which have S in accordance with Sigmund's[16] theory of collisional sputtering, with or without a change of stoichiometry due to oxygen sputtering:

$$\begin{cases} Al_2O_3, MgO, Nb_2O_5, SiO_2, Ta_2O_5, TiO_2, UO_2, ZrO_2 \\ S = S_{collisional} + (S_{oxygen}) \end{cases}$$

A second group has S values which are too large for collisional theory and which (in so far as the information is available) are temperature dependent. Both results suggest that thermal sputtering plays an important role, with or without a contribution from oxygen sputtering:

$$\begin{cases} MoO_3, SnO_2, V_2O_5, WO_3 \\ S = S_{collisional} + S_{thermal} + (S_{oxygen}) \end{cases}$$

A third group involves oxides which, whether they have S values appropriate to collisional or thermal sputtering, show highly characteristic behavior such that they change their stoichiometry during bombardment. That is, they can be said to show oxygen sputtering, with or without a thermal contribution:

$$\begin{cases} MoO_3, Nb_2O_5, TiO_2, V_2O_5, WO_3 \\ S = S_{collisional} + S_{oxygen} + (S_{thermal}) \end{cases}$$

As far as an explanation is concerned, the main problems lie with thermal and oxygen sputtering, as collisional sputtering is already (except for one feature) well understood. This feature is the behavior of binary targets, about which essentially nothing is known.

We will in Part II therefore outline a possible explanation for thermal sputtering using a modernized version of the thermal-spike approach of Thompson and Nelson.[18] In this version the damage-distribution theory of Winterbon, Sigmund, and Sanders[68] is used to specify the initial heat distribution; i.e. the oversimplification of assuming a point, line, or spherical distribution is avoided. The final result is that the four oxides showing large S (MoO_3, SnO_2, V_2O_5, WO_3) are found to have vapor pressures which meet the condition for thermal-spike induced vaporization, whereas the other eight oxides considered have vapor pressures which are orders of magnitude too small.

An explanation for oxygen sputtering must await more experimental work. For example, it is essential to decide whether the effect is collisional or thermal in origin, thence whether internal energies or free energies should be examined. (See note b at end.)

ACKNOWLEDGEMENT

The authors are grateful to D. K. Murti (McMaster University) for permission to reproduce his electron-diffraction patterns of bombarded Nb_2O_5 and Nb in advance of publication.

REFERENCES

1. R. L. Hines and R. Wallor, J. Appl. Phys. 32, 202 (1961).
2. A. I. Akishin, S. S. Vasil'ev and L. N. Isaev, Acad. Sci. USSR Bull. Phys. Sci. 26, 1379 (1962).
3. H. Schirrwitz, Beitrage zur Plasmaphysik 2, 188 (1962).
4. G. K. Wehner, C. Kenknight and D. L. Rosenberg, Planet. Space Sci. 11, 885 (1963).
5. G. V. Jorgenson and G. K. Wehner, J. Appl. Phys. 36, 2672 (1965).
6. P. D. Davidse and L. I. Maissel, J. Vac. Sci. Technol. 4, 33 (1967).

7. E. H. Hasseltine, F. C. Hurlbut, N. T. Olson and H. P. Smith, *J. Appl. Phys.* **38**, 4313 (1967).
8. H. Bach, *Nucl. Instr. Methods* **84**, 4 (1970).
9. H. Bach, *Z. Naturforsch.* **27a**, 333 (1972).
10. O. Gautsch, C. Mustacchi and H. Wahl, EURATOM Report EUR 2515.e (1965).
11. B. V. Panin and V. G. Tel'kovskii, *Sov. Phys.–Solid State* **8**, 1031 (1966).
12. R. Kelly, *Can. J. Phys.* **46**, 473 (1968).
13. N. Q. Lam (or L. Q. Nghi) and R. Kelly, *Can. J. Phys.* **48**, 137 (1970).
14. N. Q. Lam and R. Kelly, to be published.
15. E. Giani and R. Kelly, to be published.
16. P. Sigmund, *Phys. Rev.* **184**, 383 (1969).
17. O. Kubaschewski, E. L. Evans and C. B. Alcock, *Metallurgical Thermochemistry* (Pergamon Press, Oxford, 1967), p. 303.
18. M. W. Thompson and R. S. Nelson, *Phil. Mag.* **7**, 2015 (1962).
19. B. W. Farmery and M. W. Thompson, *Phil. Mag.* **18**, 415 (1968).
20. O. Almén and G. Bruce, *Trans. 8th Vac. Symp.* (Pergamon Press, Oxford, 1962), p. 245.
21. R. S. Nelson, *Phil. Mag.* **11**, 291 (1965).
22. O. Almén and G. Bruce, *Nucl. Instr. Methods* **11**, 257 (1961).
23. K. B. Winterbon, Chalk River Report AECL-3194 (1968).
24. D. P. Jackson, *Rad. Effects* (in press).
25. D. Rosenberg and G. K. Wehner, *J. Appl. Phys.* **33**, 1842 (1962).
26. R. E. Nielsen and W. B. Shepherd, *Rev. Sci. Instr.* **35**, 123 (1964).
27. P. Jespersgård and J. A. Davies, *Can. J. Phys.* **45**, 2983 (1967).
28. M. R. Arora and R. Kelly, *J. Electrochem. Soc.* **119**, 270 (1972).
29. E. F. Duffek, E. A. Benjamini and C. Mylroie, *Electrochem. Technol.* **3**, 75 (1965).
30. R. E. Pawel, *Rev. Sci. Instr.* **35**, 1066 (1964).
31. W. Mizushima, *J. Electrochem. Soc.* **108**, 825 (1961).
32. M. R. Arora and R. Kelly, *J. Electrochem. Soc.* **120**, 128 (1973).
33. R. D. Misch, *Acta Met.* **5**, 179 (1957).
34. C. E. Carlston, G. D. Magnuson, A. Comeaux and P. Mahedevan, *Phys. Rev.* **138**, A759 (1965).
35. A. E. Andrews, E. H. Hasseltine, N. T. Olson and H. P. Smith, *J. Appl. Phys.* **37**, 3344 (1966).
36. H. L. Daley and J. Perel, *AIAA J.* **5**, 113 (1967).
37. M. T. Robinson and A. E. Southern, *J. Appl. Phys.* **38**, 2969 (1967).
38. E. T. Pitkin, *Progr. Astron. Rocketry* **5**, 195 (1961).
39. S. Ya. Lebedev, Yu. Ya. Stavisskii and Yu. V. Shut'ko, *Sov. Phys.–Tech. Phys.* **9**, 854 (1964).
40. B. M. Gurmin, T. P. Martynenko and Yu. A. Ryzhov, *Sov. Phys.–Sol. State*, **10**, 324 (1968).
41. A. J. Summers, N. J. Freeman and N. R. Daly, *J. Appl. Phys.* **42**, 4774 (1971).
42. A. L. Southern, W. R. Willis and M. T. Robinson, *J. Appl. Phys.* **34**, 153 (1963).
43. M. I. Guseva, *Radio Eng. Elect. Phys.* **7**, 1563 (1962).
44. O. K. Kurbatov, *Sov. Phys.–Tech. Phys.* **12**, 1328 (1968).
45. J. F. Strachan and N. L. Harris, *Proc. Phys. Soc. London* **B69**, 1148 (1956).
46. G. K. Wehner and D. Rosenberg, *J. Appl. Phys.* **32**, 887 (1961).
47. O. C. Yonts and D. E. Harrison, *J. Appl. Phys.* **31**, 1583 (1960).
48. R. Kelly, N. Q. Lam, D. K. Murti, H. M. Naguib and T. E. Parker, *Proc. Intern. Summer School on Physics of Ionized Gases* (Split, Yugoslavia, 1972) (in press).
49. M. Meyer, P. Haymann and J.-J. Trillat, in *Growth of Crystals* (Consultants Bureau, New York, 1969), Vol. 8, p. 120.
50. M. Meyer, C. Marelle and P. Haymann, *C.R. Acad. Sci. (Paris)* **268B**, 1145 (1969).
51. J.-J. Trillat and K. Mihama, *C.R. Acad. Sci. (Paris)* **248**, 2827 (1959).
52. T. E. Parker and R. Kelly, Proc. Third Intern. Conf. on Ion Implantation in Semiconductors (Yorktown Heights, N.Y., 1972) (in press).
53. D. K. Murti and R. Kelly, to be published.
54. H. M. Naguib and R. Kelly, *J. Phys. Chem. Sol.* **33**, 1751 (1972).
55. R. J. MacDonald and D. Haneman, *J. Appl. Phys.* **37**, 3048 (1966).
56. M. R. Arora and R. Kelly, to be published.
57. G. K. Wehner, C. E. Kenknight and D. Rosenberg, *Planet. Space Sci.* **11**, 1257 (1963).
58. M. Balarin, G. Otto, I. Storbeck, M. Schenk and H. Wagner, *Thin Solid Films* **4**, 255 (1969).
59. P. W. Gilles, K. D. Carlson, H. F. Franzen and P. G. Wahlbeck, *J. Chem. Phys.* **46**, 2461 (1967).
60. M. Meyer, P. Haymann and J.-J. Trillat, *C.R. Acad. Sci. (Paris)* **261**, 4353 (1965).
61. P. Haymann, C. Waldburger and J.-J. Trillat, in *Processus de Nucléation dans les Réactions des Gaz sur les Métaux et Problèmes Connexes* (C.N.R.S., Paris, 1965), p. 135.
62. J. G. Perkins and L. E. Collins, *Thin Solid Films* **5**, R59 (1970).
63. M. Watanabe and A. Tooi, *Jap. J. Appl. Phys.* **5**, 737 (1966).
64. P. V. Pavlov and E. V. Shitova, *Sov. Phys.–Doklady* **12**, 11 (1967).
65. Hj. Matzke and J. L. Whitton, *Can. J. Phys.* **44**, 995 (1966).
66. R. S. Nelson, *J. Nucl. Mat.* **10**, 154 (1963).
67. R. E. Jones, H. F. Winters, and L. I. Maissel, *J. Vac. Sci. Technol.* **5**, 84 (1968).
68. K. B. Winterbon, P. Sigmund and J. B. Sanders, *Kgl. Danske Vid. Selsk. Mat. Fys. Medd.* **37**, No. 14 (1970).
69. J. J. Polling and A. Charlesby, *Proc. Phys. Soc.* **67B**, 201 (1954).
70. A. J. Stirling and W. D. Westwood, *Thin Solid Films* **7**, 1 (1971).
71. J. J. Trillat, L. Tertian and N. Terao, *Cah. Phys.* **12**, 161 (1958).
72. M. Meyer and P. Haymann, *C. R. Acad. Sci. (Paris)* **258**, 4690 (1964).
73. R. A. Dugdale and S. D. Ford, *Trans. Brit. Cer. Soc.* **65**, 165 (1966).

Notes added in proof

a) Important experiments due to Dugdale and Ford[73] were overlooked. They show that SiO_2 resembles WO_3, MoO_3, and V_2O_5 in having an S which increases with temperature.

This result is acceptable, for it is indeed true (see Table VI) that the apparent E_b of SiO_2 is somewhat below the thermodynamic E_b.

b) Ref. 52 includes a brief treatment of thermal sputtering.

It is shown that oxygen sputtering with TiO_2 is *not* thermal in origin, that with MoO_3 is almost certainly thermal, while that with V_2O_5 is intermediate.

SPUTTERING AND BACKSCATTERING OF keV LIGHT IONS BOMBARDING RANDOM TARGETS

R. WEISSMANN

Max-Planck-Institut für Plasmaphysik, EURATOM Association, BRD-8046 Garching, Germany

and

P. SIGMUND

H. C. Ørsted Institute, DK-2100 Copenhagen, Denmark

Range and damage distributions have been calculated for light ions slowing down in heavy targets in the energy region where electronic stopping is dominating and approximately proportional to velocity. Following Schiøtt's approach, well-known integral equations for spatial moments were approximated by differential equations, and the latter solved by numerical integration. Moments over the two distributions up to third order have been obtained, and the distributions were constructed by use of the Edgeworth expansion. From the profiles, backscattering coefficients and relative sputtering yields were determined. Comparison is made with previous theoretical results, computer simulation and experimental results. An improved calculation of the sputtering yield of light ions bombarding a thin film is also presented.

1 INTRODUCTION

Ion beams of protons, deuterons, and α particles in the upper keV and low MeV region have been exceedingly useful for a number of years as an analytic tool in investigating surface and bulk properties of solid targets.[1] More recently, the energy range of interest has been extended into the lower keV and upper eV region. Such extension is of major interest in fusion research, i.e. to predict radiation damage, sputtering and ion backscattering from the first wall of a fusion reactor.[2] Furthermore, low-energy beams have received increasing importance in the analysis of very shallow surface layers.[3]

In this contribution we present calculations on backscattering and sputtering of light ions bombarding "random" (i.e. amorphous or polycrystalline) targets. Channelling effects[1] are thus neglected.

A key question concerns the relative significance of single and multiple collisions in the problem investigated. Because of the decrease of Rutherford's cross section with increasing energy, both sputtering and backscattering are determined by *single* collisions of the ion beam at some high enough ion energies. Reversely, *multiple* scattering processes must play a dominating role at low energies. The transition region is not well known and probably not very accurately defined.

Most existent theories of backscattering[4] and sputtering[5] for light ions apply only to the single-collision region (although this is not always stated explicitly). Multiple-scattering theories of backscattering[6] and sputtering[7] have been applied mainly to the case of *heavy* ions. The latter class of theory is closely related to the problem of ion penetration[8] and energy deposition.[9] A rather comprehensive program to evaluate accurate profiles of ion penetration and energy deposition has been worked out recently by Winterbon.[10] However, for light ions the available energy range ($\epsilon \lesssim 1.5$ in dimensionless units[8]) barely extends into the low-keV region. On the other hand, Schiøtt[11] has pointed out that solutions of the integral equations of range theory can be found in a comparatively simple manner, if use is made of a number of simplifying approximations that are rather well justified in case of light-ion bombardment of heavy targets.

Starting from well-known integral equations, we evaluate the first three moments of the range and damage profiles making essentially the same approximations as Schiøtt. Sputtering yields and reflection coefficients are evaluated from the resulting profiles, and comparison is made with experimental results and some data from a computer simulation model. In Appendix B, an improved estimate is given of the single-collision sputtering yield.

Most of the general background, the basic equations, and some of the more specific techniques relevant to this paper have been reviewed in a recent series of articles by one of us.[12] Therefore, mathematical derivations have been kept to a minimum, and the equations quoted serve mainly to specify the approximations made.

2 BASIC EQUATIONS

In an infinite medium, the range distribution of an ion beam is given by $F_R(x, E, \eta)$, where x is the penetration depth, E the initial energy, and $\eta = \cos \vartheta$, ϑ being the initial angle between the beam and the x-direction. F_R is determined by the following equation,[8-13]

$$-\eta \frac{\partial F_R}{\partial x} = NS_e \frac{\partial F_R}{\partial E} + N \int d\sigma_n (F_R - F_R') \quad (1)$$

where N is the number of target atoms per unit volume, $S_e = S_e(E)$ the electronic stopping cross section, $d\sigma_n = (d\sigma_n(E, T)/dT) \, dT$ the differential cross section for energy loss (T, dT) in an elastic collision, $F_R' = F_R(x, E - T, \eta')$, and $\eta' = \cos \vartheta'$, with ϑ' being the polar angle of a deflected ion. The target has been assumed random and monatomic, and electronic stopping has been separated according to the scheme of Lindhard et al.[8] The distribution is normalized,

$$\int_{-\infty}^{\infty} dx F_R(x, E, \eta) = 1 \quad (2)$$

and the reflection coefficient $R = R(E, \eta)$ is given, to a first approximation,[6,14] by

$$R = \int_{-\infty}^{0} dx F_R(x, E, \eta) \quad (3)$$

Equation (3) has been shown[6] to give a good estimate of the reflection coefficient as long as $R \lesssim 0.1$, say. For larger values of R, a correction or multiple passage through the surface has to be made. In principle Eq. (3) is applicable in both the single and multiple-collision case. In practice, R may be so low in the single-collision case that an accurate estimate of $F_R(x, E, \eta)$ for $x < 0$ may be exceedingly difficult.

After slowing-down, the average amount of energy per incoming ion that is lost to the electronic system is given by $\eta_{(1)}(E)$, the rest, $\nu_{(1)}(E) = E - \eta_{(1)}(E)$ is spent in atomic motion.[15] The depth distribution of $\nu_{(1)}(E)$ is given by $F_{(1)}(E, x, \eta)$, and obeys the equation[9,12,13]

$$-\eta \frac{\partial F_{(1)}}{\partial x} = NS_e \frac{\partial F_{(1)}}{\partial E} +$$
$$N \int d\sigma_n \, (F_{(1)} - F_{(1)}' - F'') \quad (4)$$

where $F_{(1)}' = F_{(1)}(x, E - T, \eta')$, and $F'' = F(x, T, \eta'')$ is

the deposited-energy profile due to a target atom of energy T and polar angle ϑ'' ($\eta'' = \cos \vartheta''$). The function $F(x, E, \eta)$ follows from a similar equation.[13] Analogous to Eq. (2) we have

$$\int_{-\infty}^{\infty} dx F_{(1)}(x, E, \eta) = \nu_{(1)}(E) \quad (5)$$

Threshold and atomic binding effects have been neglected in Eq. (4).

As a first approximation, the backsputtering yield is given by[7]

$$S(E, \eta) = \Lambda \cdot F_{(1)}(0, E, \eta) \quad (6)$$

where

$$\Lambda = \frac{3}{4\pi^2} \frac{1}{NC_0 U_0} \cong \frac{0.042}{NU_0 \, \text{Å}^2} \quad (6a)$$

U_0 being the height of a planar surface potential barrier ($\approx$ sublimation energy), and C_0 a constant related to the Born–Mayer repulsion of the target atoms.

We introduce a dimensionless quantity α by

$$F_{(1)}(0, E, \eta) = \alpha N S_n(E) \quad (7)$$

where $S_n = \int T d\sigma_n$ is the nuclear stopping cross section. In general we have $\alpha = \alpha(E, \eta)$ for a given ion-target combination. For perpendicular incidence, $\eta = 1$, α may be larger or smaller than 1, dependent on the importance of wide-angle multiple scattering. For single collisions, i.e. large energy and $\eta = 1$, α is always smaller than 1.

3 LIGHT IONS IN HEAVY TARGETS

From Eq. (1), we obtain moment equations[9]

$$nl F_{Rl-1}{}^{n-1} + n(l + 1) F_{Rl+1}{}^{n-1}$$
$$= (2l + 1) NS_e \frac{dF_{Rl}{}^n}{dE} + (2l + 1) N \int d\sigma_n$$
$$\times (F_{Rl}{}^n(E) - P_l(\cos \vartheta') F_{Rl}(E - T)) \quad (8)$$

where the $F_{Rl}{}^n = F_{Rl}{}^n(E)$ are coefficients in the Legendre polynomial expansion,

$$\int_{-\infty}^{\infty} dx \, x^n F(x, E, \eta) = \sum_{l=0}^{n} (2l + 1) \, F_{Rl}{}^n(E)$$
$$\times P_n(\eta) \quad (9)$$

and ϑ' the laboratory scattering angle.

We have $n = 1, 2, 3, \ldots$, and $l = 0, 1, 2, \ldots, n$.

For light ions in heavy targets, we have

$$T \leqslant T_m = \frac{4M_1M_2}{(M_1 + M_2)^2} E \ll E$$

$$M_1 = \text{ion mass} \qquad M_2 = \text{target mass} \qquad (10)$$

so that

$$F_{Rl}{}^n(E - T) \approx F_{Rl}{}^n(E) - T\frac{d}{dE}F_{Rl}{}^n(E) \qquad (10a)$$

The integral on the r.h.s. of Eq. (8) then takes the form

$$\approx (2l + 1)F_{Rl}{}^n(E)N \int d\sigma_n(1 - P_l(\cos\varphi')) \qquad (11)$$

while the subsequent term, which contains $(d/dE)F_{Rl}{}^n(E)$, is small compared to the first term on the r.h.s. of Eq. (8), as long as electric stopping is the dominating mechanism of energy loss of the ion. For light ions, this holds to a very good approximation[11] for $\epsilon \gtrsim 1$ where

$$\epsilon = \left(\frac{M_2E}{M_1 + M_2}\right) \bigg/ \left(\frac{Z_1Z_2e^2}{a}\right) \qquad (12)$$

Equation (4) can be brought into similar form as Eq. (8). In fact, the (additional) recoil term

$$-(2l + 1)N \int d\sigma_n P_l(\cos\varphi'')F_l{}^n(T) \qquad (13)$$

where φ'' is the laboratory scattering angle for a recoiling atom, will be neglected so that exactly the same set of differential equations determines range and damage profiles, the sole difference being the normalization

$$F_{Rl}{}^0(E) = \delta_{l0} \qquad (14a)$$

$$F_{(1)l}{}^0(E) = \delta_{l0}\nu_{(1)}(E) \qquad (14b)$$

Neglecting the recoil term (13) is justified within the present scheme (i.e. within moment expansion) since the maximum ranges of recoil atoms are very small compared to the length of the trajectory of the ion. Note that not only is the maximum recoil energy small as compared to E (Eq. (10)), but even at the same initial energy, proton ranges are much larger than corresponding heavy-particle ranges. The physical argument is the same as that of Brice,[16] although Brice's approach—which does not go over moment equations—does not appear to be suited for evaluation of $F_{(1)}(0, E, \eta)$ in Eq. (6) (see also Ref. 17).

Evaluation is done in dimensionless units, ϵ (Eq. (12)) and ρ according to[18]

$$\rho = xN\pi a^2 \gamma \qquad (15)$$

where $\gamma = 4M_1M_2/(M_1 + M_2)^2$, and by use of the Thomas–Fermi expressions[8,11,18,19] for S_e and $d\sigma_n$, i.e.

$$S_e(\epsilon) = \left(\frac{d\epsilon}{d\rho}\right)_e = k \cdot \epsilon^{1/2} \quad \text{for}$$

$$E < Z_1{}^{4/3}M_1 \cdot 25 \text{ keV} \qquad (16)$$

Here, for $M_1 \ll M_2$,

$$k \cong 0.0793\left(\frac{Z_1}{A_1}\right)^{1/2} \cdot \frac{A_2}{A_1}$$

$$(A_1, A_2 = \text{mass numbers}) \qquad (16a)$$

and

$$d\sigma_n = \pi a^2 \frac{dt}{2t^{3/2}}f(t^{1/2}) \qquad (17)$$

with[9]

$$f(\xi) = \lambda'\xi^{1/3}[1 + (2\lambda'\xi^{4/3})^{2/3}]^{-3/2} \qquad (17a)$$

and $\lambda' = 1.309$.

With the notations

$$f_{Rl}{}^n(\epsilon) = \left(\frac{\rho}{x}\right)^n F_{Rl}{}^n(E) \qquad (18a)$$

$$f_{(1)l}{}^n(\epsilon) = \frac{\epsilon}{E}\left(\frac{\rho}{x}\right)^n F_{(1)l}{}^n(E) \qquad (18b)$$

Eqs. (1) and (4) reduce to

$$nlf_{l-1}{}^{n-1} + n(l + 1)f_{l-1}{}^{n-1}$$

$$= (2l + 1)S_e \frac{df_l{}^n}{d\epsilon} + (2l + 1)f_l{}^n \cdot \frac{M_2}{4M_1}$$

$$\times \int \frac{dt}{2t^{3/2}}f(t^{1/2})(1 - P_l(\cos\varphi')) \qquad (19)$$

This set of differential equations has been solved numerically for $n \leqslant 3$ on the IBM 360-91 computing system of the IPP. The zero moment $\nu_{(1)}(\epsilon)$ is evaluated in Appendix A.

Profiles have been constructed from the moments by use of the Edgeworth expansion,[9,20,21]

$$f(\rho, \epsilon, \eta) = \frac{\bar{f}}{\langle\Delta\rho^2\rangle^{1/2}}$$

$$\times \left(\varphi_0(\xi) - \frac{\Gamma_1}{6}\varphi_3(\xi)\dots\right) \qquad (20)$$

where

$$\bar{f} = \begin{cases} 1 \qquad\qquad f_R \\ \qquad \text{for} \\ \nu_{(1)} \qquad\quad f_{(1)} \end{cases} \qquad (20a)$$

$$\xi = \frac{\rho - \langle \rho \rangle}{\langle \Delta \rho^2 \rangle^{1/2}} \qquad (20b)$$

$$\Gamma_1 = \frac{\langle \Delta \rho^3 \rangle}{\langle \Delta \rho^2 \rangle^{3/2}} \qquad (20c)$$

and $\varphi_0(\xi)$ and $\varphi_3(\xi)$ the gaussian and its third derivative.

4 PENETRATION PROFILES AND REFLECTION COEFFICIENTS

Our calculated values of the average projected range, $\langle \rho \rangle_R$, and straggling, $\langle \Delta \rho^2 \rangle_R$, agree well with those of Schiøtt.[11] Figure 1 shows penetration profiles for protons of different energies, calculated within Schiøtt's gaussian approximation, and from Eq. (20). It is seen that in the energy range $1 < \epsilon < 10$, the

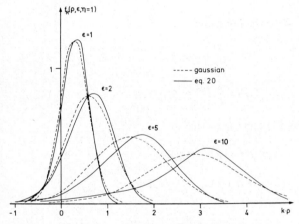

FIGURE 1   Calculated projected range distributions for protons in heavy targets. In dimensionless units (Eqs. (12), (15), (16a)).

skewness introduced by the third order moment is noticeable, but not prohibitively great. This gives some justification to our neglecting higher-order terms in the Edgeworth expansion.

Figure 2 shows reflection coefficients calculated from Eqns. (3) and (20), for protons, deuterons, tritons, and α-particles. As a function of the dimensionless energy ϵ, all four curves are rather similar in shape and magnitude. In the region $\epsilon > 1$ where the neglect of nuclear stopping is justified, reflection coefficients turn out to be smaller than 20%. Thus, a surface correction would have to be made at the lowest energies. Within the accuracy of the scheme, Figure 8 of Ref. 6 can be used. For more accurate calculations, the surface

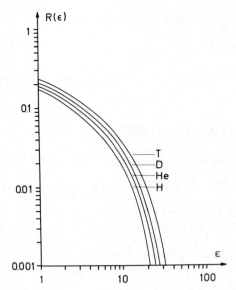

FIGURE 2   Reflection coefficients for normally incident ions (H, D, T, He) from heavy targets as a function of the dimensionless energy ϵ (Eq. (12)).

correction would have to be re-evaluated on the basis of realistic energy and angular distributions, neither of the latter being well known.

Figure 3 shows the reflection coefficients calculated for protons in three different ways. Curve (a) is the one from Figure 2, and curve (c) follows from the gaussian approximation to the range profile. Curve (b) follows from the assumption of a single Rutherford collision and velocity-proportional stopping power,[4] where, in our notation

$$R = 1.05 \, \epsilon^{-3/2} \qquad (21)$$

For several obvious reasons, this expression over-estimates R at low energies. However, at high energies it must give a realistic value. The fact that curve (a) drops far below curve (b) for $\epsilon > 20$ is due to increasing deviation from gaussian shape of the penetration profile[10] with increasing ϵ. For qualitative orientation, we included a feasible interpolation between the two curves. It follows that the transition from clear multiple-scattering to clear single-scattering type of reflection occurs around $\epsilon = 10$–20. Also included are the results of Ishitani et al.[22] that have been obtained by Monte-Carlo simulation of the H^+–Cu system, with rather similar input quantities. The agreement is satisfactory in view of our neglect of the surface correction.

We are not aware of any measured values of total reflection coefficients for light ions in the energy range considered.

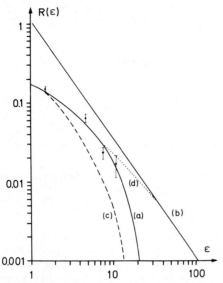

FIGURE 3   Reflection coefficient for protons. (a) curve from Figure 2; (b) single collision model Eq. (21); (c) gaussian approximation of the range profile and computer simulations (error bars) Ref. 22. Dotted line is an interpolation by eye between the curves (a) and (b).

5 DAMAGE PROFILES AND SPUTTERING YIELD

Figure 4 shows damage (= deposited energy) profiles corresponding to the penetration profiles of Figure 1. In Figures 5 and 6 first, second and third order averages over damage distributions are compared with those over range distributions obtained by Schiøtt[11] and our own calculations. The damage profiles show a very similar behaviour as the depth profiles. The skewness is consistently smaller than in the case of the penetration profiles (Figure 6). Figures 7a–c show the energy deposited at the surface ($\rho = 0$) for perpendicular incidence, $\eta = 1$, as a function of energy ϵ.

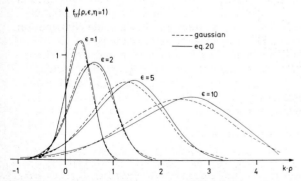

FIGURE 4   Damage profiles corresponding to Figure 1. Normalized to $\nu_1(E)$.

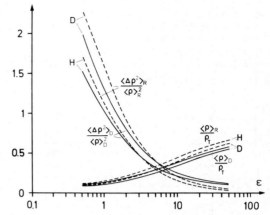

FIGURE 5   First and second order averages over damage and projected range distributions for protons and deuterons as a function of the dimensionless energy ϵ. The path length $\rho_T = (2/k)\epsilon^{1/2}$. Solid lines: Damage. Dashed lines: Projected ranges obtained by Schiøtt.[11]

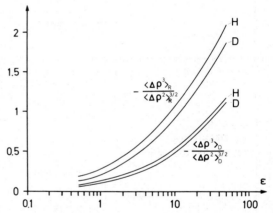

FIGURE 6   Third order averages over damage and projected range distributions.

According to Eq. (6), this quantity determines the dependence of the backsputtering yield on ion type and energy. Experimental sputtering yields[23-30] have been divided by Λ (Eq. (6a)) and included. Also included is a zero-order approximation, $f_{(1)}(\epsilon) = s_n(\epsilon)$, where the deposited energy is set equal to the nuclear stopping power. This simplifying assumption has been made in many previous sputtering theories, especially those listed in Ref. 5. It corresponds to setting $\alpha = 1$ in Eq. (7) (see also Figure 8 below). According to Andersen and Bay,[31] experimental sputtering yields may show a significant dependence on ion dose, and the type of variation may depend on the bombarding ion. Since no dose dependence has been reported on the light-ion yield data of Refs. 23–30, both the

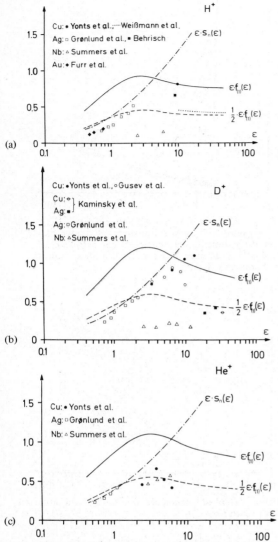

FIGURE 7   Deposited energy at the target surface for normally incident protons (a), deuterons (b) and helium ions (c) compared with reduced experimental sputtering yields by Yonts et al.,[23] Grønlund and Moore,[24] Summers et al.,[25] Behrisch,[26] Furr and Finfgeld,[27] Kaminsky,[28] Gusev,[29] Weissmann and Behrisch.[30] See text $\epsilon\ f(\epsilon)$ instead of $f(\epsilon)$ has been plotted for convenience.

absolute magnitude (i.e. the value of the factor Λ) and even the variation with the type of ion has to be considered with some reservation. The least uncertain factor seems to be the variation with energy, which indeed appears to be very well represented for both protons (Figure 7a), deuterons (Figure 7b), and He⁺-ions (Figure 7c). The only significant exception is the D⁺-

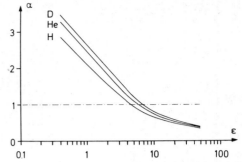

FIGURE 8   Factor α (Eq. (7)) for protons, deuterons and helium ions defined as a function of energy. The straight line $\alpha \equiv 1$ follows from the single-collision sputtering theories quoted in Ref. 5.

data of Yonts et al.,[23] that do not follow the theoretical energy dependence nor that observed by Gusev et al.[29]

Inspection of all three graphs indicates that the energy dependence of the sputtering yield is represented better by the $f_{(1)}(\epsilon)$ curves than by the nuclear stopping power (or $\alpha \equiv 1$). The absolute magnitude is in most cases equal to about half the value calculated from Eqs. (6) and (6a). However, exceptions in both directions are found. To describe those deviations in more detail would require some information on the dependence of the yield on ion dose. However, special attention is paid to the fact that exceptionally low yields are observed for Nb in case of proton and deuteron bombardment (Figures 7a–b), while the same authors[25] reported He⁺–Nb yields that follow the general picture quite well (Figure 7c). The reason for this behaviour is not understood. Note that Eckstein et al.[32] reported H⁺–Nb yields still lower than those of Ref. 25.

Figure 8 shows the factor α as defined by Eq. (7) as a function of ϵ for perpendicular incidence. Contrary to the case of heavy keV ions,[7,13] α depends sensitively on ion energy. Note especially that for an ion beam penetrating straight into the target, α cannot be greater than $\frac{1}{2}$.[7] The rather large values of α at low and intermediate ϵ-energies indicate a considerable contribution to the sputtering yield of deflected ions. This is consistent with the experimental findings.[30,33] This point is elaborated in Appendix B, and discussed in the subsequent paper.[30] It is pointed out that Figure 8 replaces Figure 15 of Ref. 7 that was derived by a less quantitative argument.

ACKNOWLEDGEMENTS

We thank Rainer Behrisch for his continuous encouragement and enthusiastic criticism. One of us (P.S.) would like to thank the Garching group for their kind hospitality on several occasions. Also we wish to thank Mrs. Marianne Walter for setting up the computer program.

REFERENCES

1. For example: J. W. Mayer, L. Eriksson and J. A. Davies, *Ion Implantation in semiconductors* (Academic Press Inc., New York, 1970).
2. For example: R. Behrisch, *Nuclear Fusion* 12, 695 (1972); H. Vernickel, *Nuclear Fusion* 12, 386 (1972); W. Eckstein and H. Verbeek, *J. Vac. Sci. Tech.* 9, 612 (1972).
3. D. P. Smith, *Surface Science* 25, 171 (1971); T. S. Buck and G. H. Wheatley, *Surface Science* 33, 35 (1972); W. Heiland and E. Taglauer, *J. Vac. Sci. Tech.* 9, 620 (1972).
4. G. McCracken and N. J. Freeman, *J. Phys. B.* 2, 661 (1969); R. Behrisch, IPP-Report 2/68 (1968).
5. D. T. Goldman and A. Simon, *Phys. Rev.* 111, 383 (1958); R. S. Pease, in *Proceedings of the International School of Physics "Enrico Fermi" Course* (Academic Press Inc., New York, 1964), Vol. 13, p. 158; Yu. V. Bulgakov, *Zh. Tekhn. Fiz.* 33, 500 (1963) (English transl.: *Soviet Phys.-Techn. Phys.* 8, 369 (1963)); W. Brandt and R. Laubert, *Nucl. Instr. Methods*, 47, 201 (1967).
6. J. Bøttiger, J. A. Davies, P. Sigmund and K. B. Winterbon *Rad. Effects* 11, 69 (1971).
7. P. Sigmund, *Phys. Rev.* 184, 383 (1969).
8. J. Lindhard, M. Scharff and H. E. Schiøtt, *Kgl. Danske Videnskab. Selskab Mat.-Fys. Medd.* 33, No. 14 (1963).
9. K. B. Winterbon, P. Sigmund and J. B. Sanders, *ibid* 37, No. 14 (1970).
10. K. B. Winterbon, *Rad. Effects* 13, 215 (1972).
11. H. E. Schiøtt, *Kgl. Danske Videnskab Selskab, Mat.-Fys. Medd.* 35, No. 9 (1966).
12. P. Sigmund, *Rev. Roum. Phys.* 17, 823, 969, 1079 (1972).
13. P. Sigmund, M. T. Matthies and D. L. Phillips, *Rad. Effects* 11, 39 (1971).
14. P. Sigmund, *Can. J. Phys.* 46, 731 (1968).
15. J. Lindhard, V. Nielsen, M. Scharff and P. V. Thomsen, *Kgl. Danske Videnskab. Selskab., Mat.-Fys. Medd.* 33, No. 10 (1963).
16. D. K. Brice, *Rad. Effects*, 6, 77 (1970).
17. P. Sigmund, in *Proc. of the VIth International Summer School on the Physics of Ionized Gases,* edited by M. Kurepa, in press.
18. J. Lindhard and M. Scharff, *Phys. Rev.* 124, 128 (1961).
19. J. Lindhard, V. Nielsen and M. Scharff, *Kgl. Danske Videnskab. Selskab., Mat.-Fys. Medd.* 36, No. 10 (1968).
20. J. B. Sanders, *Can. J. Phys.* 46, 455 (1968).
21. E. M. Baroody, *J. Appl. Phys.* 36, 3565 (1965).
22. T. Ishitani, R. Shimizu and K. Murata, *Jap. J. Appl. Phys.* 11, 125 (1972).
23. O. C. Yonts, C. E. Normand and Don E. Harrison, Jr., *J. Appl. Phys.* 31, 447 (1960).
24. F. Grønlund, W. J. Moore, *J. Chem. Phys.* 32, 1540 (1960).
25. A. J. Summers, N. J. Freeman and N. R. Daly, *J. Appl. Phys.* 42, 4774 (1971).
26. R. Behrisch, *Diplomarbeit München* (1960).
27. A. K. Furr and C. R. Finfgeld, *J. Appl. Phys.* 41, 1739 (1970).
28. M. Kaminsky, *Phys. Rev.* 126, 1267 (1962).
29. V. M. Gusev, *Jzvestiya ANSSR Sov. Fiz.* 24, 6, 689 (1960).
30. R. Weissmann and R. Behrisch, submitted to *Rad. Effects* (subsequent paper).
31. H. H. Andersen and H. Bay, *Rad. Effects* 13, 67 (1972).
32. W. Eckstein, B. M. U. Scherzer and H. Verbeek, Int. Conf. Ion-Surface Interaction, Garching (1972).
33. R. Behrisch and R. Weissmann, *Phys. Lett.* 30A, 506 (1969).

Appendices

A EVALUATION OF $v_{(1)}(E)$

According to Lindhard *et al.*,[15]

$$O = NS_e(E) \frac{dv_{(1)}(E)}{dE} + N \int d\sigma_n$$

$$\times (v_{(1)}(E) - v_{(1)}(E - T) - v(T)) \qquad (A1)$$

where $v(T)$ refers to a moving target atom. Applying Eq. (11), and approximating $v(T) \approx T$ (T is in the eV or lower keV range in all cases considered here), Eq. (A1) reads

$$\frac{dv_1(E)}{dE} = \frac{S_n(E)}{S_e(E)} \qquad (A2)$$

In ϵ-units

$$v_1(\epsilon) = \frac{1}{k} \int_0^\epsilon d\epsilon' \frac{S_n(\epsilon')}{(\epsilon')^{1/2}} = \frac{1}{k} g(\epsilon) \qquad (A3)$$

where $g(\epsilon)$ is a universal function that has been evaluated numerically.

B SPUTTERING YIELD OF UNDEFLECTED IONS

In Ref. 7, a formula was derived for the sputtering yield of an ion beam that penetrates into the target without any deflection. In the Rutherford collision range, this is essentially equivalent to assuming that only the first collision undergone by the ion contributes to the yield (or that the target is a thin film). The result was

$$S_I = \Lambda \cdot N \int d\sigma_n T\gamma(T_1 \cos \varphi'') \qquad (B1)$$

for perpendicular incidence, where γ is the sputtering efficiency[14] and φ'' the laboratory scattering angle of a recoiling target atom, i.e. $\cos \varphi'' = (T/T_m)^{1/2}$. Equation (B1) was evaluated in Ref. 7 by use of the expression

$$\gamma(E, \eta) \cong \gamma(\eta) = \gamma_p + (\tfrac{1}{2} - \gamma_p)(1 - \eta)^2 \qquad (B2)$$

and $\gamma_p = 0.028$, and a power cross section for $d\sigma_n$.

With the same assumption, but $d\sigma_n$ given by Eqs. (17) and (17a), we can define α_I ($<\alpha$) by means of Eqs. (6), and (B1). The result is shown in Figure 9.

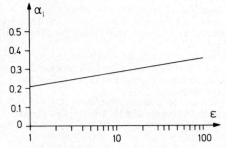

FIGURE 9   Factor α_I defining the backsputtering yield (Eq. (7)) in the case of undeflected ions (thin-film case).

Because of the screening of the Thomas–Fermi cross section function, the asymptotic value[7] $\alpha_I = \frac{1}{2}$ is reached only at ϵ-values that are substantially higher than anticipated.

The difference

$$S_{II} = S - S_I \tag{B3}$$

can be interpreted as the sputtering yield due to deflected or backscattered ions. It will be compared with experimental results in the subsequent paper.[30]

CONTRIBUTIONS OF BACKSCATTERED IONS TO SPUTTERING YIELDS DEPENDING ON PRIMARY ION ENERGY

R. WEISSMANN and R. BEHRISCH

Max-Planck-Institut für Plasmaphysik, EURATOM Association, 8046 Garching, Germany

Sputtering yields for 33–150 keV protons normally incident on Cu-films deposited on different backing materials have been measured. The decrease of the Cu-film thickness was determined by Rutherford-backscattering.

As the used base materials have different backscattering intensities, the contribution to sputtering yield by incident ions and by the ions backscattered to the surface within the material could be separated. It was found that the relative contribution to total sputtering yield by the backscattered ions increases slowly for protons in the energy range from 50 to 150 keV. For energies lower than 50 keV the sputtering yields are determined by the special geometry of the target sandwich used here.

1 INTRODUCTION

Sputtering, i.e. the erosion of solid surfaces due to the bombardment with energetic ions, is caused by collision cascades initiated in the solid by the incident ions. When the collision cascades reach the surface with an energy larger than the surface binding energy, one or more lattice atoms can leave the surface. A measure of the erosion is the sputtering yield S, which is defined as the ratio of the number of sputtered atoms to the number of incident ions.[1-6]

The collision cascades in the surface layers can be initiated by the incoming ions in two ways[7] (Figure 1).

Sputtering mechanism I: When the ions enter the surface layer.

Sputtering mechanism II: When the ions are back-scattered from the interior of the solid to the surface.

The sputtering yield accordingly consists of these two components

$$S = S_I + S_{II} \tag{1}$$

The magnitude S_{II} is governed by the intensity of the ions backscattered from the deeper layers of the solid as well as their energy and angular distribution. This depends strongly on the target material, becoming larger with increasing atomic number z. Hitherto sputtering has been regarded as essentially caused by mechanism I, while mechanism II has been considered as minor.[8,9] In the case of 100 keV protons, however, it was shown that S_{II} is of the same order of magnitude as S_I.[7] The purpose of this paper is to investigate mechanism II and its dependence on the bombardment energy in more detail.

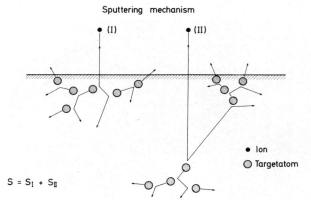

FIGURE 1 Sputtering mechanism for light ions. (I) sputtering by incoming ions, (II) sputtering by ions reflected from the interior of the solid to the surface.

2 MEASURING METHOD

a Detection of Mechanism II

In order to determine the influence of the back-scattered ions on the sputtering process, sputtering process, sputtering measurements were made on thin Cu-films deposited on different base materials. The thickness of the Cu-film of ≈ 230 Å was much smaller than the range of the protons used. The base materials Be, V, Nb and Ta chosen backscatter the protons with different intensities. The yield of the Cu-film due to sputtering mechanism I should be independent of the base material, while sputtering mechanism II should give very different yields for the different base materials.

In the experiments only the total sputtering yields (mechanism I plus mechanism II) could be determined. However, as Be due to its small z has a very low backscattering intensity, the yield measured for the Cu-films on Be as base material have been taken for S_I. The difference between the total yield and S_I was put equal to S_{II}.

b Measurement of the Sputtering Yield

Because of the very low sputtering yields of 10^{-2} to 10^{-3} atoms/ion, expected in the case of proton bombardment it is not possible to measure the sputtering yield by weight loss of the target. Due to the entrapment of hydrogen the weight will rather increase after the ion bombardment.

If the hydrogen is not soluble in the target material[10,11] e.g. Cu an additional problem arises. The injected gas agglomerates to gas bubbles[12-16] which finally develop blisters on the surface (Figure 2).

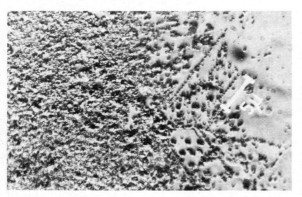

FIGURE 2   Blistering of polycrystalline Cu caused by the bombardment with 100 keV protons (total dose $\approx 10^{20}$ ions/cm^2) Right side is unbombarded. At the grain boundaries smaller blisters can be seen, while they are much larger in the center of the grains.

Blistering could, however, be completely avoided for the ion doses of $\approx 10^{20}$ ions cm^{-2} used in the experiments in selecting V, Nb and Ta as base materials. In these metals hydrogen is well soluble and diffuses well at room temperature.[10,11,17] In case of Be as base material some large blisters have always been observed starting at doses of about $5 \cdot 10^{15}$ ions/cm^2 (Figure 3). However, since the Be was a sintered material the surface showed no major erosion but mainly long cracks which should not change the sputtering yield according to mechanism I.

In order to measure the sputtering yields, the thickness of the Cu-films evaporated onto the different base

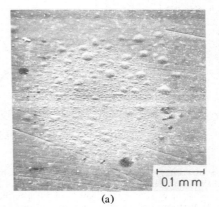

(a)

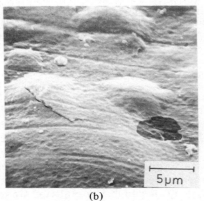

(b)

FIGURE 3   Scanning electron micrographs of a Be–Be–Cu sandwich target bombarded with 100 keV H_3^+ (total dose 2.10^{19} ions/cm^2) (a) overall picture of the spot sputtered. (b) detail enlarged. It can be seen that one of the blisters broke away. A careful check of the whole spot showed that this was less than 2% of the whole spot area.

materials was determined by the Rutherford backscattering technique.[7] Compared to earlier light ion sputtering experiments done by others[18-21] this method has the advantage that small changes of the Cu-film thickness could be measured and the time or dose dependence of the sputtering yields could be followed *in situ.*

The Rutherford backscattering technique has recently found considerable interest as a practically destruction free method to determine depth profiles in solids.[22,23] This information is obtained from the energy spectrum of the backscattered light ions as H, D or He. Light ions at energies larger than 100 keV mostly penetrate deep into a solid and only a few are backscattered. This takes place predominantly by a single head-on collision with a target atom, which can be described by the Coulomb potential of the nuclei.[24,25]

On its path in and out of the solid the ion loses energy to electrons. Thus the energy of the backscattered ions is a measure of the depth where the backscattering took place. The intensity of the backscattered ions at a certain energy gives information on the density and atomic number of the atoms at this depth.[22–25]

Figure 4 shows an energy spectrum of the ions backscattered under an angle of 30° to the normal from a sandwich used in the experiments. The base

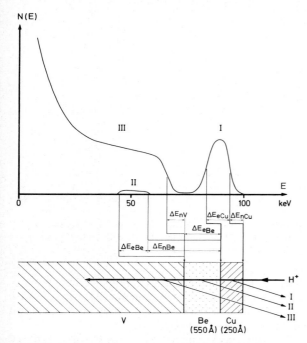

FIGURE 4   Energy spectra of protons backscattered from a V–Be–Cu–target. ΔE_n is the energy loss of the protons to the target atoms during backscattering. ΔE_e is the electronic energy loss of the protons in the matter. The parts I, II, III of the energy spectrum of backscattered protons correspond to the Cu-film (I), to the Be-film (II) and to the base material V (III).

material, V in this case, contributes to the ions back-scattered at the lowest energy. The broad peak at high energy separated from the rest is due to the protons backscattered from the Cu-film on the surface. In order to separate in the energy spectrum the backscattering from the Cu-film on the surface and from the V base material, a small film of Be has been evaporated in between. Due to the low z of Be this contributes only negligibly to backscattered intensity. Additionally due to the large energy loss of the protons in the binary collision at the Be atoms, the small Be-peak is shifted to lower energies as seen in Figure 4.

The thickness d of the Cu-film can be calculated from the integral intensity N_r of the particles back-scattered in the high energy peak for a given solid angle $\Delta\Omega$ and primary dose N_i. Actually, Rutherford back-scattering does not give the thickness d, but the number of atoms per unit area $N \cdot d$, were N is the number of target atoms per unit volume. In a first approximation we have

$$\frac{N_r}{N_i} = N \cdot d \cdot \frac{d\sigma_R}{d\Omega} \cdot \Delta\Omega \qquad (2)$$

The electronic energy loss of the protons in Cu means that the Rutherford cross section $d\sigma_R$ is not constant, but depends on the thickness. To allow for this fact, not the primary energy $E_0 = 100$ keV was used, but as an approximation the energy E_0' which the proton has half way through the film. For a film thickness of 230 Å and a differential energy loss $dE/dx \cong 20$ eV/Å for 100 keV protons this yields $E_0' = 97.7$ keV.

From the energy spectra before and after sputtering with a certain ion dose N_z it is possible to determine the decrease Δd of the film thickness d and hence, for given bombarded area F the sputtering yield

$$S = N \cdot \gamma \cdot \frac{F \cdot \Delta d}{N_z}$$

$$= \gamma \cdot \frac{N_{r_1} - N_{r_2}}{N_i} \cdot \frac{F}{\frac{d\sigma_R}{d\Omega} \cdot \Delta\Omega} \cdot \frac{1}{N_z} \qquad (3)$$

where γ is a correction factor to allow for the fact that the current density over the bombarded area may not be completely uniform. For the beam geometry of these measurements γ was 0.97 (see Appendix).

At $2.0 \cdot 10^{16}$ protons/cm² the dose of N_i used for the thickness measurement was so small that the resulting sputtering was negligible.

c Experimental Apparatus

The sputtering measurements were made on the "Pharao", an accelerator for light ions (H, D, He) with energies between 33 and 150 keV described elsewhere.[26] The ion current measurement for determining the bombardment dose was possible in two ways. A rotating diaphragm could be used to scatter part of the proton beam onto a reference counter. The number of scattered particles was proportional to the total current calibrated with a faraday cup. The current could also be measured directly at the target using a slit diaphragm at negative potential placed around the target to suppress the secondary electrons. The beam diameter used was 400 μm, the currents were 0.3–0.5 μA. In one set of measurements up to 4 targets and on every target three different spots could be sputtered. The vacuum during the measurement was typically $\approx 5 \cdot 10^{-8}$ torr.

The energy spectrum of the protons backscattered by the target was recorded with a Si surface barrier detector placed at 30° to the direction of incidence of the proton beam. The energy resolution of the detector was 3.5 keV F.W.H.M.[27] Both charged and neutral hydrogen atoms down to about 10 keV could be counted. The electrical pulses from the detector are proportional to the particle energy. They were amplified and analysed according to height in a multichannel analyser Didac 800 Intertechnique.

Before evaporation the targets Be, V, Nb and Ta were mechanically polished with diamond-paste (final grain diameter 0.5 μm) and subsequently electro-polished. The Be and Cu-films were evaporated in a standard evaporation chamber with a background pressure of the order of $\approx 10^{-5}$ torr. The film thickness measured before starting the sputtering measurements by means of Rutherford backscattering agreed within the margins of errors with the thickness measurement during evaporation where a crystal oscillator had been used. The sputtering measurements were performed by alternately determining the thickness of the Cu-film by means of Rutherford backscattering with 100 keV protons, and sputtering of the film with the same ion beam at high intensity usually at a different energy. The bombardment direction was always perpendicular to the surface. The total ion bombardment dose was 0.64–4.0 10^{19} ions/cm². In the measurements generally a total of about 10–20% of the film thickness was sputtered away. The measuring errors set the lower detection limit for variations of the film thickness at 5 Å. This yields a relative experimental error of 15% for the sputtering yield.

For the low ion energy sputtering experiments mostly 100 keV H_3^+ and 100 keV H_2^+ beams were used instead of 33 and 50 keV H^+ in order to have higher beam currents and shorter measuring times. Test measurements with 75 keV H^+ and 150 keV H_2^+ have shown that within the errors of measurements the sputtering yields are the same in both cases.

3 EXPERIMENTAL RESULTS

The thickness of the Cu-film after several bombardments is plotted in Figure 5 as a function of ion dose. Within the experimental error a linear decrease is obtained for the 4 different base materials. From the slope of these lines the sputtering yield was calculated.

Before final measurements many test measurements with different targets vapor deposited at different conditions were performed. They all showed the same trend in yield dependence on base materials. But for targets vapor deposited at different times the measured sputtering yield deviated about a factor of two or three. This is consistent with observations by others, that the sputtering yield depends on evaporation conditions.[28] Therefore one set of targets, all deposited in one evaporation, have been used for the final

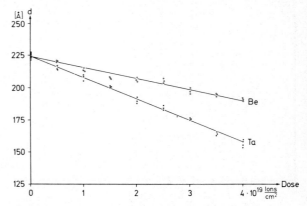

FIGURE 5   Decrease of the Cu-film deposited on the base materials Be and Ta during sputtering with 75 keV protons. Thickness measurements with 100 keV protons.

measurements. This means that all sputtering yields have been measured on the same Cu-film with the same degree of contamination and they may not be identical to those for very clean Cu-films.

The total sputtering yields for the Cu-film measured for the four base materials at proton energies of 33–150 keV are presented in Figure 6. At proton energies greater than 50 keV the sputtering yield of the Cu-film increases with atomic number z of the base material as expected. From this z dependence the sputtering yields of solid Cu ($z = 29$) is expected to lie between the curves for V ($z = 23$) and Nb ($z = 41$).

For energies below 50 keV the sputtering yields for the Cu-film converge to the same value. Here, the sandwich structure of the target has to be taken into account. As the range of the protons decreases less and less protons backscattered from the base material to reach the Cu-surface and the sputtering yield becomes representative of solid Cu.

For the three base materials V, Nb and Ta the contribution S_{II} of the backscattered ions to sputtering was determined by substracting from the measured total yields S the yields $S = S_I$ for Be as base material. The results obtained are shown in Figure 7a. At energies

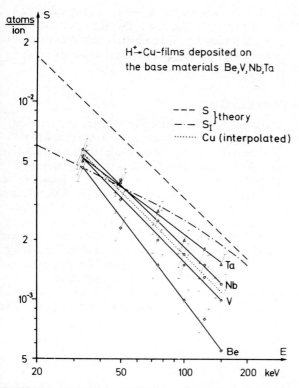

FIGURE 6   Sputtering yields of proton bombarded Cu-films deposited on the base materials Be, V, Nb and Ta compared with theoretical results for solid Cu. From the z-dependence of the sputtering yields on the base materials the case of Cu as base materials is interpolated (dotted line) between the measured curves of V and Nb as base materials.

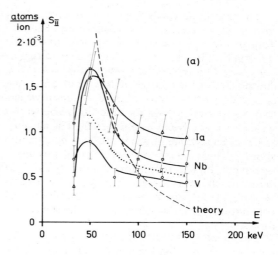

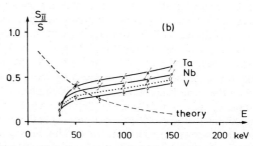

FIGURE 7   The total (a) and relative (b) sputtering yield S_{II}/S of Cu-films for protons backscattered from the base materials V, Nb and Ta. Dotted line: interpolated curve for Cu as base material. Dashed line: theoretical values for solid Cu.

greater than 50 keV the sputtering yields decrease with increasing bombarding ion energy. In Figure 7b the relative contribution of the backscattered ions to the sputtering yields S_{II}/S is plotted. This stays nearly constant or slightly increases with ion energy for the energy range investigated.

4 COMPARISON WITH THE THEORY

Most of the sputtering theories developed for light ions took into account sputtering mechanism I only.[29-31] A comprehensive theory for amorphous materials and for all ion-target combinations which in principle also includes mechanism II has been formulated by P. Sigmund.[32] This theory describes the sputtering process as a multi particle problem by means of the Boltzmann transport equation. It yields a general formual for the sputtering yield S

$$S = \Lambda \cdot F(x = 0, E, \eta) \qquad (4)$$

Where $F(x, E, \eta)$ is the deposited energy function, i.e. the energy which an incident particle with the energy E and the cosine η of the angle of incidence to the normal deposits in nuclear motion at a depth (x, dx) as a result of collision cascades. Λ is a constant depending on the target material and given in first approximation by

$$\Lambda = \frac{0.042}{NU_0} \; [\text{Å}^{-2}] \qquad (5)$$

where U_0 is the surface binding energy. Also from Sigmunds theory we can obtain the yield S_I which is given by

$$S_I = \Lambda \cdot \alpha \cdot S_n(E) \qquad (6)$$

Where $N \cdot S_n(E)$ is the nuclear stopping power and α a factor which is known from theory and smaller than one.[32,33] The difference of the two yields $S - S_I$ gives the yield S_{II} which was defined at the beginning of this work.

The sputtering yields S and S_I have been calculated[33] for the case of solid Cu and are plotted in Figure 6. Though a comparison with the experimental values for the thin Cu-films on the different base materials can be done only with reservations, we see that the energy dependence of the theoretical total sputtering yield agrees very well with the measured values if we assume that the sputtering yield for solid Cu lies between the yields for V and Nb as base materials. The absolute values calculated are however too high by a factor of about 2. There are two main reasons for this disagreement. Firstly, the factor Λ in Eqs. (4) and (5) is hardly much better known than by a factor of two.[32] Secondly, as already mentioned the Cu-film probably contains impurities which may alter the surface binding forces.

The theoretical values of S_{II} and S_{II}/S for solid Cu have been included in Figures 7a and 7b. In Figure 7a it can be seen that the calculated S_{II} are of the same order of magnitude as the experimental results and also decrease towards higher bombarding energies. However for the relative contribution S_{II}/S of the backscattered ions to sputtering as shown in Figure 7b the energy dependence of theory and experiment are opposite. The theoretical values definitely show a decrease with energy, while the experimental values slightly increase. The discrepancy may well be real and demonstrate some weakness in theory.[33] However also several experimental reasons may account for it. The measurements have been performed with thin Cu-films with a structure and density probably different from solid Cu. Also some blistering had been taken into account in determining

S_I. However theoretical values have been calculated for solid Cu only, and the collision cascades leading to sputtering may not be completely identical for both cases though the basic effects should be the same.

CONCLUSIONS

A new method was applied to measure the very low sputtering yields which occur on bombardment with light ions. It consists in measuring the thickness of thin films before and after sputtering by means of Rutherford backscattering of protons. It proves to be very sensitive, allowing variations of the film thickness as small as 5 Å to be measured. Thus more detailed investigations of sputtering mechanism in the case of light ions became possible. For sputtering by protons it was shown that the ions backscattered to the surface from the interior of the bombarded material substantially contribute to the sputtering yields in the investigated energy range of 33–150 keV. This result is in qualitative agreement with the sputtering theory of Sigmund.

ACKNOWLEDGEMENTS

The authors would like to thank Dr. P. Sigmund and Dr. B. M. U. Scherzer for many clarifying discussions. We also thank H. Wacker and A. Eicher for their technical assistance. The targets were prepared in the engineering division. The scanning electron micrographs were made in the Laboratory for Scanning-Electron-Microscopy of Dr. H. Klingele, Munich.

Appendix

INFLUENCE OF THE CURRENT DENSITY DISTRIBUTION ON THE SPUTTERING RATE

When the current densities j_i and j_z during the thickness measurement or sputtering process are not constant over the bombarded area, the sputtering yield is given by the equation

$$S = \frac{N_{r_1} - N_{r_2}}{\frac{d\sigma_R}{d\Omega} \cdot \Delta\Omega \cdot \frac{1}{e^2} \cdot t_i \cdot t_z \int_F j_i \cdot j_z \cdot df} \quad (7)$$

where t_i and t_z are the ion bombardment times of the thickness measurement and sputtering respectively and e is the elementary charge. All that can be measured

are the bombardment doses

$$N_i = \frac{1}{e} t_i \cdot \int j_i df; \qquad N_z = \frac{1}{e} t_z \cdot \int_F j_z df$$

Comparison of Eqs. (5) and (9) yields for the correction factor

$$\gamma = \frac{\left(\int j_i df\right) \cdot \left(\int j_z df\right)}{F \cdot \int j_i \cdot j_z df} \quad (8)$$

In the Pharao device the current density distribution is governed by the arrangement of the diaphragms and their diameters. It was shown that the current density was essentially constant over the bombarded area and strongly decreased at the boundary only, hence a trapezoidal current density distribution could be assumed. The calculation for this geometry yielded $\gamma = 0.97$.

REFERENCES

1. G. K. Wehner, *Advances in Electronics and Electron Physics*, **VIII**, 239 (1955).
2. R. Behrisch, *Erg. Exact. Naturw.* **35**, 295 (1964).
3. M. Kaminsky, *Atomic and Ionic Impact Phenomena on Metal Surfaces* (Springer Verlag, Berlin, 1965).
4. G. Carter and J. S. Colligon, *Ion Bombardment of Solids* (Elsevier Publishing Co., Inc., New York, 1968).
5. N. V. Pleshivtsev, *Cathode Sputtering* (in Russian) (Atomizdat Moscow, 1968).
6. R. J. MacDonald, *Adv. Phys.* **19**, 457 (1970).
7. R. Behrisch and R. Weißmann, *Phys. Lett.* **30A**, 506 (1969).
8. E. B. Henschke, *Phys. Rev.* **106**, 737 (1957).
9. D. E. Harrison Jr., N. S. Levy, I. P. Johnson III and H. M. Effron, *J. Appl. Phys.* **39**, 3742 (1968).
10. E. Waldschmidt, *Metall.* **8**, 749 (1954).
11. G. Alefeld, *Phys. Stat. Sol.* **32**, 67 (1968).
12. W. Primak, *J. Appl. Phys.* **34**, 1342 (1964).
13. M. Kaminsky, *Adv. Mass. Spectrometry* **3**, 69 (1964).
14. W. Primak and J. Luthra, *J. Appl. Phys.* **37**, 2287 (1966).
15. L. H. Milacek and R. D. Daniels, *J. Appl. Phys.* **39**, 2803 (1968).
16. G. J. Thomas, W. Bauer and J. B. Holt, *Rad. Effects* **7**, 269 (1971).
17. G. McCracken, *J. Vac. Sc. Techn.* **9**, 361 (1972).
18. C. E. KenKnight and G. K. Wehner, *J. Appl. Phys.* **35**, 322 (1964).
19. A. Benninghoven, *Z. angew. Phys.* **27**, 51 (1969).
20. A. K. Furr and C. R. Finfgeld, *J. Appl. Phys.* **41**, 1739 (1970).
21. A. J. Summers, N. J. Freeman and N. R. Daly, *J. Appl. Phys.* **42**, 4774 (1971).
22. E. Bøgh, *Can. J. Phys.* **46**, 653 (1968).
23. J. W. Mayer, L. Eriksson and J. A. Davies, *Ion implantation in semiconductors* (Academic Press Inc., New York, 1970).
24. A. B. Brown, C. W. Snyder, A. W. Fowler and C. C. Lauritsen, *Phys. Rev.* **82**, 159 (1951).
25. S. Rubin, *Nucl. Instr. Meth.* **5**, 177 (1959).
26. R. Behrisch, *Vak. Technik* **10**, 250 (1967).
27. H. Schmidl, IPP 9/3 Laborbericht (1971).
28. H. H. Andersen and H. Bay, *Rad. Effects* **13**, 67 (1972).
29. D. T. Goldman and A. Simon, *Phys. Rev.* **111**, 383 (1968).
30. R. S. Pease, *Proceedings of the Int. School of Phys. "Enrico Fermi" Course* (Academic Press Inc., New York, 1964), Vol. 13, p. 158.
31. Yu. V. Bulgakov, *Sov. Phys.-Techn. Phys.* **8**, 369 (1963).
32. P. Sigmund, *Phys. Rev.* **184**, 383 (1969).
33. R. Weißmann and P. Sigmund, submitted to *Rad. Effects* (previous paper).

SPUTTERING-YIELD STUDIES ON SILICON AND SILVER TARGETS

H. H. ANDERSEN and H. L. BAY

Institute of Physics, University of Aarhus DK-8000 Aarhus C, Denmark

The sputtering yield of vacuum-deposited silicon and silver targets has been measured with 15 different 45-keV ions throughout the periodic system. The dependence of the yield on the projectile atomic number follows rather closely Sigmund's prediction, especially for silicon. The self-sputtering yield of silver was measured in the 30–500 keV energy range. Similar to what is known to be the case for other heavy particles on silver, the maximum of the yield was found to be much more pronounced than predicted by Sigmund's theory. This effect, together with small systematic deviations in the Z_1 dependence, is explained as being caused by non-linear effects in very dense collision cascades. This point of view is strongly supported by a comparison of the sputtering yield per atom for irradiation with atomic and molecular ions.

INTRODUCTION

The study of sputtering yields of amorphous and poly-crystalline targets has, throughout the years, produced a confusing amount of apparently uncorrelated data.[1] The lack of a comprehensive theory describing the effect of different projectile-target combinations and projectile energies was strongly felt. The recent publication of Sigmund's theoretical paper on sputtering[2] has, to a great extent, changed this situation, and his theoretical results constitute a convenient starting point for new experimental studies.

Sigmund's theory is based upon the collision-cascade concept and thus connects sputtering to other cascade phenomena such as range distributions, radiation damage, etc. The theoretical treatment assumes the target to be random and infinite. Through moment expansion of a linearized Boltzmann transport equation, it is shown that the sputtering yield is directly proportional to the value at the target surface of the function, describing the final spatial distribution of the projectile energy. The thick-target sputtering yield (number of sputtered atoms per incoming ion) is, in the elastic-collision region, found to be

$$S = \lambda \frac{S_n(E) \cdot \alpha(M_2/M_1)}{U_0} \qquad (1)$$

for perpendicular incidence of the ions. Here, λ is a calculable constant, $S_n(E)$ the nuclear stopping power,[3] and $\alpha(M_2/M_1)$ is a numerically calculated function of the ratio between the target mass (M_2) and projectile mass (M_1). For $M_2/M_1 < 1$, this function is nearly constant, and it increases rather sharply for high mass ratios. (The theoretical α function may be seen in Figure 10.) Finally, U_0 is a surface-binding energy, usually set equal to the sublimation energy of the target material.

The proportionality of the yield to $S_n(E)$ has been argued before[4] in a much less detailed manner. As is seen from Sigmund's comprehensive comparisons with existing experimental results, this proportionality is found experimentally for relatively light ions (mainly Ne and Ar) on all targets and also for intermediate ions (e.g. Kr) on light targets. All other ion-target combinations yield a more pronounced maximum than predicted by theory. Furthermore, the maximum is apparently also found at lower energies than predicted.

Sigmund points out (Ref. 2, p. 397) that studies of the dependence of sputtering yield on the projectile mass for a fixed target are a better check of his basic assumptions than are comparisons with measured energy dependences. Furthermore, the dependence on angle of incidence should also be sensitive to critical parameters in the theory. As our experimental equipment allows no adjustment of the incidence angle, we shall concentrate on the variations of Z_1, Z_2, and E. Extensive studies of the variation of yield as a function of Z_1 have been performed by Almén and Bruce[5] for copper, silver, and tantalum targets. All these measurements were carried out at constant energy. For a fixed target, these authors found the yield to increase with increasing Z_1. Strong oscillations following the chemical nature of the projectiles were superimposed on this increase. In an earlier publication[6] we showed that for copper, these oscillations were a consequence of the large doses used by Almén and Bruce. The α function deduced from our measurements on copper was in good agreement with Sigmund's predictions except in the case of large mass ratios, where α was found to increase more slowly with M_2/M_1 than calculated.

The purpose of the present study is partly to find whether a common α function may be deduced from measurements on silicon, copper, and silver, and partly

63

to look for explanations of the systematic differences
between theory and experiment found in Ref. 6. More
generally, the data presented here have been obtained
in order to make it possible to investigate the influence
of some of the basic assumptions in Sigmund's theory,
viz. the use of a linear transport equation as well as the
assumption concerning an infinite target medium.
Finally, the influence of the neglect of electronic-energy
losses will be touched upon.

EXPERIMENTAL

In sputtering experiments, contamination of the target
is a serious problem and may be caused by the background gas in the sputtering chamber as well as by the
irradiating particles. Hence it is essential to be able to
work in a good vacuum as well as with high sensitivity.
The former condition will keep the number of gas
molecules hitting the target low, while the latter allows
yields to be measured by means of small doses. The
above aims are partly reached by the equipment used
to obtain the data presented here.

The equipment is shown in Figure 1. The chamber
is made from stainless steel with metal sealings. It is
pumped by a standard 4″ diffusion pump with liquid-
nitrogen cooling trap. During irradiation, the pressure

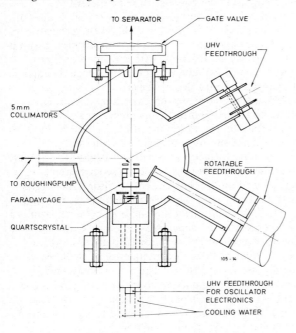

FIGURE 1   The stainless-steel sputtering chamber viewed
from above. Diffusion pumps are placed below the section
shown and the ionization pressure gauge above.

is 4×10^{-8} to 10×10^{-8} torr. Prior to a run, the target
material has been vacuum-deposited on one side of thin
quartz-crystal discs. These discs are part of an oscillator
circuit, and their eigenfrequency depends on their mass.
The mass changes during irradiation are small as compared to the total mass of the crystal, and the frequency
change will thus be directly proportional to the mass
change of the target. The system therefore constitutes
an *in situ* determination of the sputtering weight loss
of the target with a sensitivity of approximately
6×10^{-9} g. As the irradiated area is 5 mm in diameter,
this sensitivity corresponds to the removal of less than
one atomic layer for all elements. The absolute
accuracy of the weight-loss measurements (for
$\Delta f > 50$ Hz) was found to be better than 2%. The dose
in each run is determined by measuring the current
before and after the run and the irradiation time.
Measurements of the current are performed by means
of a removable Faraday cup provided with secondary
electron suppression. The absolute accuracy of the
current measurements was found to be 8% (one standard
deviation).[6] For a fixed projectile-target combination at
a given energy, the scatter in the yield data is of the
order of 2–3% when the beam current is stable.

From the measured frequency changes Δf and the
number N of the projectile atoms, the sputtering yield
S is obtained as

$$S = \frac{k_0 N_0}{m_2} \times \frac{\Delta f}{N} + \frac{m_1}{m_2}. \qquad (2)$$

Here, m_1 and m_2 are the atomic weights (g/mole) of the
projectile and target material, respectively. N_0 is
Avogadro's number (6.02×10^{23} atoms/mole), and k_0
the calibration constant of the instrument (5.97×10^{-9}
g/Hz). The term m_1/m_2 corrects for the mass of the
incident ions. In Eq. (2), direct reflection of the incoming ions is neglected. This could be taken into
account by a further correction term $-R \times m_1/m_2$,
where R is the reflection coefficient. Using the reflection coefficients of Bøttiger *et al.*,[7] this term was found
to be less than 1% of S in all cases where S was larger
than 1. The highest value ($\sim$4%) was reached for C^+
irradiation of Cu, Ag, and Au. As 4% is not very
significant in our measurements, the term was disregarded. Equation (2) is thus a good approximation at
the beginning of the irradiation of a clean target, which
may be ensured by prolonged self-sputtering. During
irradiation with an alien projectile, a concentration of
this projectile builds up within the target. If an equilibrium is reached, as many projectile atoms leave as
arrive on the target. Equation (2) without the term
m_1/m_2 then gives the yield of atoms of the original

target material. If the presence of projectile species within the target influences the yield, saturation is reached when the yield stops changing with dose. In Ref. 6 it was found that equilibrium in the above sense was not often reached in the case of copper as target material. In these intermediate situations, it appears to be appropriate to use Eq. (2) to *define* sputtering yields as measured by weight losses. For copper, silver, and gold, we always find yields higher than 10 when m_1/m_2 is higher than one. As seen below, however, the correction term m_1/m_2 constitutes 80% of the total yield in the case of lead irradiation of silicon, and the data will necessarily be less accurate in such cases.

When a light target is irradiated with a heavy projectile (e.g. $Pb^+ \rightarrow Si$), the implanted lead atoms might enhance the reflection coefficient for subsequent irradiations with lighter ions. As discussed below, argon irradiation was used to clean our silicon target between irradiations with different heavier ions. All irradiations of silicon with ions heavier than argon were analysed in an attempt to find systematic influences on the silicon–argon reflection coefficient. Within our measuring accuracy, no systematic variation was found.

Finally, the target-preparation procedure is part of the experimental technique and will briefly be commented on. The metal targets (V, Mn, Fe, Cu, and Zn in Ref. 6, Ag in this publication, and Zr and Au in stock for future work) are all prepared by electron-beam evaporation in an UHV evaporation unit. Usually, evaporation does not start until the pressure is well down in the 10^{-8} torr region. During evaporation, the pressure often rises to 10^{-6} torr. 1 micron material is deposited within a few minutes; the fast evaporation in a good vacuum ensures clean targets. All metal targets were found to be polycrystalline and texture-free with a grain size of the order of 1 micron. The silicon targets were prepared by low-energy, argon-sputtering deposition also in an UHV unit. These targets were amorphous (or at least extremely fine-grained). Hence they are particularly well-suited as targets for experiments aimed at a comparison with theories assuming a random target.

RESULTS AND DISCUSSION

The results presented in this section will concentrate on the influence of three different parameters of the incident projectile on the sputtering yield of silver and silicon: The atomic number Z_1 and energy E of the incoming ions, and the possibility that they arrive as atomic or molecular ions. For all three parameters, we shall see deviations from the dependence predicted by theory and try to give a common, qualitative explanation of these deviations.

In Ref. 6, a strong dose dependence was found for the sputtering yield of several elements incident on copper targets. Two cases with a particularly strong dose dependence were selected for a further, more detailed study.[8] Here we shall only briefly mention the dose dependence found for silver and silicon targets. For silver, the self-sputtering yield at 45 keV decreased rather strongly with dose, as was the case for copper. The extrapolation back to the original surface (zero dose) yielded a value of 50 for the self-sputtering yield. This may be compared with the value of 26, obtained by Almén and Bruce.[5] It is, however, more reasonable to compare our average yield for a dose similar to that used by Almén and Bruce with their value. This comparison shows good agreement (see Figure 5). For silver, P, Zn, Xe, Cs, Pb, and Bi ions showed pronounced dose effects. Note the contrast to copper (Ref. 6), where P, Zn, and Pb showed no dose effects. As was generally the case also for copper, elements corresponding to minima in Almén and Bruce's high-dose yield versus Z_1 showed a decrease with increasing dose. For silicon targets, Zn and Ag ions gave moderate dose effects, while all projectiles heavier than silver showed strong increases in yield with dose. Similar, but less pronounced effects were found for copper. This is probably in many cases caused by the assumption that all sputtered atoms have mass m_2 (Eq. 2).

The strong dose effects prevent us from using absolute yield data when comparing with theory. As in Ref. 6, we use normalized yields. We irradiate alternatingly with a normalizing ion (45 keV Ag in the case of silver targets, 45 keV Ar in the case of silicon targets) and with the ion in question. Four to six runs are normally made with a specific combination before we switch back to the normalizing case for which another four–six runs are made. At the moment of switching, the target is exactly the same for the two combinations, and we may compare the measured yield ratio with the theoretical prediction

$$\frac{S_{Z_1 \rightarrow Z_2}}{S_{norm}} = \frac{S_n(Z_1, Z_2) \times \alpha(M_2/M_1)}{S_n(Z_{norm}, Z_2) \times \alpha(M_2/M_{norm})}, \quad (3)$$

obtained from Eq. (1). If self-sputtering is used for normalization, the denominator reduces to $S_n(Z_2, Z_2) \times \alpha(1)$. This normalization is the most convenient as the target is then simultaneously cleaned for alien elements. For silicon, this procedure was not possible as it is difficult to obtain a clean silicon beam from an isotope separator (mass number 28 is mostly N_2^+ and CO^+). Hence argon was chosen for normalization, which was fortunate as no dose effects were seen for this combination.

The results of relative sputtering-yield measurements at 45 keV on silicon, copper (from Ref. 6), and silver are shown in Figures 2, 3, and 4, where also the theoretical predictions are shown (Eq. 3). The numerical values were calculated by means of the α function of Sigmund[2] and the nuclear stopping power of Lindhard et al.[3] The fact that the measured points lie on a smooth curve (scatter less than ±10%) shows that the nuclear stopping power varies smoothly as a function of Z_1. This conclusion has been reached recently also for other targets by means of measurements of multiple scattering[9] and energy loss.[10]

The theory is seen to predict the general trend of the data well. For silicon, the agreement is nearly perfect, but the heavier the target, the more pronounced

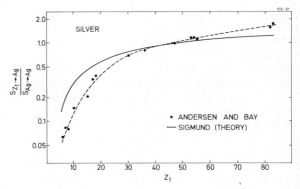

FIGURE 4   The sputtering yield of 15 different 45 keV ions in silver, normalized to silver self-sputtering. The fully drawn line represents the theoretical prediction of Sigmund.[2]

the tendency of the experimental points to increase faster with Z_1 than predicted theoretically. This appears to be connected to tendencies in the energy dependence of the yield which, as mentioned in the introduction, tends to show a more pronounced maximum for heavy targets than expected from the behaviour of S_n. We shall thus resume the discussion of Figures 2–4 when data on energy dependence and the influence of molecular ions have been presented.

Our equipment is not particularly well-suited for measurements of the energy dependence of sputtering yields because of the need for normalization. In Figure 5 we present the result of an investigation of the self-sputtering yield of silver. The data are given as absolute yields, where we have chosen the average value over all measured points of 45 keV self-sputtering as the normalization value. Hence the data correspond to large doses, and their absolute magnitude agrees rather well with the data of Almén and Bruce.[5] As shown by Sigmund,[2] the maximum may be nearly a factor of two

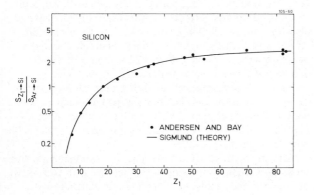

FIGURE 2   The sputtering yield of 15 different 45 keV ions in silicon normalized to 45 keV Ar⁺ yields in silicon. The fully drawn line represents the theoretical prediction of Sigmund.[2]

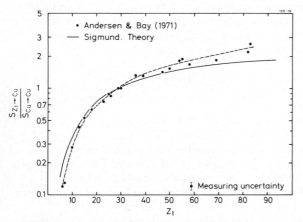

FIGURE 3   The sputtering yield of 21 different 45 keV ions in copper normalized to copper self-sputtering. The fully drawn line represents the theoretical prediction of Sigmund.[2]

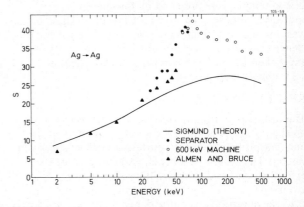

FIGURE 5   The energy dependence of silver self-sputtering. Also shown are data of Almén and Bruce[5] and the theoretical prediction.[2]

higher than predicted, while the absolute agreement is rather good at lower energies. Although the position of the maximum is not deduced with high precision from the data, the experimental maximum is clearly found at lower energies than the theoretical one.

We have also attempted to measure the energy dependence for a very heavy projectile (Pb) in silicon. The data points scatter considerably as the frequency often changes by 10 Hz only during one run. Although measurements were performed over the energy range from 30 to 500 keV, no definite conclusion as to the position of the maximum could be reached.

From the data presented here, both concerning Z_1 dependence and energy dependence, it is clear that deviations from the theoretical predictions of Sigmund are seen in the case of high nuclear-stopping powers. Sigmund[11] proposed that the deviation from proportionality with $S_n(E)$ was a non-linear effect arising in very dense cascades. We looked for possibilities of isolating this effect and concluded that the use of molecular ions probably would serve this purpose. By irradiation with molecular ions, a high degree of overlapping of the individual cascades created by the members of the molecule will be obtained. This increases the energy density in the overlapping region. A comparison with the yield *per atom* for atomic and molecular ions with the same energy *per atom* will clearly show the possible effect of the overlapping cascades as it constitutes the only difference between the two cases.

Measurements were performed at energies slightly below the maximum of the nuclear-stopping power. Results for silver targets irradiated with Cl_2^+, Se_2^+, and Te_2^+ are presented in Figures 6 through 8. These particular projectiles have been chosen because the ion source conveniently delivers sufficiently large currents

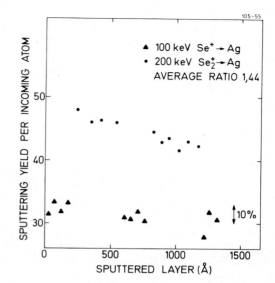

FIGURE 7   The sputtering yield *per atom* of silver irradiated with 100 keV Se^+ and 200 keV Se_2^+ ions. Non-linear effects are clearly seen.

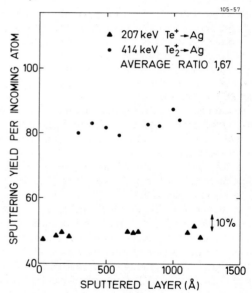

FIGURE 8   The sputtering yield *per atom* of silver irradiated with 207 keV Te^+ and 414 keV Te_2^+ ions. Very pronounced non-linear effects are seen.

of the molecular ions. The figures show absolute yields, and it is seen that we have alternated between irradiations with the atomic and molecular ions to be able to correct for dose effects. The average molecular-to-atomic-yield ratio is stated in the figures. We also tried to irradiate silicon with Se_2^+ and Te_2^+. The results for $Te^+ - Te_2^+$ are shown in Figure 9, where stronger

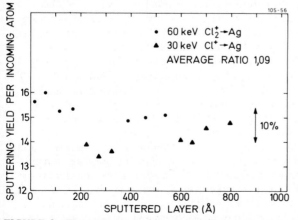

FIGURE 6   The sputtering yield *per atom* of silver irradiated with 30 keV Cl^+ and 60 keV Cl_2^+ ions. Non-linear effects are not very significant.

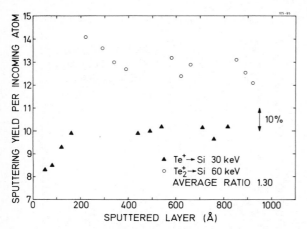

FIGURE 9   The sputtering yield *per atom* of silicon irradiated with 30 keV Te^+ and 60 keV Te_2^+ ions. Although rather strong dose effects are present, non-linear effects are also clearly seen.

systematic dose effects are seen, probably caused by tellurium trapped in the target. The tellurium concentration increases during irradiation with the atomic beam. As discussed in connection with Eq. (2), more tellurium is then resputtered, corresponding to an increasing weight loss per incoming atom with increasing dose. Switching to molecular ions, we have a considerably higher yield. This will cause the tellurium concentration to decrease and hence the weight loss per atom to decrease with increasing dose.

The general trend of the results shown in Figures 6 through 9 is the same as that for the energy dependence. The heavier the projectile and target, the more pronounced the deviation from the expected behaviour. As we do not have additivity of the effect of the individual members of the molecule, there is a clear evidence of non-linear effects.

The influence of irradiation with molecular ions has been investigated earlier.[12–14] With one exception, no evidence of non-linear effects was found, but in the light of our results, it is plain that they were made with either too light ions (N_2^+ being the heaviest) or at too low energies (below 1 keV). Among other cases, where the above reservations hold, Rol, *et al.*,[14] however, studied the yield of 30 to 50 keV KI^+ ions impinging on copper and compared it with the result calculated from the yield of the individual K^+ and I^+ ions at the appropriate energies. The yield of the molecular ions was 20% higher than calculated. No corrections for dose effects are possible in this case. The result has, apart from being mentioned very briefly in Behrisch's review article,[1] never been discussed in the literature.

We shall here outline a qualitative explanation of the observed non-linear effects within the scheme

originally proposed by Sigmund.[11] Within a dense cascade, the density especially of atoms moving with low energies will be high as the energy distribution of the moving atoms within the cascade is roughly proportional to E^{-2} (see e.g. Ref. 2). At some low energy, the linear transport theory must thus break down as it requires one partner in each collision to lie still in the laboratory system prior to the collision. At some point in the energy spectrum we must therefore have an enhanced number of recoils as they cannot distribute their energy to other atoms through the mechanism working at higher energies. If this critical energy rises above the binding energy at the surface, we shall find an enhancement in sputtering. Within this picture, non-linear effects for molecular ions may be seen also in cases where no non-linear effects are seen in the energy dependence. If non-linear effects exist in the energy dependence, they will be most pronounced below the maximum in S_n, as is also observed experimentally.

In a qualitative way, the outlined model is connected to the concept of thermal spikes used to explain the energy spectra of sputtered atoms at very low energies.[15] If the explanation of our experimental results is correct, then an enhancement in the low-energy end of the spectra must be seen in the case of irradiation with molecular ions.

As the non-linear effects discussed above are not contained in Eq. (1), we may try to correct for them when comparing the Z_1 dependences shown in Figures 2 through 4 with the theoretical predictions. Firstly, it seems reasonable to explain the rise above the theoretical curve for large Z_1 by this effect. Secondly, if large, non-linear effects are significant at the normalization point, this will give too high a normalization yield. As the normalization point must coincide with the theoretical curve, points for lower Z_1 will be forced below the curve. Only in the case of silver are non-linear effects considered important in this context.

Figure 10 shows $\alpha(M_2/M_1)$ extracted from Z_1-dependence measurements on Si, Cu, and Ag, using the nuclear-stopping powers, $S_n(E)$, of Lindherd *et al.*[3] The non-linear effect for the silver normalization has been estimated from the energy dependence of silver self-sputtering (Figure 5), and the arrows for large mass ratios (low Z_1) indicate the approximate size of corrections for this particular non-linear effect. Taking this correction into account, a very good agreement between α functions deduced from measurements on different targets is found, indicating that Eq. (1) with a universal α function is a good description of the (Z_1, Z_2) variation of sputtering yields. If we disregard the possibility of compensating errors, Figure 10 also shows that the nuclear-stopping powers of Lindhard *et al.*[3]

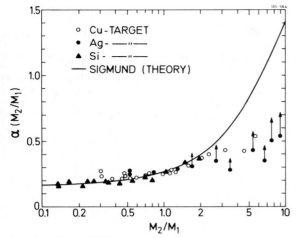

FIGURE 10 $\alpha(M_2/M_1)$ extracted from measurements with 15 different ions on silicon, 21 on copper, and 15 on silver. The height of the arrow for the silver points at large mass ratios indicates tentative corrections for non-linear-effect influence on the normalization point.

describe the (Z_1, Z_2) variation of S_n well. No evidence of systematic deviations for heavy targets, which is indicated to exist from multiple-scattering measurements,[9] is seen here.

The deduced α function is nearly constant for $M_2/M_1 < 1$, as predicted theoretically. The silicon results, where non-linear effects are expected to be very small, show that even the detailed behaviour follows the theory closely. This is not the case for large mass ratios. Theory predicts α to increase by a factor of four by increasing M_2/M_1 from 1 to 5. Experimentally we found an increase of a factor of two only. In Ref. 6 we have tentatively explained this difference as being caused by electronic-energy losses. In the light of the agreement between α functions deduced from measurements on copper and silver, this explanation is not very probable. As an example, mass ratio 5.3 corresponds to $C^+ \rightarrow Cu$ and $N^+ \rightarrow Ag$. The points lie close to each other, but in the former case, the ratio between electronic and nuclear stopping is 2.0, while it is 0.5 in the latter case.

More recent calculations[16] indicated some influence of electronic-energy losses at $M_2/M_1 = 1$. For very light projectiles, Weissmann and Sigmund[17] found α to be a function of projectile species and energy only, i.e., not of the target. The numerical calculations were performed for hydrogen and helium isotopes only, and it is not clear whether the approximations involved will hold for, e.g. carbon and nitrogen at sufficiently high energies. If values belonging to the above region are plotted as a function of M_2/M_1 at a fixed energy, they

will decrease with increasing M_2/M_1. We do not find such a behaviour, even for the lightest projectiles on copper, but it would be interesting to measure $\alpha(M_2/M_1)$ at considerably higher energies than used here.

The most probable cause of the difference between theoretical and experimental α values at large mass ratios is the assumption of an infinite target used in the calculations. In the model, a light projectile once scattered out through the surface may be scattered back into the target, and it may, in particular, pass along close to the surface, depositing large amounts of energy in this region, which is favourable for sputtering. As backscattering is most important for light ions, this source of error will be most important for large M_2/M_1. Surface corrections have been developed for the simpler case of calculating backscattering coefficients,[7] and theoretical progress along similar lines must be awaited with interest as far as sputtering is concerned.

CONCLUSIONS

1) The sputtering yield of polycrystalline and amorphous samples is well described by Sigmund's theory.[2]

2) Non-linear effects modify conclusion (1) for the case of intermediate and heavy projectiles with energies close to the maximum in nuclear stopping.

3) Non-linear effects are most strongly in evidence when the yield *per atom* obtained for irradiations is compared with atomic and molecular ions. These effects increase with increasing Z_1 and Z_2.

4) Sigmund's α function is high by at least a factor of two for large mass ratios. This is possibly caused by a lack of surface corrections in the theory.

ACKNOWLEDGEMENTS

All our metal targets were prepared by Arne Nordskov and Hans Sørensen of the Danish Atomic Energy Commission. The silicon targets were supplied by Dr. H. Garvin, Hughes Aircraft Laboratories, Malibu, California. We are very grateful for these significant contributions to our experiments. We are also indebted to Dr. P. Sigmund for his taking part in numerous discussions during the preparation of this work.

REFERENCES

1. K. Behrisch, *Ergeb. exakt Naturw.* **35**, 295 (1964).
2. P. Sigmund, *Phys. Rev.* **184**, 383 (1969).
3. J. Lindhard, M. Scharff and H. E. Schiøtt, *Kgl. Danske Videnskab. Selskab, Mat.-Fys. Medd.* **33**, No. 14 (1963).
4. P. V. Rol, J. M. Fluit and J. Kistemaker, *Physica* **26**, 1009 (1960).
5. O. Almén and G. Bruce, *Nucl. Instr. Methods* **11**, 279 (1961).

6. H. H. Andersen and H. Bay, *Rad. Effects* **13**, 67 (1972).
7. J. Bøttiger, J. Davies, P. Sigmund and B. Winterbon, *Rad. Effects* **11**, 69 (1971).
8. H. H. Andersen; Proc. this Conf., *Rad. Effects* (to be published).
9. H. H. Andersen, J. Bøttiger and H. Knudsen, *Rad. Effects* **13**, 203 (1972).
10. G. Högberg and R. Skoog, *Rad. Effects* **13**, 197 (1972).
11. P. Sigmund, Int. Conf. on Atomic Collisions in Solids, Gausdal, Norway (1971) (unpublished).

12. F. Grønlund and W. J. Moore, *J. Chem. Phys.* **32**, 1540 (1960).
13. M. Bader, F. C. Witteborn and T. W. Snouse; NASA Techn. Report R-105 (1961).
14. P. K. Rol, J. M. Fluit, and J. Kistemaker, *Physica* **26**, 1000 (1960).
15. M. W. Thompson and R. S. Nelson, *Phil. Mag.* **7**, 2015 (1962).
16. P. Sigmund, M. T. Matthies and D. L. Phillips, *Rad. Effects* **11**, 34 (1971).
17. R. Weissmann and P. Sigmund; Proc. this Conf., *Rad. Effects* (to be published).

THE DOSE DEPENDENCE OF 45 keV V^+ AND Bi^+ ION SPUTTERING YIELD OF COPPER

HANS HENRIK ANDERSEN

Institute of Physics, University of Aarhus DK-8000 Aarhus C, Denmark

The sputtering yield of 45 keV V^+ ions on copper is found to decrease with increasing dose even for very large doses, whilst the yield of 45 keV Bi^+ ions on copper increases with increasing dose until approximately ten times the projected range of the Bi^+ ions has been sputtered away. The fate of the incoming V^+ and Bi^+ ions was investigated. A radiotracer technique showed that vanadium accumulated on the target surface, whilst bismuth diffused into the target, as was seen by means of Rutherford-scattering of light ions.

INTRODUCTION

Until ten years ago, the dose dependence of the heavy-ion sputtering yield has been only sporadically studied. At the time, Almén and Bruce[1] published some spectacular examples of 45 keV V^+-ion irradiation of copper and tantalum targets, where they found a marked decrease in the sputtering yield as a function of dose. For 45 keV Ca^+ ions on the same targets, they even found a weight gain in their targets. They had to use doses of the order of 10^{17} ions/cm^2 to obtain measurable weight changes and to sputter several hundred atomic layers off their target before obtaining the first point on a dose curve.

In the same article, the authors also reported measurements of the Z_1 variation of the 45 keV sputtering yield on copper, silver, and tantalum targets. In the Z_1 variation they found peculiar oscillations following the chemical properties of the incident ions and ascribed these variations to the change in target material caused by the projectile atoms embedded in the target. This assumption has recently been shown to be correct,[2] as the periodic Z_1 variations are not found for much smaller doses (10^{15} ions/cm^2). Hence the periodic Z_1 variations are also an evidence of dose effects in heavy-ion sputtering yields of metals.

A few cases have been studied in more detail. Thackery and Nelson[3] irradiated aluminum with heavy ions. They studied their target foils in the electron microscope and found cases, most notable copper in aluminum, where the irradiated species precipitated as a separate phase. This happened where the projectile atoms were known to have a very low solubility in the target material. Among the many studies of surface topography, Wehner and Hajicek[4] found that molybdenum had a curious influence on the sputtering process of copper. The presence of molybdenum strongly favoured the creation of erosion cones on the copper surface, cones that were otherwise known to mainly originate beneath dust and oxide particles on the surface.[5] The formation of a high concentration of erosion cones on a target surface may also give rise to dose effects in the sputtering yield.

Dose effects may also occur if gas bubbles are formed during irradiation. Such bubbles are seen in many cases where irradiation with large doses of noble gases is performed (e.g. Ar on Cu[6]). Gas-bubble formation and, eventually, blistering of the surface will also be found in the cases of deuterium and proton irradiation.[7] The influence of these phenomena on the sputtering yield has not been studied in any detail.

Sigmund[8] predicted the back-sputtering yield to be given by the formula

$$S = \lambda \frac{S_n(E_1, Z_1, Z_2) \times \alpha(M_2/M_1)}{U_0} \qquad (1)$$

for perpendicular incidence. Here, λ is a calculable constant in the elastic-collision region, S_n the nuclear stopping power, which is a function of projectile energy, atomic number, and target material, and α a function depending on the ratio of target (M_2) and projectile (M_1) mass only. U_0 is a surface-binding energy, mostly set equal to the sublimation energy of the target material. Equation (1) is a convenient starting point for a discussion of dose effects in sputtering. The equilibrium concentration of projectile atoms within the target will be of the order of the inverse of the sputtering yield, i.e., only for targets with a very low sputtering yield may we expect to find an appreciable influence on S_n or α. In most cases, dose effects must be expected to proceed via a change in U_0. This change may, in turn, be caused by changes in the electronic structure as well as in the surface topography. If the

projectiles neither precipitate nor diffuse to any considerable extent, saturation must be reached when a
layer corresponding to the depth distribution of the
projectiles has been sputtered away. Only changes in
the surface topography may, under the above conditions, cause exceptions from this rule, but then they
will hardly be correlated with the specific chemical
properties of the implanted ions.

The work of Andersen and Bay[2] showed several
examples of such simple behaviour, where saturation
is reached before twice the projected range of the
incident particle is sputtered away. Examples of this
are seen for 45 keV N^+, Ar^+, Kr^+, Cs^+, Ce^+, and Al^+
irradiations of copper. Strong decreases with no sign
of rapid saturation were found for Sn^+, V^+, and Y^+—
the latter two cases in complete agreement with the
results of Almén and Bruce.[1] In Ref. 2, a few examples
of strong increases in yield have been found. This has
not previously been reported in the literature. The
cases occurred for irradiation with Tm^+ and Bi^+ and
did not reveal saturation until after the erosion of
approximately ten times the ion range in the target
material.

We have selected two cases from Ref. 2 for a more
detailed study, i.e., sputtering of copper with 45 keV
V^+ and Bi^+ ions. As mentioned above, both cases show
strong dose dependence in connection with a slow
approach toward saturation. As discussed below, there
is evidence for both cases that the dose dependence is
caused by the specific behaviour of the implanted ions
in the target, and the remainder of this paper is devoted
to a particular study of the fate of vanadium and
bismuth ions implanted into copper.

VANADIUM IRRADIATIONS

The targets used for the yield studies in Ref. 2 were
evaporated onto silver electrodes, which had been
previously deposited on quartz discs, constituting the
sensitive elements in a vacuum micro-balance. The size
of the quartz-crystal discs made them unsuitable for the
present studies. To simulate the targets used for yield
measurements as closely as possible, vacuum-deposited
samples were prepared from the same charge in the
same vacuum unit and under the same conditions as
the targets used for the yield studies, but small pieces of
0.2 mm thick copper foils were used as substrates
instead of the silver electrodes of the quartz crystals.
As was the case for the yield-study targets, no texture
was found in the targets deposited on copper substrates.

Figure 1 shows the yield data of Ref. 1, replotted on a
double-logarithmic scale, together with our data. The
absolute agreement between our results and those of

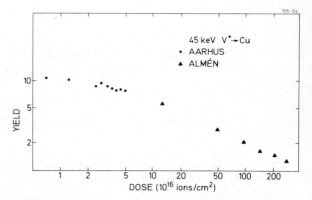

FIGURE 1   Dose dependence of 45 keV V^+ sputtering yield
of copper. Data of Almén and Bruce[1] are also shown.

Almén and Bruce is surprisingly good. It is seen that
saturation is still not reached after a dose of 3×10^{18}
ions/cm^2. At this point, several thousand Ångströms of
target material have been sputtered away, and the yield
has decreased by more than one order of magnitude. In
Ref. 2 we switched to irradiation with 45 keV Cu^+ ions
after irradiation with 5×10^{16} V^+ ions/cm^2 and found
an increase in yield as a function of the copper-ion dose.

As the copper ions are rather close in mass to those of
vanadium, the collision cascade will be very similar in
the two cases, and the different saturation behaviour
thus indicates that the decrease in the case of vanadium
ions is specifically caused by the presence of the
vanadium atoms in the copper matrix.

The equilibrium behaviour of vanadium in copper is
not very well known,[9] but several authors claim the
solubility of vanadium in copper to be low.[10] It is of
special interest here that Guillet[11] maintains the
existence of a copper-rich phase of a characteristic
bluish colour. We observed this colour for all specimens
irradiated with more than 10^{16} V^+ ions/cm^2.

To reveal the fate of the incoming vanadium atoms,
trace amounts of ^{48}V radioactive nuclei were implanted
into a number of identical copper samples. These
samples were then sputtered with non-radioactive, 45
keV ^{51}V in doses of 4×10^{15}/cm^2 to 4×10^{17}/cm^2.
Figure 2 shows the fraction of the ^{48}V activity which
remains as a function of sputtering dose. Arrows marked
$2 \times R$, etc., indicate the doses calculated to sputter away
different multiples of the range of the ^{48}V atoms. To
calculate the sputtered layers, the data in Figure 1 have
been used. It is seen that the removal of ten times the
tracer range still leaves 15% of the original activity
within the target.

Figure 3 shows the depth distribution of the origin-
ally implanted ^{48}V atoms after sputtering with 4×10^{14},

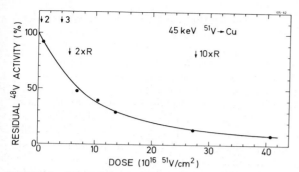

FIGURE 2   The residual activity of a tracer implantation of 45 keV ^{48}V after sputtering with different doses of ^{51}V. Arrows 2 x R show doses corresponding to the removal of twice and ten times the projected range of the tracer-implanted ions.

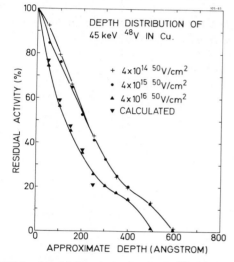

FIGURE 3   Integral depth distribution of ^{48}V in copper after sputtering with ^{50}V at the doses indicated. The calculated curve was obtained from that of the low dose by a 300 Å shift to the surface and renormalization at the resulting surface.

4×10^{15}, and 4×10^{16} ^{50}V/cm^2, respectively. Thin layers were stripped from the copper targets by means of the corrosion-film technique of Andersen and Sørensen.[12] The 4×10^{14} curve shows a median penetration of ~220 Å, in good agreement with the theoretical prediction of 240 Å.[13] Sputtering with 3.6×10^{16} ^{50}V/cm^2 corresponds to the removal of approx. 300 Å of target material before the chemical stripping is performed. The curve marked "calculated" is obtained from the 4×10^{14} curve shifted 300 Å to the left and normalized at the resulting surface. The good agreement between the experimental and the calculated curves indicates that diffusion plays only a minor role. Corrosion-film stripping of the targets sputtered with very high doses indicates that for these doses, all the ^{48}V activity was contained within the outermost 50 Å

of the target and most probably situated at the surface.

The experimental facts emerging from this investigation are thus as follows: The sputtering yield of vanadium-irradiated copper decreases steadily as a function of the vanadium-ion dose. As the process proceeds, more and more vanadium atoms are concentrated on the surface, where they remain in spite of the average corrosion of the surface. No evidence is found of a deeply penetrating tail, growing as a function of dose. Hence bulk diffusion plays no important role. The surface attains a bluish colour characteristic of a specific, copper-rich, copper–vanadium phase.

The following model may offer an explanation of the above experimental observations: The vanadium atoms introduced in the copper matrix through irradiation precipitate in a separate copper–vanadium phase. This phase has a much lower sputtering yield than the matrix within which it is embedded. As the retreating surface passes the precipitate, an erosion cone is formed beneath it. As the process goes on, more and more cones are formed on the surface, a higher fraction of the surface is covered with low-yield phase, and the yield decreases steadily. At a certain concentration, the surface attains the colour of the precipitate. The size of the cones will depend on the size of the precipitates and the difference in sputtering yield between precipitates and matrix.

We tried to observe the postulated cones in a scanning electron microscope. The negative result of the attempt is shown in Figure 4. The target was irradiated with 26×10^{16} ^{51}V ions/cm^2. Figure 1 shows that this dose causes approx. 2500 Å to be sputtered away. This

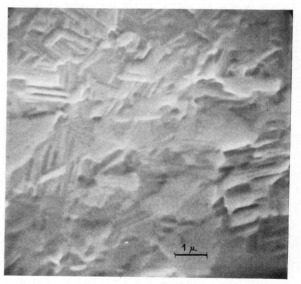

FIGURE 4   Scanning electron micrograph of a copper surface irradiated with 2.6×10^{17} 45 keV V$^+$ ions/cm^2.

will, in any case, be the maximum height of possible cones. To make them observable in the microscope we had access to, at least the lower 500 Å of the cone must be more than 500 Å wide. This corresponds to the half-angle of a cone with a maximum height of more than 15°, which is unlikely.[5] The chances of observing any cones with the scanning microscope are therefore small unless considerably larger sputtering doses are used, and even then cones may be eroded away before reaching an observable size. Figure 4 clearly indicates differential sputtering of differently oriented grains. Further it illustrates the development of a wavy surface structure for some orientations of the grains.

In conclusion we may claim that the model explains the experimental observations, but further studies with a high-resolution scanning microscope would be welcome. Furthermore, transmission electron microscopy of thinned samples would be most interesting for the revelation of the presence of precipitates.

BISMUTH IRRADIATIONS

The targets used for bismuth irradiations were prepared similarly to those described above. The dose dependence of the sputtering yield of 45 keV Bi^+ in copper is shown in Figure 5. These data have been replotted from Figure 4 of Ref. 2, here as a function of bismuth–ion dose. Arrows 2 x R, etc., indicate the doses corresponding to the removal of the respective multiples of the projected bismuth–ion ranges. Arrows at the top denote doses used in the studies described below. Saturation is seen to be reached only after the sputtering away of

much more than the trivial amount of once or twice the projected range. Figure 5 of Ref. 2 shows that a similar behaviour is not found in the case of lead–ion irradiation. As lead is the neighbouring element to bismuth, the slow approach toward saturation in the bismuth case appears to be caused by the presence of bismuth atoms and not by surface-topography changes generally appearing for heavy-ion irradiations.

The case of bismuth in copper is particularly well suited for analysis of the depth distribution by means of Rutherford back-scattering (see e.g. Ref. 14). 45 keV Bi^+ implantations were made at doses marked 1–4 in Fig. 5. The depth distribution was obtained through 90° scattering of 500 keV He^+ ions. The scattered ions were momentum-analyzed in a high-resolution magnet. Figure 6 shows the momentum analysis of scattered alpha particles from sample no. 1 (Figure 5). Known stopping powers have been used to convert the He^+

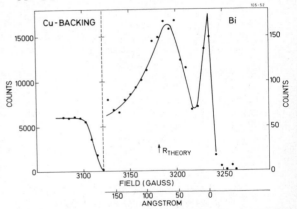

FIGURE 6   Differential depth distribution of 45 keV bismuth ions implanted into copper. The dose corresponds to sample no. 1 of Figure 5.

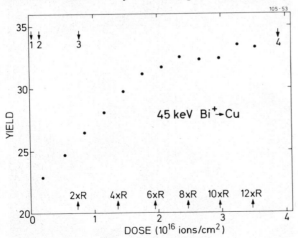

FIGURE 5   Dose dependence of the sputtering yield of 45 keV bismuth ions on copper. Different multiples of the bismuth range in copper are indicated at the bottom of the figure. Numbered arrows at the top indicate implantation doses used for further analysis.

energies obtained from the magnet field to an approximate depth scale. The broad peak corresponds well to the theoretical range prediction.[13] Furthermore, a pronounced surface peak is seen together with a tail extending into large depths. This tail is much more pronounced than predicted by theory. A corresponding analysis of sample no. 4 ($\sim$4 x 10^{16} Bi^+/cm^2) is shown in Figure 7. Now the surface peak has disappeared, and a broad distribution with a concentration of 2.5 atomic percent bismuth is seen. This corresponds well to a saturation yield of 35 (Figure 5), as the saturation concentration is expected roughly to equal the inverse of the saturation yield.

The solubility of bismuth in copper is known to be very low.[15,16] The implanted bismuth ions must thus either precipitate within their range distribution or else diffuse until they get trapped at defects. The former

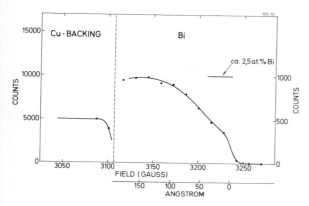

FIGURE 7   Differential depth distribution of 45 keV bismuth ions implanted in copper. The dose corresponds to sample no. 4 of Figure 5.

possibility will not give rise to a slow approach toward saturation of the sputtering yield, and thus diffusion appears to be the most probable explanation. Furthermore, no large dose effects are found in the case of lead irradiation of copper; this indicates that the dose effects for the bismuth–copper combination are specifically connected to the presence of bismuth in the copper matrix. The mechanism postulated for the vanadium–copper system only works for cases where the yield of the precipitate is less than of the substrate. This combination causes the yield to decrease with increasing dose and thus does not work for the bismuth–copper system. We must thus conclude that the dose effects for the bismuth–copper system are caused by a decrease of the surface-binding energy due to the presence of bismuth. The slow approach toward saturation will then be caused by diffusion, which is entirely in agreement with the data presented in Figures 6 and 7. Diffusion terminating at defects will give rise to a surface peak and a tail. For high doses, the valley in front of the main peak will disappear (simply by the sputtering away of the corresponding layer), and the peak may no longer be discerned. Diffusion into the depth will proceed,[17] broadening the distribution more and more, but in our case, the bismuth concentration can be followed no further down than 200 Å, due to the large scattering yield from the copper matrix. Analysis with, for example, 2 MeV protons could be used to extend the depth scale at the expense of depth resolution.

CONCLUSIONS

The analysis of the fate of implanted vanadium and

bismuth ions in copper results in peculiarities, which may well explain the strong dose dependences of the sputtering yield of the above combinations.

Vanadium is concentrated on the surface as the sputtering process proceeds. This concentration results in a decrease in the sputtering yield, which has not reached saturation even at doses of 3×10^{18} ions/cm^2.

Bismuth diffuses into the copper matrix. This in itself will not result in sputtering yields different from those obtained without diffusion, but it takes much longer time to reach saturation.

ACKNOWLEDGEMENTS

To a large extent, this work consists of the piecing-together of work made by others. G. Sørensen and H. Jensen from this Institute performed the stripping analysis of the vanadium-implanted targets, while E. Bøgh and H. Wolder Jørgensen, also from this Institute, performed the Rutherford-backscattering measurements on the bismuth-implanted samples. The vacuum deposition of the high-purity copper samples was performed by A. Nordskov and H. Sørensen at the Danish Atomic Energy Commission. Finally, I am grateful to the Mineralogy Institute of the University of Copenhagen for giving me access to their scanning electron microscope.

REFERENCES

1. O. Almén and G. Bruce, *Nucl. Instr. Methods* **11**, 279 (1961).
2. H. H. Andersen and H. Bay, *Rad. Effects* **13**, 67 (1972).
3. H. Thackery and R. S. Nelson, *Phil. Mag.* **19**, 169 (1969).
4. G. K. Wehner and P. J. Hajicek, *J. Appl. Phys.* **42**, 1145 (1971).
5. A. D. G. Stewart and M. W. Thompson, *J. Mat. Sci.* **4**, 56 (1969).
6. R. S. Nelson, *Phil. Mag.* **9**, 343 (1964).
7. M. Kaminsky, *Advan. Mass Spectrometry* **3**, 69 (1964).
8. P. Sigmund, *Phys. Rev.* **184**, 383 (1969).
9. M. Hansen, *The Constitution of Binary Alloys,* 2nd ed., (McGraw-Hill, New York 1958) p. 648.
10. See e.g. C. O. McHugh, Ph.D. Thesis. Pennsylvania State University (1959).
11. L. Guillet, *Rev. mét.* **3**, 174 (1906).
12. T. Andersen and G. Sørensen, *Rad. Effects* **2**, 111 (1967).
13. J. Lindhard, M. Scharff and H. E. Schiøtt, *Kgl. Danske Videnskab. Selskab, Mat.-Fys. Medd.* **33**, No. 10 (1963).
14. S. Rubin, *Nucl. Instr. Methods* **5**, 177 (1959).
15. D. Hanson and G. W. Ford, *J. Inst. Metals* **37**, 169 (1927).
16. E. Raub and A. Engel, *Z. Metallkunde* **37**, 76 (1946).
17. H. J. Smith (*Phys. Letters* **37A**, 289 (1971)) bombarded Zn with 39 keV Ni$^+$ and measured the sputtering yield and the amount of Ni retained in the Zn target. The Ni retention showed no saturation as a function of dose at room temperature, but rapid saturation at 77 K. Unfortunately, the sputtering yields are not presented as a function of dose, but apparently similar diffusion effects as those for the bismuth–copper system occur here.

VARIATION OF THE SPUTTERING YIELD OF GOLD WITH ION DOSE

J. S. COLLIGON, C. M. HICKS and A. P. NEOKLEOUS

*Department of Electrical Engineering, University of Salford,
Salford M5 4WT, Lancashire, England*

In an earlier publication[1] it was noted that the sputtering yields of gold varied considerably over an initial bombardment period during which approximately 6×10^{16} ions/cm^2 struck the target surface; above this dose yields became constant and agreed reasonably well with other sources of data. The present work describes a modified version of the original apparatus which allows up to one hundred experiments to be carried out in one evacuation under better U.H.V. conditions than previously attained. 0-600 eV Nitrogen (N$_2$) ions have been used and, as the present technique allows measurement of total yields as low as 10^{12} atoms, a careful study of the sputtering yield-dose pattern has been possible. Scanning electron micrographs of the gold surface at various stages of the bombardment indicate only slight surface roughening. A model is proposed based on the development of dislocation loops which could explain the relatively long recovery time in the 600 eV yield which follows a short bombardment at a lower energy.

1 INTRODUCTION

In an earlier investigation of the sputtering yield of gold targets at energies near threshold an extremely sensitive radio-active technique was developed, capable of detecting minimum total yields of 10^{12} atoms.[1] During these measurements it was found that the sputtering yield at a higher energy of 600 eV (used for cleaning the target) varied with the total ion dose received by the gold surface; approaching an apparent steady value after approximately 6×10^{16} ions/cm^2 had bombarded the target. This final value of the sputtering yield compared favourably with yields obtained by less sensitive techniques where ion doses were considerably in excess of 6×10^{16} ions/cm^2. Due to the limited number of results obtainable at the time and the relatively poor vacuum in the target chamber, it was not clear how much of this change in yield was due to a removal of surface contamination and how much was due to a change in target structure induced by the bombardment. It was decided therefore to modify the original apparatus, both to improve the ultimate pressure attainable in the target chamber and also to allow many more observations of yield during one evacuation.

In the meantime experiments in other laboratories have been performed in which sputtering yields for 5 keV ions of various species on copper have been measured with a sensitive oscillating quartz microbalance.[2] These again indicate a marked dependence of sputtering yield on the initial ion dose as the bombarding ion is changed from one species to another.

Both this and the radio-active gold experiments suggest therefore that, as ions bombard a target, an apparent steady yield condition is approached which is probably dependent on the damage state of the target induced by the particular bombarding ion, and on composition changes and metallographic effects which occur in the near surface region where ions come to rest. A similar effect should occur for one particular ion species bombarding a target at different energies where, at each new energy, a new damage-equilibrium must be established before the sputtering yield reaches a constant value and it is this particular aspect of the phenomenon that is discussed here.

2 EXPERIMENTAL DETAILS

Full details of the ion gun, which is shown schematically in Figure 1, and also the radio-active measuring technique, have been presented in an earlier publication.[1] Two alterations to this original apparatus have been made. First, a new section has been introduced at the position of the first deflector plates which contains a non-bakeable straight-through valve followed on the U.H.V. side by a liquid-nitrogen-cooled cold finger. This valve was closed when changing ion source filaments so that the U.H.V. section remained evacuated and was maintained closed whenever the ion beam was not required, to prevent diffusion of contaminants from the analyser section into the U.H.V. part of the apparatus. With the new section included ultimate pressures of order 2 ntorr were attainable. However,

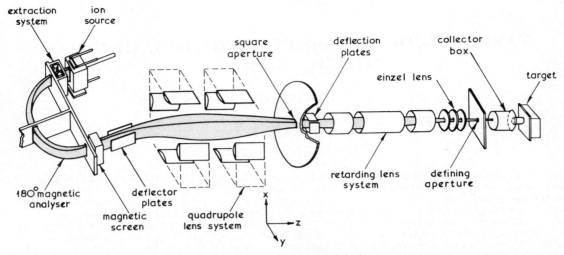

FIGURE 1   A schematic diagram of the ion gun.

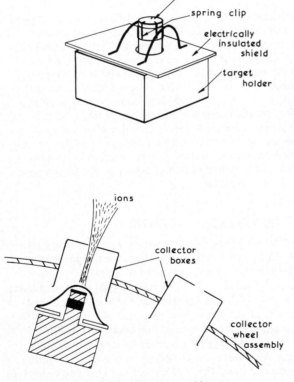

FIGURE 2   Target assembly.

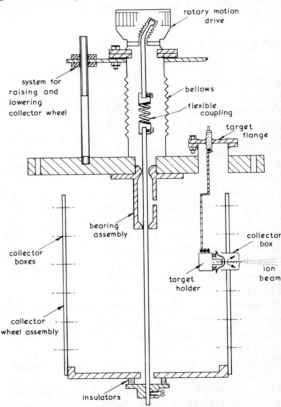

FIGURE 3   Collector and target assembly.

in the particular experiment reported here, the sublimation pumps failed to operate and a base pressure of
only 8 ntorr was obtained. Apart from the vacuum

advantages of including this section, calculations of
the optical properties of the quadrupole system indicated than an improved focus would be obtained were

the path length of the ion beam increased in this region and, in fact, a five times larger current (now 100 nA) can be focused onto the target which has an area of about 0.1 cm². The present bombardment rate corresponds therefore to about 6 x 10¹² ions/cm²-sec.

The second alteration to the previously described apparatus was the construction of a new target and collector-wheel assembly. The target mounting arrangement is shown in Figure 2 and allowed much easier insertion of the radio-active gold disc than in the earlier design. Furthermore, the target could be introduced into the vacuum system without demounting the collector-wheel flange; by using the smaller sub-flange (see Figure 3). Electrically insulated springs on the line of the front face of the target allowed the target to ride over protruding edges of collector boxes; and a heating system and thermocouple will enable sputtering measurements to be made at elevated temperatures in later work. By attaching the central support shaft of the collector-wheel to a bakeable rotary drive mechanism which itself was mounted from the main flange via a bellows (see Figure 3) it was possible to rotate the collector-wheel or move it vertically up or down with respect to the fixed target and ion beam axes. Thus any collector in the 6 layers, each of 18 boxes, could in turn by brought into line with the ion beam so that a maximum 108 experimental observations could be made during a single evacuation.

On arrival from the reactor, the radio-active gold was mounted into the vacuum system, the apparatus pumped and the U.H.V. section baked for 12 hours at 300°C. When reasonably cool a beam of N_2^+ ions was focused through the $\frac{5}{32}$ inch dia aperture of one selected collector box onto the target, and the target current–time integral recorded (from which the number of ions striking the target could be determined). At the end of the required bombardment, another collector was brought into line and a further experiment initiated. When all boxes had been utilized they were removed from the vacuum and, from the radio-activity on each box, the number of gold atoms on it was determined. Occasionally, boxes were then segmented, and each portion counted, to obtain the spatial distribution of the sputtered gold.

3 EXPERIMENTAL RESULTS AND DISCUSSION

In the present study no correction has been made for secondary electron emission from the gold target. Sputtering yields quoted therefore are $S_m = S/(1 + \gamma)$ where S is the correct value of yield and γ is the number of electrons emitted by the gold target per incident ion. Data for the electron emission coefficient of N_2^+ on gold is not available. However, at these energies, the emission is a potential process so that the value of γ

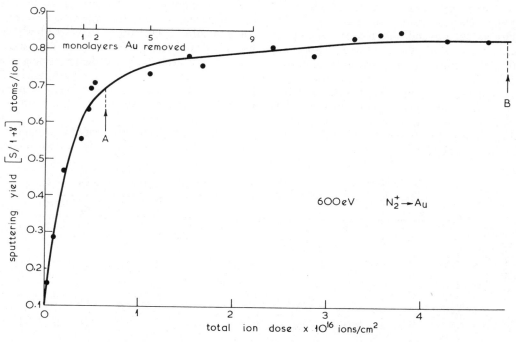

FIGURE 4   Variation of sputtering yield with ion dose for 600 eV N_2^+ ions on gold.

for N_2^+ is expected to be somewhat less than that for Ar^+ where a value of γ of 0.15 electrons per ion is quoted.[3] The error in the results will therefore be small if the electron emission correction is neglected.

In Figure 4 the variation of sputtering coefficient S_m with ion dose is shown for 600 eV N_2^+ ions on gold and it is clear that there is an initial rapid increase in S_m from 0.1 to 0.7 atoms/ion as the first 5×10^{15} ions/cm^2 bombard the target. Alternatively, we can define this initial interval by measuring the total number of gold atoms removed during the 5×10^{15} ions/cm^2 bombardment; which is 2.25×10^{15} atoms/cm^2. Assuming that the ion beam is uniform over the whole $\frac{1}{10}$ cm^2 target area, this corresponds to a removal of about $1\frac{1}{2}$ monolayers of gold and the change could therefore be attributed to the removal of an adsorbed or contaminant layer. However, in a later experiment the target was re-exposed to air for 30 minutes and the yield recorded for subsequent bombardment. Again the yield was low initially and increased rapidly to a steady value, this time before a further monolayer of gold had been removed (see Figure 5). This tends to suggest that the initial change in yield is predominantly due to the removal of an adsorbed layer from the surface rather than the removal of a contaminant layer.

In the present work the gold targets were not subjected to any cleaning procedure apart from degreasing in carbon tetrachloride and de-ionized water. However, in other work[4] where gold was etched in aqua regia, subsequent Auger electron analysis revealed oxygen, nitrogen, chlorine, sulphur and carbon contaminants on the surface, which persisted even after a 20 hour bake at 400°C in a vacuum of 20 ntorr. Although it is likely that some of these contaminants arise from the contact with aqua regia, the experiment indicates that the adsorbed contaminant layer on gold is very tenacious and one would expect continued adsorption of at least oxygen and nitrogen on the present gold samples even after the vacuum anneal of 300°C.

At higher ion doses (see Figure 4 again) there is a further gradual increase in yield up to an apparent steady value of about 0.85 atoms/ion after a bombardment of 50×10^{15} ions/cm^2 and, as this corresponds to a removal of 17–20 monolayers of gold, the effect is difficult to explain in terms of surface contamination.

Scanning electron micrographs of an identical gold target before bombardment, after 6×10^{16} ions/cm^2 and after 50×10^{15} ions/cm^2 had interacted with the target (corresponding to points 0, A, and B on Figure 4) have been taken, but show only slight roughening in certain regions; no macroscopic surface topographical changes appear to be developing which could explain this change in sputtering yield.

One possibility considered was that the re-contamination rate from the residual gases in the target chamber may be such that the surface was only slowly being cleaned after the initial monolayer of contamination had been removed. However, if this were the case, a prolonged interruption in the bombardment would be expected to drastically alter the yield due to re-contamination and Figure 6 illustrates that, even after a 12 hour rest, there is no apparent change in the measurements. Further conformation that the ion bombardment rate is not competing closely with residual gases has been obtained by varying the dose-rate from 0.05–0.8 $\mu A/cm^2$. No change in measured yield was observed over this interval.

The most significant indication that the variation in yield, which occurs after the initial monolayer has been removed, is due to the target itself rather than contamination is given in Figures 7 and 8 which show the effect of changing the ion energy for a short time from the normal 600 eV and then returning to 600 eV for further measurements of the sputtering coefficient. In all cases the subsequent 600 eV yield begins at a value higher than normal and decreases with increasing dose towards the original level. It appears that a layer of material, re-ordered or damaged in a manner characteristic of the new ion energy, is created, and has to be

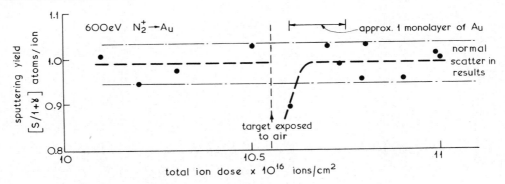

FIGURE 5   Effect on sputtering yield of 30-minute exposure of target to air.

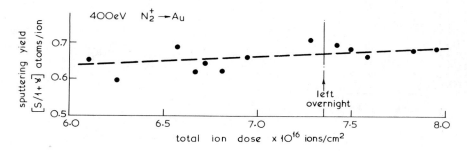

FIGURE 6   Variation of sputtering yield with dose of 400 eV N_2^+ after interruption in bombardment overnight (~12 hours)

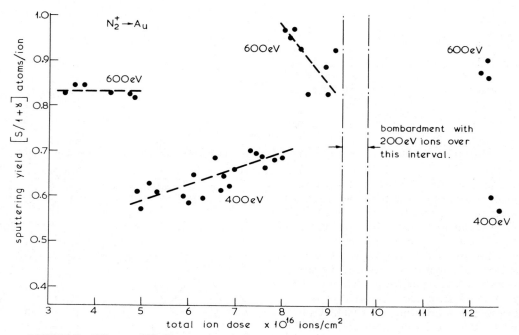

FIGURE 7   Effect on 600 eV sputtering yield of changing beam energy to 400 eV for a short time.

re-sputtered away before the yield returns to its steady 600 eV value. The length of the recovery interval is thus a measure of the depth of the damaged layer which would be expected to be of the same order as the most probable range of the bombarding ions.

However, present results indicate that the thickness of gold which has to be removed is, at the very least, 20 Å which is an order of magnitude larger than the ion range. A possible explanation for this could be based on a similar model to that proposed by Hermanne et al.[5] to explain the development of typical surface topo-

graphical features during ion bombardment of single crystals. The basis of this model, which is confirmed by the experimental work of Ogilvie et al.[6] and Venables et al.,[7] is that, during bombardment, defects are continuously being created within the target; the rate of production being governed by the type of ion and its energy. These defects will diffuse in the material to a suitable sink, such as a grain boundary, and will also agglomerate into line defects or loops. At the same time, however, the surface is receding due to the sputtering process at a rate again determined by the

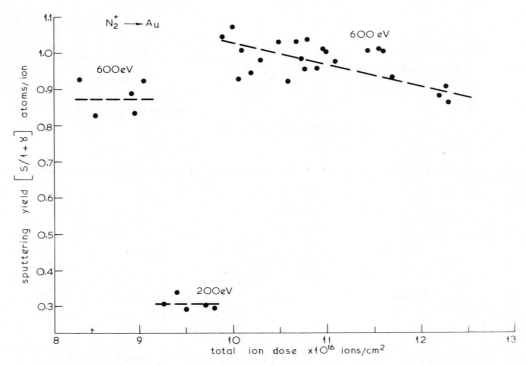

FIGURE 8   Effect on 600 eV sputtering yield of changing beam energy to 200 eV for a short time.

type of ion and its energy. The number and size of
dislocation loops formed thus depends on the compe-
tition between defect migration rate on the one hand
and erosion rate, and consequent liberation of defects,
on the other. Whilst Hermanne et al.[5] conclude that it
is the intersection of the defects with the surface which
give rise to local variations in sputtering and hence a
development of surface topography, we now propose
that the presence of these loops accounts for the
measured variations in total sputtering yield. Ogilvie
et al.[6] observed that the damage associated with 500
eV ion bombardment extended to depths of order
100 Å which is compatible with the observed minimum
erosion of 20 Å which must take place before the yield
approaches its steady value after the ion energy has
been changed. Presumably a similar recovery interval
would be required were the ion species changed.

Although results reported here refer to a nominally
polycrystalline gold target, the distribution of radio-
activity on sectioned collector boxes showed that
there were marked peaks in the intensity of the deposit
consistent with the usual spot patterns obtained in
single crystal sputtering.[9,10] Figure 9 shows the distribution of sputtered material around the cylindrical
segment of the collector which subtends an angle of
$\phi = 45°\,(\pm 5°)$ to the target normal and it is clearly

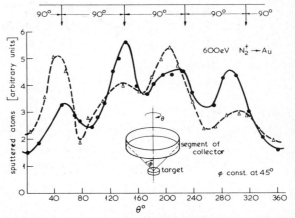

FIGURE 9   Angular distribution of sputtered gold around
a cylindrical section of the collector which collects all atoms
sputtered at $\phi = 45°$ to the target normal. (Values shown •
and △ refer to different collector boxes.)

non-uniform; having intense spots separated from each
other by 90°. These deposits are also seen to be a maximum when sectioning along strips of collector which
are parallel to the cylinder axis, and thus represent
true local maxima or "spots", which are characteristic
of a sputtering from a (100) surface of a fcc material.
It thus appears that the gold target is relatively well

ordered during the sputtering and it could be that the extent of this ordering also contributes to the variation in sputtering yield with ion dose at a given energy.

4 CONCLUSIONS

From the results reported here it is clear that the sputtering yield of gold is dependent on the bombarding ion dose. The major variation in yield occurs with a target recently exposed to air and takes place whilst about one monolayer of gold is removed, but a further smaller ($\sim$20%) and much slower variation occurs which is apparently due to a change in the target structure; characteristic of the bombarding ion and its energy. Experimental evidence suggests that this latter change in yield is not due to contamination nor to macroscopic changes in surface topography, but is consistent with formation of dislocation loops in the target material. Furthermore, because the target current remained constant throughout each experiment it is unlikely that a change in secondary electron emission coefficient is responsible for the observed

effects. Further results must be obtained before a final explanation of the process can be attempted, but with the effect representing 20% of the final yield, it could be a significant factor in the sputtering phenomenon.

REFERENCES

1. J. S. Colligon and R. W. Bramham. *Atomic Collision Phenomena in Solids* edited by D. W. Palmer, M. W. Thompson and P. D. Townsend (North-Holland Pub. Co., 1970), pp. 258–265.
2. H. H. Andersen and H. Bay. *Rad. Effects* **13,** 67 (1972).
3. A. R. Bayly. PhD thesis. University of Salford (1970).
4. D. L. Olson, H. R. Patil and J. M. Blakely. *Scripta Metallurgica* **6,** 229 (1972).
5. N. Hermanne and A. Art. *Fizika* **2** (Suppl. 1) 73 (1970).
6. G. J. Ogilvie, J. V. Sanders and A. A. Thomson. *J. Phys. Chem. Solids* **24,** 247 (1963).
7. J. A. Venables and R. W. Balluffi. *Bull. Amer. Phys. Soc.* **9,** 295 (1964).
8. J. A. Venables and R. W. Balluffi. *Phil. Mag.* **11,** 1021 and 1039 (1965).
9. G. S. Anderson and G. K. Wehner. *J. Appl. Phys.* **31,** 2305 (1960).
10. M. Koedam and A. Hoogendoorn. *Physica (letters)* **26,** 351 (1960).

SPUTTERING YIELDS OF NIOBIUM BY DEUTERIUM IN THE keV RANGE

W. ECKSTEIN, B. M. U. SCHERZER and H. VERBEEK

Max-Planck-Institut für Plasmaphysik, EURATOM-Association, BRD-8046 Garching, Germany

The sputtering yields of ~600 Å niobium films bombarded by 3–8 keV D^+ ions have been measured. Rutherford back-scattering of 150 keV H^+ ions has been used as a probe to measure the decrease of film thickness due to sputtering. A sputtering yield of $S = 4 \times 10^{-3}$ atoms/ion was found which is only weakly dependent on the energy. The result is about a factor of two lower than expected from extrapolating measurements made by Summers *et al.*[2] at higher energies. The discrepancy may be due to differences in film structure, contamination by oxygen, and/or differences in the measuring methods, but cannot at the moment be clearly accounted for.

1 INTRODUCTION

The sputtering yield of refractory metals with hydrogen, deuterium, tritium and helium is of importance for assessing wall erosion in fusion reactors. At present niobium seems to be the favourite first wall material.[1] Sputtering yield measurements of this material with D^+ ions have been made by Summers *et al.*[2] at energies above 10 keV. However data at lower energies are of even greater importance, since (1) the sputtering yield maximum is expected in this range, and (2) the mean energy of plasma particles hitting the wall will probably be less than 10 keV.

Sputtering yield measurements with light ions are difficult because the weight loss method normally used is not applicable in this case. Since sputtering yields are low, the weight loss due to sputtering may be partly or totally compensated by the weight gain due to trapped primary ions. A number of methods have been used to avoid this difficulty.[2,3,4] In this paper we used a new method first introduced by R. Behrisch and R. Weissmann.[5] The decrease in the number of atoms per unit area given by the product $N_{Nb} \cdot d$ of the number of niobium atoms per unit volume times the film thickness d of a niobium film, is measured by Rutherford backscattering of 150 keV H^+ ions. For these measurements the film is evaporated onto a low-Z material which does not contribute to backscattering in the energy range of interest. The sputtering yield is proportional to the decrease in the number of Nb-atoms per unit area Δn_{Nb} due to the bombardment with D^+ ions:

$$S = \Delta n_{Nb} \cdot F/J$$

where J is the total number of incident D^+ ions and F is the area of the bombarded spot.

It is essential for these measurements that the primary D^+ beam be evenly distributed over the bombarded spot.

2 EXPERIMENTAL

Measurements were performed on ~600 Å thick Nb films evaporated onto mechanically polished Be substrates. The Be surface is covered with a thick oxide layer. The residual gas pressure during electron beam evaporation was 10^{-7} torr. The niobium has a purity of 99.95%, the main impurity being ~500 ppm Ta. The thickness of the film was chosen such that no backscattered deuterons from below the Nb layer can penetrate to the surface and contribute to the sputtering yield. The mean projected range of 10 keV D^+ ions in niobium as calculated by theory[6–8] is $\bar{R}_p = 610$ Å with a range straggling of ~90%.

The targets were bombarded with D_2^+ ions at normal incidence and room temperature in the BOMBARDON accelerator which is described elsewhere.[9] The beam diameter was defined by a 0.6 mm diaphragm mounted at a distance of 2 mm from the target. Typical bombardment parameters are given in Table I. For evaluation of the sputtering yield S it is assumed that a D_2^+ ion of energy $2E$ sputters like two D^+ ions of energy E, as was also assumed by KenKnight and Wehner[3] in the case of H_2^+ and H_3^+.

Thickness measurement: After D^+ bombardment the targets were removed from the low energy accelerator target chamber and mounted in the PHARAO 150 keV accelerator[10] for thickness measurements by backscattering. A 150 keV proton beam with a diameter of 0.14 mm at the target was used at normal incidence. The beam current was ~6×10^{-8} A, and the total dose

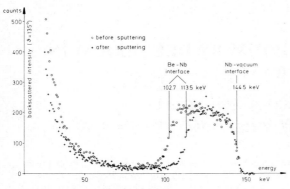

FIGURE 1   Energy distributions of protons backscattered at an angle of $\theta = 135°$ from a 600 Å Nb film on a Be substrate before and after sputtering with D^+ ions.

TABLE I

List of energies, beam-currents, and total doses per unit area of sputtering measurements.

D_2^+-ion energy [keV]	Beam-current [A]	Total number of D^+ ions per unit area [cm^{-2}]
6.16	$1.7 \cdot 10^{-7}$	$1.45 \cdot 10^{19}$
8.23	$1.7 \cdot 10^{-7}$	$1.59 \cdot 10^{19}$
10.27	$1.75 \cdot 10^{-7}$	$1.42 \cdot 10^{19}$
13.3	$7 \cdot 10^{-7}$	$1.33 \cdot 10^{19}$
15.76	$5.5 \cdot 10^{-7}$	$1.56 \cdot 10^{19}$

The total amount of niobium atoms per unit area is calculated by the formula

$$n_{Nb} = \frac{N}{(i \cdot t/e)_{H^+} \cdot \sigma(\bar{E}, \theta, \Delta\Omega)}$$

where N is the total number of counts in the Nb peak, $(i \cdot t/e)_{H^+}$ the number of primary H^+ ions and $\sigma(\bar{E}, \theta, \Delta\Omega)$ the cross section for scattering of H^+ ions with a mean energy of $\bar{E}$ at an angle θ into a solid angle $\Delta\Omega$. Rutherford scattering was assumed and the mean energy of H^+ penetrating the Nb film was calculated from

$$\bar{E} = E_1 - \frac{\Delta E}{2(1 + \cos^{-1}\beta)}$$

where β is the angle of the scattered beam to the surface normal of the target, and E_1 is the primary energy. Constant electronic stopping is assumed over the energy range of ΔE. This is very well fulfilled for the energy range of interest (100–150 keV) in niobium.[11]

3 RESULTS AND DISCUSSION

Measured sputtering yields are shown in Figure 2. For comparison, values measured by Summers et al.[2] at higher energies are also presented. From the sputtering yield measurements of He on Nb by Rosenberg and Wehner[12] and by Summers et al.,[2] the maximum value is expected to be somewhere near 5 keV. Our measurements by themselves corroborate the assumption of a

per spectrum 10^{-6} A·s, corresponding to a current density of 4×10^{-4} A·cm^{-2} and a fluence of 6.7×10^{-3} A·s·cm^{-2}. Sputtering by the probe beam is negligible since the sputtering yield of 150 keV protons is $<10^{-3}$. This was also confirmed experimentally. The targets were scanned over the deuterium-bombarded spot in order to obtain backscattered energy spectra inside and outside the sputtered region. From these scans it could also be assessed whether the deuterium beam had been homogeneous and the thickness variations over the bombarded spot were small. This condition was not very well fulfilled in some of the measurements reported here. The calculated sputtering yields give mean values over the bombarded area.

Typical backscattered energy spectra are shown in Figure 1. Protons scattered at an angle of 135° from the target surface appear at an energy of 144.5 keV, whereas the energy of protons backscattered from the lower Nb surface depends on the thickness of the film.

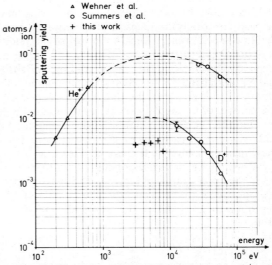

FIGURE 2   Sputtering yields of Nb bombarded by D^+ and He^+.

flat maximum in this energy range, but compared with the measurements at higher energies they are about a factor of two lower than expected. There may be several reasons for this discrepancy:

1) It is well known that sputtering yields of poly-crystalline materials depend to a large extent on the structure of the material used. There is not much information on the films used by Summers *et al.*, which were deposited by sputtering, whereas our films were thermally evaporated. The structure of Nb-films evaporated in the same way onto $BaCl_2$ covered glass substrates was shown by electron diffraction to be nearly amorphous.[13]

2) Nb is oxidized very easily and the oxide may well penetrate to a depth of several 100 Å. Since our films were exposed to air after evaporation and also after sputtering, before thickness measurements were taken, the question arises to what extent oxygen is present in the films and what influence it exerts on measured sputtering yields. Oxygen cannot be directly seen in our backscattering measurements with H^+, but its presence may be induced from energy loss considerations. From the energy difference between the leading and trailing edges of the Nb peak in our backscattered energy spectra, Figure 1, a mean electronic stopping power of 5.3 eV · cm^2/atom is calculated. The corresponding value for 100–150 keV H^+ on clean solid Nb which was measured in the same accelerator by R. Behrisch and B. Scherzer[14] is 3.2 eV cm^2/atom. If the difference in stopping power is ascribed to oxygen atoms, a concentration somewhat less than 2 oxygen atoms per niobium atom results, assuming that the stopping power of the bound oxygen atoms in the film is equal to the free atom value of 1.2 eV cm^2/atom.[11]

The most common oxide of Nb is Nb_2O_5 and our result indicates that the oxygen concentration in our film is lower than in this oxide. It was shown by Nghi and Kelly[15] that sputtering yield of Kr on Ta_2O_5 is lower than the pure metal value, by a factor of 3 and of Kr on WO_3 is lower by a factor of 1.5 if only the number of sputtered metal atoms is regarded. In this respect a reduction of sputtering yield in our Nb-films by about a factor of 2 due to oxygen contamination appears reasonable. On the other hand the Nb-films of Summers *et al.*[2] have also been exposed to air before bombardment. So, either there exists a difference in take up of oxygen between the r.f. sputtered and the evaporated films, or the oxygen is already incorporated in the film during evaporation. This may be the case if niobium oxide is more volatile than the metal itself. In this case very thorough cleaning of the niobium prior to evaporation will be necessary.

4 CONCLUSION

Sputtering yield measurements of niobium films with D^+ show only a weak energy dependence between 3 and 8 keV. They are about a factor of two lower than would be expected from measurements by Summers *et al.*[2] at higher energies. The difference may be due to film structure and/or oxygen contamination. Measurements on well defined clean targets have to be made in order to obtain more reliable data.

ACKNOWLEDGEMENTS

We thank Dr. H. Vernickel for his constant interest in this work. Dr. R. Behrisch, H. G. Schäffler, and R. Weissmann helped us with many discussions. The Nb films were prepared by H. Kukral. R. Hippele, H. Schmidl, S. Schrapel, and H. Wacker were responsible for the electronics and the mechanical setup of the experiment and assisted in the measurements. We take pleasure in gratefully acknowledging all of these contributions to this work.

REFERENCES

1. D. J. Rose, ORNL-TM-2204 (1968).
2. A. J. Summers, N. J. Freeman and N. R. Daly, *J. Appl. Phys.* **42**, 4774 (1971).
3. C. E. KenKnight and G. K. Wehner, *J. Appl. Phys.* **35**, 322 (1964).
4. O. C. Yonts, B.N.E.S. Nuclear Fusion Reactors Conference, Culham, Sept. (1969) 424.
5. R. Behrisch and R. Weissmann, *Phys. Letters* **30A**, 506 (1969).
6. J. Lindhard, M. Scharff and H. E. Schiøtt, *Mat. Fys. Medd. Dan. Vid. Selsk.* **33**, no. 14 (1963).
7. H. E. Schiøtt, *Mat. Fys. Medd. Dan. Vid. Selsk.* **35**, no. 9 (1966).
8. K. B. Winterbon, AECL-3194 (1968).
9. W. Eckstein and H. Verbeek, IPP 9/7 (1972).
10. R. Behrisch, *Vak. tech.* **10**, 250 (1967).
11. L. C. Northcliffe and R. F. Schilling, *Nucl. Data Tables* **A7**, no. 3–4 (1970).
12. D. Rosenberg and G. K. Wehner, *J. Appl. Phys.* **33**, 1842 (1968).
13. H. G. Schäffler, private communication.
14. R. Behrisch and B. M. U. Scherzer, to be published.
15. L. Q. Nghi and R. Kelly, *Can. J. Phys.* **48**, 137 (1970).

CURRENT TRENDS IN ION SCATTERING FROM SOLID SURFACES

E. S. MASHKOVA and V. A. MOLCHANOV

Institut of Nuclear Physics, Moscow State University, Moscow, 117234, USSR

Several current trends in the study of ion scattering from solid surfaces are discussed, especially in respect to the conditions necessary for reproducible measurements and to interpretation of the experimental results.

INTRODUCTION

This paper is not a review, but merely an attempt to trace on the basis of recent publications some current trends in the study medium ion scattering from solid surfaces. In the study of this problem three stages may be distinguished (see Figure 1).

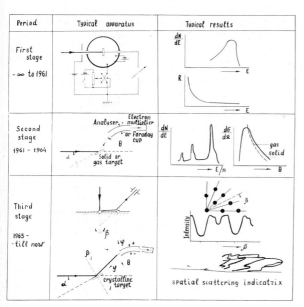

FIGURE 1 Typical apparatus and results in investigations of ion scattering.

The first stage (which ended in 1961) may be referred to as "experimental data accumulation". The main attention of scientists was paid to the study of the "total" characteristics of ion scattering from solid surfaces (such as the total ion reflection ratios and the mean energy of the scattered ions). Two more features of this stage will be emphasized here. First, at that time this field of physics was not included in solid state physics, but in electronics. For this reason most investigations were performed by specialists in gas discharge studies. Second, the scientists working in this field were not yet divided into theorists and experimentalists, so that every scientist tried to develop his own "theory" for explaining the observed experimental results. These circumstances should be taken into account, otherwise several explanations made at that time now look extremely naive.

The second stage in the study of ion scattering from solid surfaces began approximately at the end of 1961 and continued to the end of 1964. The main attention of scientists was paid to studying the differential characteristics of ion scattering i.e. scattered ions angular distributions; energy distributions of ions scattered in chosen directions, and so on. During this stage the analogy between the ion-atom collisions in gas and ion-solid collisions was traced in detail and main differences were found. The investigations were strongly influenced by the experiments made by Everhart and co-workers and Fedorenko and co-workers, who studied ion-atom collisions in gas (see Ref. 1).

The third stage began in 1965 when the first directional effects in ion scattering from ordered media were observed. During this stage the main attention of scientists was paid to the search and analysis of various correlation effects of atomic collisions on solid surfaces caused by regular arrangement of target atoms. As a result it was established beyond all doubt that under some conditions correlation effects played an important role in the ion scattering process. In recent years successful attempts have been made to detect such effects caused by target lattice irregularities, in particular by radiation damage produced under ion bombardment.

SURFACE CONDITIONS DURING ION BOMBARDMENT

Thus, key attention is now being concentrated on ion scattering from crystals. Here it is necessary to emphasize the following circumstances. As early as 1955, Günterschulze[2] wrote that the target surface should be "clean and well-defined". When we say, for example: "As target, the (100) face of a copper crystal was used", this statement is much more definite than when we say "As target, polycrystalline copper was used". In fact, when a polycrystalline target is used it is necessary to know (among other parameters) grain sizes, grain arrangement and so on. Besides, in a number of cases target recrystallization as well as target topography transformation may occur during the experiment. So, there is preferential sputtering of some grains due to the strong dependence of the sputtering yield on the grain orientation of the polycrystalline target with respect to the ion beam. By way of example we quote data obtained ten years ago by Southern, Willis and Robinson.[3] These authors found that the sputtering yields of polycrystalline targets from different sources may be markedly different owing to differences in the preferred orientation of the individual grains, (texture), see Figure 2. The problem of the structure

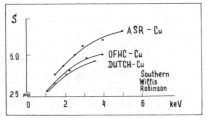

FIGURE 2   Energy dependence of the sputtering yield for different polycrystalline copper samples, data.[3]

of the polycrystalline surface has become much more important in recent scattering studies, because narrow ion beams are usually used. For this reason the dimension of the irradiated spot of the target surface may be comparable to the individual grain size, so that the ion scattering is studied not from a polycrystal, but rather from either a single grain or a few grains. Of course, with monocrystal targets some unknown parameters also exist. In particular, only the purity of the target substance is usually stated, but not the degree of crystal perfection (i.e. a concentration of structure defects). On the whole, however, the term "monocrystalline target" is more definite than the term "polycrystalline target".

The studies carried out during the last decade have shown that it is also necessary to take into account the

change of the surface target layer caused by ion bombardment. This refers as well to usual surface "etching" under ion bombardment as to more essential changes (radiation damage, ion implantation). One of the principal requirements for experiments is that the target surface has to be clean. Usually, the cleanness of the target is achieved by using intensive ion beams: the ion current density is chosen such that the number of sputtered particles ejected from the target surface appreciable exceeds the number of residual gas molecules incident on the target surface. Under typical experimental conditions this leads one to use ion current densities of about 0.1–1.0 mA cm^{-2}, i.e. converting to particle number, $\sim 10^{15}$ part. cm^{-2} sec^{-1}, which is comparable with the surface atom density ($\sim 10^{15}$ atoms cm^{-2}). This leads to at least two results. Firstly, in the surface target layer a large number of displaced atoms is produced per second (about $E/2E_d$ per incident ion). In some cases radiation damage is largely annealed, depending on the target temperature. In other cases radiation damage is accumulated, thus causing the target to become amorphous. Secondly, the target surface boundary moves rather quickly owing to sputtering (up to a few times ten atomic layers per second). These processes are demonstrated in Figure 3 and Figure 4. From the start of irradiation

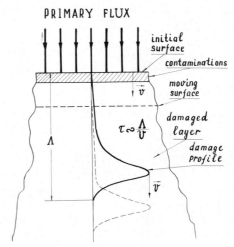

FIGURE 3   Change of surface conditions during ion bombardment. Λ is the mean range of the ions in the solid and v the velocity of the moving surface due to loss of surface material.

the properties of the target surface are generally modified. The time of equilibration due to these processes (relaxation time) is expected to be of the order of the ion penetration depth divided by the velocity of the

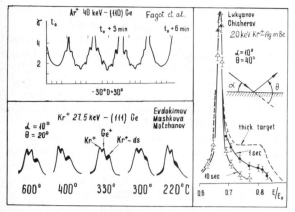

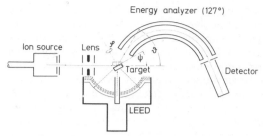

FIGURE 5 Scheme of an experimental setup for low energy ion scattering (100 to 2000 eV) and low energy electron diffraction (LEED) under ultrahigh vacuum conditions.

FIGURE 4 At the top—dependence of the anistropy of the secondary ion-electron emission coefficient on time of irradiation Data.[4] At the bottom—dependence of the scattered ion energy distribution on the target temperature. Data.[12]

On the right—dependence of the scattered ion energy distribution on the time of irradiation. The thin silver film was deposited on a solid beryllium sample. Data.[5]

irradiated surface motion due to target sputtering. Simple estimates show that the relaxation time is of the order of minutes or seconds, depending on the experimental conditions (Figure 4 top). After this time the number of bombarding particles inside the target proves to be practically independent of the ion current density. The relaxation process may be illustrated by data of French authors[4] obtained in the study of ion-electron emission or by data of Lykyanov and Chicherov,[5] in the study of ion scattering from thin layers deposited on massive targets of light substances, see Figure 4 on the right.

In our opinion, the study of the effect of surface layer properties (or, more exactly, the change of surface layer properties) on ion scattering regularities is a very interesting and essential problem that has so far been investigated only slightly. However, experiments in this field call for a great deal of methodical work. In particular, effective control of surface conditions such as LEED or Auger is needed. It has been demonstrated by Heiland and Taglauer that LEED together with a high vacuum and low primary ion density afford essentially new prospects in ion scattering studies. Figure 5 shows a scheme of their experimental setup.[21]

MECHANISM OF ION SCATTERING PROCESS

The general objective of the study of ion scattering from solids is to find the mechanism of ion scattering process, i.e. to find the main regularities of elementary collision events and the sequences by which the process

is realized. Ion scattering from solids is a rather complicated process that may be represented as the sum of several simpler processes, which may be considered as its "elementary components". Among these processes are uncorrelated scattering i.e. either single scattering or ordinary multiple scattering typical of random media and the various types of correlated atomic collision sequences, e.g. correlated collisions of ions with target atoms, which form either separate atomic rows or a crystalline plane. All the above mentioned simple processes are typical not only of ion scattering from solids, but also of the other radiation processes which take place in the bulk of the target. However, there is a specific type of correlated atomic collision which is typical of ion scattering from solid surfaces only, such a type is the ion motion governed by the field of incomplete channels formed by close-packed atomic rows lying just on and near the target surface. Both the experiments and the calculations show that such "half-channels" may under specific conditions play an important role in the formation of the scattered ion flux.[18]

The role of all the above mentioned "simple processes" may be different under different circumstances. However, all of them should be taken into account when analysing real ion scattering from solids. In our opinion, it is of particular interest to study the mechanism of ion scattering from solids at low ion energies when one can expect to detect experimentally such effects as the influence of the non-binary character of ion-atom interactions, the influence of the binding energy of target atoms and in particular its directional dependence. Experiments at low ion energies are extremely complicated because in this energy region perfect experimental equipment and excellent vacuum conditions are required. Besides, the experimentalist must be extremely unbiased, otherwise he may jump to speculative conclusions. Some interesting results in the low-energy range have already been obtained[21,22] (Figure 7).

In this paper, however, we have confined ourselves to examining only the "medium-energy" interval, in which the experiments are much easier and the obtained experimental data are numerous. As the "medium energy range" is usually believed to be one in which strictly speaking both energy loss mechanisms, elastic and inelastic, must be taken into account. By tradition, the limiting energy demarcating the intervals with different dominant energy loss mechanisms is usually expressed in keV and is equal to the atomic number of the incident ions, irrespective of the target material (so-called "simplified criterion of Seitz"; see monograph by Dienes and Vineyard.[6] Thus, "medium-energies" are the energy interval of the order of a few keV—a few times ten keV, depending on the sort of ions used. Analysis of the experimental investigations performed in this energy interval shows that most of them were carried out using gas ions (as a rule, noble gas ions). This is reasonable because such ions do not contaminate the target surface. On the other hand, most of the medium-energy gas ions are neutralized during the scattering process so that the ion component is only a small part of the scattered particle flux. In most of the experiments hitherto only the charged component of the scattered particle flux has been investigated. For this reason the serious problem arises to what extent the results obtained are characteristic of the whole scattered particle flux. In other words, the important question is as follows: In which cases may one extrapolate the results inherent in the ion component of the scattered particle flux to the whole scattered particle flux. To answer this question, it is necessary to perform comparative measurements of both the angular and energy distributions of the particles scattered in various charge states, including the neutral state. This problem is not a new one, but it recently became extremely important because several essential experimental data proved to be assumed alternative interpretations.

DECOMPOSITION OF THE SCATTERING PROCESS†

To decompose the real scattering process into a set of simple processes, the various correlation effects are usually employed. As far as we know, the first experiment of this kind was performed by Datz and Snoek in 1964.[7] To single out the peak due to elementary collisions between primary ions with target atoms, these authors oriented the crystalline target with respect to the primary ion beam in such a manner that most primary ions were channelled into the target. For this reason multiple ion scattering was suppressed and the observed peaks in the scattered ion energy distribution became narrow compared with the random orientation. Some time ago this method seemed most promising for obtaining the ion-atom interaction potential. However, soon after it was admitted that, with the exception of several specific cases (such as target materials with a high melting point), a much more reliable method for obtaining the ion-atom interaction potentials is ion scattering from evaporated target atoms.† This statement may be confirmed by the experimental data, obtained by Dutch scientists[13] see Figure 6. One can see that the general character of the dependences obtained, including such a delicate effect as inelastic energy loss, are similar to these for the solid and evaporated target material. An essential difference was found only in the average number of missing electrons: this number was found to be smaller in the case of ion scattering from solid targets than from gas ones. The authors believe that the cause of this phenomenon is Auger neutralization of scattered particles near the target surface. It is interesting to note that Dutch authors accepted this concept (the important role of ion neutralization near the target surface) as a guidance line in their studies of ion scattering from crystals; see below.

The use of other correlation effects also allows effects due to ion scattering from close-packed atomic rows to be singled out. Let us assume that the (100) face of a b.c.c. crystal is used as the target, and that the incidence plane of the primary ions is parallel to the (100) faces of the target. One can see that in such a case a considerable part of the primary ions will be channelled into the channels formed by the (100) crystalline planes of the target. For this reason one can expect that a considerable part of the scattered ion flux will consist of ions scattered from the atoms which form the close-packed rows (100) lying on the target surface. Finally, one can expect to detect effects

---

† Actually, for a number of reasons the ion-atom interaction potential cannot be reliably obtained from the ion-solid scattering data. For one thing, when the ion-atom interaction potential is obtained from the scattering data one must be sure that only single scattering takes place. This fact is experimentally tested when ion scattering in gases is investigated. On the other hand, ion scattering from solids may never be treated as a purely single collision process. Besides, the key interest is to investigate the interaction potential distortion caused by the fact that the atom is not free, but is embedded in the lattice. However, this delicate effect cannot be investigated in principle using keV-energy ions. In fact, to obtain the interaction potential at a reasonably far distance ($\sim d/2$, d is the distance between lattice atoms), it is necessary to study small angle scattering. But it is known that small angle ion scattering from solids is a plural or multiple process.

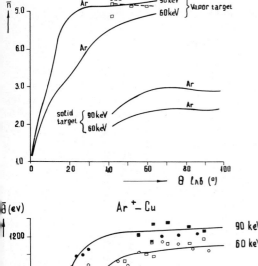

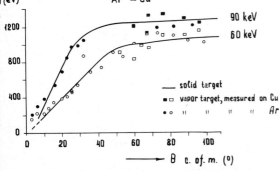

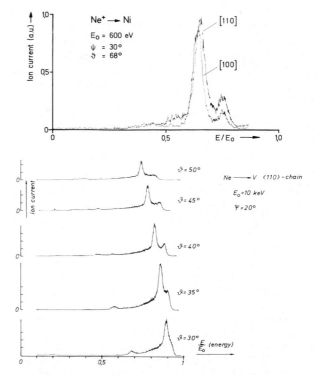

FIGURE 6   Comparative measurements of argon ion scattering using vapor and solid copper targets. At the top—the average number of missing electrons; at the bottom—the mean inelastic energy loss. Data.[13]

due to correlation of the ion collisions, using various faces of the crystals as targets and varying the incidence planes of the primary ions.

PLURAL SCATTERING

We now consider several results obtained during the study of ion scattering from crystals. In 1965 the first experimental data (Mashkova and co-workers)[23] concerned with the so-called "double-scattering effect" were published. It was found that under specific conditions in the energy distribution of ions scattered from crystals one more sharp peak (as compared with random solids) could be observed. Both the position and height of the peaks proved to be dependent on such parameters as the type and energy of the primary ions, type of target atoms, distance between the neighbouring target atoms, and also upon the angles which characterized the conditions of the experiment. An example of recent results on the double-scattering effect is reproduced in Figure 7.[20,21]

FIGURE 7   Top—Backscattering of 600 eV Ne$^+$ ions from a Ni (110) single crystal surface. The plane of scattering is perpendicular to the surface, the impact angle between impinging beam and the surface is $\psi = 30°$, and the laboratory scattering angle $\vartheta = 68°$ (see Figure 5). Spectrum [110] is taken with the plane of scattering parallel to the [110] direction in the surface, the spectrum [100] with the plane of scattering parallel to the [100] direction.

Bottom—Energy spectra of 10 keV Ne-ions backscattered from a vanadium single crystal for scattering angles ϑ between 30° and 50°. Angle of incidence to the − ⟨110⟩ chain was $\psi = 20°$.

In 1968 a similar effect was observed in ionized recoils,[16] and in 1969 for GeV-energy protons scattered from deuteron nuclei;[15] see Figure 8. To explain the main principles of the ion double scattering effect, a so-called "atomic pair" model was suggested.[8] However, before long it was found[17] that the presence of the high-energy peak in the scattered ion energy distributions may also be explained in the framework of a more complex model ("infinite atomic row"). According to calculations based on two different models the position of the peaks must be slightly different. It must be borne in mind that the calculated (in the framework of both models) positions of the peaks depend on the magnitude of the inelastic energy loss, which cannot be correctly calculated at present. Besides, the difference

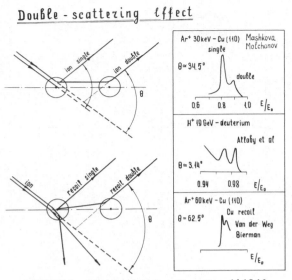

FIGURE 8   Double scattering effect. Data.[14,15,16]

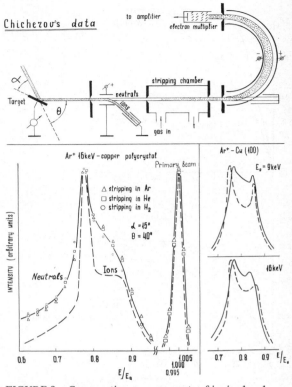

FIGURE 9   Comparative measurements of ionized and neutralized components of scattered ion flux (see text). Data.[5]

caused by using different models is of the same order of magnitude as the experimental accuracy (±1%). For these reasons the choice between two theoretical models could not be made until now. One may assume that the data obtained quite recently by Chicherov[5] will be useful for this purpose. Chicherov performed comparative measurements of the energy distribution of scattered ions and neutrals. His experimental technique as well as several experimental data are shown in Figure 9. One can see that the distance between peaks, which correspond to neutralized scattered particles, is somewhat smaller than those which corresponds to scattered ions. This fact (if, of course, it is a fact and not a systematic error in the experiment!) is a very interesting one. Actually, according to the atomic pair model the low-energy peak is practically[†] caused by single collision of an ion with a target atom. For this reason its position is practically the same as in the case of ion scattering from a gas target. The high-energy peak is caused by two collisions of the ions with the target atoms. According to the infinite atomic row model not only the high-energy peak, but also the low-energy one is caused by a number of successive collisions of the ions with the atoms in the row. Owing to this the low-energy peak lies at a somewhat higher energy compared with "pure single scattering". Because collision sequences include a number of weak (small-angle) collisions, the scattered ions have a good chance

---

[†] Generally speaking, this peak position is also caused by two collisions namely strong collision with one of two atoms and a weak collision with another atom; see, for example.[14]

of picking up the electrons during such collisions. For this reason one can expect that the scattered particles which interact with the atomic row leave the target mostly as neutral particles, whereas the particles which interact with atomic pairs have a somewhat greater chance of leaving the target in the ionized state. In this sense Chicherov's data are in qualitative agreement with the data (concerning the dependence of the peak's shape on the scattered ion charge) obtained in 1964 by Datz and Snoek.

SPATIAL DISTRIBUTION OF SCATTERED IONS

In this section we consider spatial measurements. As far as we know, spatial measurements in the high-energy region have been made since 1965. The main result obtained in this energy region was the observation of a number of phenomena associated with the "blocking effect". In the medium energy region spatial measurements were started somewhat later (1968–1969) These experiments were performed in two different ways. In Amsterdam (Kistemaker's Institute, Van der

Weg and Bierman)[9] the scattered ion analyzer was mounted mostly at a fixed angle relative to the primary ion beam, which was directed at the crystalline target usually along the normal to its surface. In the course of the experiment the target was rotated about the azimuth (i.e. around the ion beam direction). One can see that in this case the target rotation is equivalent to the azimuthal rotation of the analyzer with respect to the immobile target, so that the dependences obtained correspond to the dependences of the scattered ion intensity on the azimuthal scattering angle. In Moscow (Institut of Nuclear Physics, Mashkova and co-workers)[10,14,18] the primary ion beam was directed at the crystalline target at a small angle to its surface (so that only the ion forward scattering was studied). In the course of the experiment all the angles which characterized the ion scattering process were varied in certain intervals. Because of the great difference in the experimental conditions the data obtained in Amsterdam and in Moscow cannot be directly compared. However, some important conclusions concerning the spatial distributions of the scattered particles can be made.

First of all a strong dependence of the scattered ion intensity on the azimuthal scattering angle may be observed under some conditions; see Figure 10. Second, these strong oscillations of the scattered ion intensities are observed when the scattered ion outlet angles (i.e.

the ejection angle read from the target surface) are not too large, while in other cases the oscillations of the scattered ion intensity are much smaller, see also Figure 6 in Ref. 9. Third (hitherto observed in Moscow only) under some experimental conditions the specific effect of "ion focusing" may be observed. The crystalline target turned out to act on the scattered particles like focussing a mirror and may perform strong concentration of the scattered ions in a certain direction. At present there is no complete consensus of opinion concerning the origin of the observed regularities. In Moscow these effects are believed to be mainly caused by processes which are practically charge independent, while according to the Amsterdam group the directional dependence of the neutralization of the scattered ion will play an important role. The crucial experiments (the comparative measurements of both ionized and neutralized components of the scattered particle flux) have only just been started; see Refs. 5 and 14. Nevertheless, some of the experiments carried out with ionized components of the scattered particle flux could, in our opinion, hardly be explained in the framework of the mechanism suggested by van der Weg and Bierman.[9] In particular, it was observed that, when the ion scattering plane is parallel to the close-packed atomic rows, not only would the minima (as in the van der Weg and Bierman experiments), but also

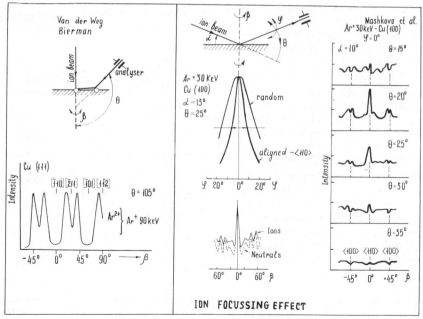

FIGURE 10   The dependence of the intensity of the scattered particles on the azimuthal angle of target rotation β and on the azimuthal scattering angle φ. On the left—data[9] obtained for charged components of scattered beam. On the right—data.[10,18]

the maxima of the scattered particle number (ions and neutrals—ion focusing effect) be observed; see Figure 10.

PROBLEM OF INTERPRETATION

For the time being the experimental study of ion scattering forestalls progress in the theory of this phenomenon. One can see from publications that the purpose of the theorist seems to be to fit (using various assumptions) formulae to experimental results. Such a situation is not surprising it is typical in early stages of all fields of science. Meanwhile, detection of the disagreement between theory and experiment is much more important for the progress of the science than confirmation of the agreement between the concrete theory and concrete experiment.

Current theories of medium-energy ion scattering from solid surfaces are essentially based on the assumption that the ion scattering process can be treated as a sequence of binary collisions of ions with target atoms placed in space, either randomly or regularly. The analysis of numerous publications shows that two asymptotic approaches to the problem of ion scattering from solids have been most developed.

In the first approach ion scattering from solids is treated as a result of single violent collisions only. Usually it is assumed that it takes place either just on the target surface (single-atom model of ion scattering) or inside the target, at some distance from the target surface. In the last instance it is believed that before and after collision the ion moves on a straight trajectory and loses energy continuously (single-collision model or model of continuous stopping).

In the second approach ion scattering from solids is treated as an essentially multiple collision process, so that the scattering process is described with kinetic equations (Firsov[11]). It should be noted that, strictly speaking, for medium ion energies neither approach is to be regarded as justified. Actually, the medium energy is high enough so that the scattering process cannot be considered as a surface process only. On the other hand, the ion energies are not yet sufficiently high to fulfil the assumption about the pure direct character of the ion trajectory. One can hardly accept unconditionally the assumption that the considered process is completely a multiple one. It is likely that for medium energies of the primary ions it is rather a plural process.

Unfortunately, the theory of plural ion scattering in solids in cases where the problem is not symmetrical is poorly developed. The existing situation may be illustrated by the following example. It is known that under some conditions the energy distributions of scattered ions have the shape of a wide cupola with a narrow peak in its high energy part. Such distributions have mostly been studied for helium ion scattering. Recently, similar data were also published for proton scattering; see Figure 4 of Ref. 20. Such results may be explained in terms of the single-collision model as well as by the Firsov multiple collision theory. In Figure 11 we present the energy distribution of the scattered particles (for helium scattering from a copper polycrystal) calculated according to the single-collision model, to the Firsov multiple collision theory and also the corresponding distribution of the scattered ions. The single collision model predicts a smooth rise of the number of scattered particles as their energy decreases. This fact is in disagreement with the experiment. The only chance of fitting the calculations to the experiment is to take into account the energy

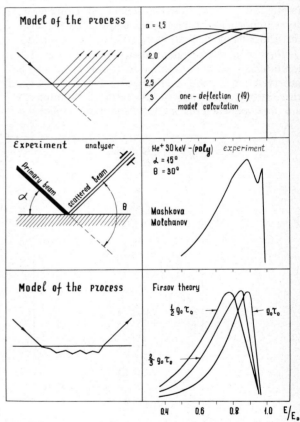

FIGURE 11   In the centre—experimental data.[19] At the top—calculations according to the single—deflection model.[19] At the bottom—calculations according to Firsov's multiple theory, $g_0\tau_0$—value characterizing the average square of scattering angle.[11]

dependence of the ion neutralization process (see Figure 11 top). The multiple collision theory suggested by Firsov describes the measured energy distributions of the scattered ions rather well (see Figure 11, bottom), although it completely ignores the process of ion charge exchange. Thus, the charge exchange process in particle scattering is one of the key problems.

CONCLUSION

On the whole, we must conclude the following. Up to now significant progress has been achieved in understanding ion scattering from solids. However, the situation is such that some principal experimental facts allow alternative theoretical explanations. To arrive at the truth comparative measurements (both extensive and precise) of the ions and the neutralized components of the scattered beam must be made.

REFERENCES

1. J. B. Hasted, *Physics of Atomic Collisions* (London, Butterworth, 1964).
2. A. Günterschulze, *Vacuum* **3**, 360 (1965).
3. A. L. Southern, W. R. Willis and M. T. Robinson, *J. Appl. Phys.* **34**, 153 (1963).
4. B. Fagot, N. Colombie, R. Thiry and Ch. Fert, *C. R. Acad. Sci.* **B 262**, 173 (1966).
5. S. Yu. Lukyanov and V. M. Chicherov, *JETP* **60**, 1399 (1971); V. M. Chicherov, *JETP Letters* **16**, wb, p. 328-331 (1972).
6. G. J. Dienes and G. H. Vineyard, *Radiation Effects in Solids* (Academic Press, New York, 1957).
7. S. Datz and C. Snoek, *Phys. Rev.,* **A134**, 347 (1964).
8. E. S. Mashkova and V. A. Molchanov, *Phys. Tverd. Tela* **8**, 1517 (1966).
9. W. F. v. d. Weg and D. J. Bierman, *Physica,* **44**, 177 (1969).
10. E. S. Mashkova, V. A. Molchanov and Yu. G. Skripka, *Doklady Acad. Nauk.* **190**, 73 (1970).
11. O. B. Firsov, *JETP* **61**, 1452 (1967).
12. I. N. Evdokimov, E. S. Mashkova and V. A. Molchanov, *Phys. Letters* **25A**, 619 (1967).
13. C. Snoek, W. F. v. d. Weg, R. Geballe and P. K. Rol in *Proc. of 7th Conference on Phenomen in Ionized Gases* (Belgrade, 1966), p. 145.
14. E. S. Mashkova and V. A. Molchanov, *Rad. Effects* **13**, 183 (1972); V. A. Molchanov and V. A. Snisar, *Rad. Effects* **12**, 105 (1972).
15. I. V. Allaby, A. N. Diddens *et al., Phys. Rev. Letters* **30B**, 549 (1969).
16. W. F. v. d. Weg and D. J. Bierman, *Physica,* **38**, 406 (1968).
17. V. M. Kivilis, E. S. Parilis and N. Yu. Turaev, *Doklady Acad. Nauk.* **173**, 805 (1967).
18. E. S. Mashkova, V. A. Molchanov and Yu. G. Skripka, *Doklady Acad. Nauk.* **198**, 809 (1971).
19. E. S. Mashkova, V. A. Molchanov and A. H. Rakhmatulina, *Doklady Acad. Nauk.* **181**, 49 (1968); E. S. Mashkova and V. A. Molchanov, *Rad. Effects* **13**, 131 (1972).
20. W. Eckstein and H. Verbeek, *J. of Vacuum Science and Technology* **9**, 612 (1972) and private communication.
21. E. Taglauer and W. Heiland, *Surface Science* **33**, 27 (1972).
22. W. Heiland, H. G. Schäffler and E. Taglauer, *Surface Science* **35**, 381 (1973).
23. E. S. Mashkova, V. A. Molchanov, E. S. Parilis and N. Yu. Turaev, *Phys. Letters* **18**, 7 (1965).

ESTIMATION OF ION SCATTERING BY ATOMIC CHAINS ON SINGLE CRYSTAL SURFACES

S. D. MARCHENKO, E. S. PARILIS and N. Y. TURAYEV

Institute for Electronics, Uzbek Academy of Sciences, Tashkent, USSR.

Estimation of ion movement close to atomic chains taking into account the elastic and inelastic energy losses and vacancy type imperfections was performed on the basis of a model of binary interactions of ions with the atoms of a crystal.

An analysis of results shows that at scattering by atomic chains a phenomenon which makes a substantial contribution to scattering by the single crystal surface the inelastic energy losses exceed the single scattering ones by a factor of 5–6.

Previously a method of successive binary collisions was applied with the purpose of clarifying the role of surface atomic chains in reflecting fast ions from a single crystal face at glancing angles of incidence.[1-5]

Estimation of ion movement close to atomic chains taking into account the elastic and inelastic energy losses was performed on the basis of a model of binary interaction of ion with the atoms of a crystal.

It was shown that scattering of ions by atomic chains is a phenomenon which makes a substantial contribution to scattering by the surface of a single crystal. The energy and angular distributions depend to a great extent on the type of the crystal axis. The angular distributions of reflected ions at glancing angles of incidence are limited by minimum and maximum angles of escaping the complete losses of energy are sharply reduced along the low index atomic chains.

The scattered ions possess an energy which exceeds that of single and even double scattering at the same angle.[1,2]

A calculation of the inelastic energy losses is of considerable interest. Although this energy constitutes a small part (10%) of overall energy losses within the range of 10–30 keV, it plays an important role in processes of ionization and electron emission which follows ion scattering and to a considerable extent account for the charge state of the scattered particles.

Calculations were performed with the help of an M-220 electronic computer applying the same technique as before.[1,2,5] This technique makes it possible at the given value of the glancing angle of incidence ψ and angle with chain axis ξ to make a juxtaposition of the impact point on the elementary area with the trajectory of the ion and the direction of its escaping which is determined by the angle with the surface of the crystal β and the azimuthal angle φ. The coordinates of the impact point are r (along the bombarded chain in units

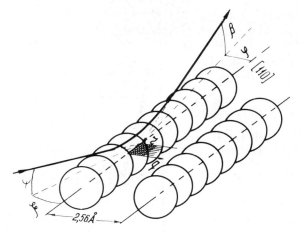

FIGURE 1   Geometry of calculations: angle of incidence ψ, angle with chain axis ξ, angle of escape β, azimuthal angle φ, impact coordinates r and q.

0.01 of the atomic spacing) and q (in normal direction in units 0.005 of the atomic spacing (Figure 1).

To each value of q in the scattered beam there corresponds a loop (Figures 2 and 3). In contrast to previous calculations the chain was taken not as a solid, but with a "defect" which was simulated by the absence of one atom in the chain. This vacancy was located either in the point preceding the first point of the impact area (Figure 2) or in the first point (Figure 3).

In the first case the picture of scattering differed little from scattering an a defectless chain. Only in Figure 2 a branch was observed corresponding to ions passing through the chain (negative values of β).

In the second case passage through the chain is dominant. Scattering occurs mainly downward and only individual branches for each value of q emerge to the surface.

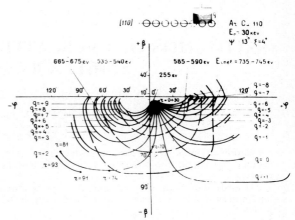

FIGURE 3   Results of calculations for the scattering of 30 keV Ar⁺ from a Cu-surface with a defect in the impact area.

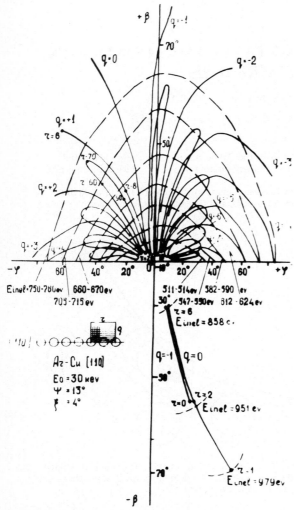

FIGURE 2   Results of calculations for the scattering of 30 keV Ar⁺ from a Cu-surface with a defect preceding the impact area.

scattering β and φ corresponds to inelastic energy losses within the range 650–760 eV. In Figure 3 a similar curve corresponds to 735–745 eV.

Attention is drawn to the large values of inelastic energy losses which exceed corresponding energy losses at single scattering of the same angle at a factor of 5–6 or more.

When the scattering angle is reduced the energy losses also diminish and on the shadow boundary they amount to about 500 eV (Figure 2) and 225 eV (Figure 3).

The inelastic energy losses were estimated on the basis of the Firsov formula[6] for collisions with each atom individually depending on the energy of the ion and the impact parameters:

$$\epsilon(P, E) = \frac{0.35(Z_1 + Z_2)^{5/3}\dfrac{\hbar}{a_0}\sqrt{\dfrac{2E}{m_1}}}{\left[1 + 0.16(Z_1 + Z_3)^{1/3}\dfrac{R}{a_0}\right]^5}$$

where $\epsilon(P, E)$—inelastic energy loss in one collision;
              P—impact parameter;
              E—ion energy prior to given collision;
    Z_1, Z_2—atomic numbers of ions and atoms of the crystal respectively;
              a_0—Bohr radius;
              m_1—mass of the ion;
   $R = R(P, E)$—distance of closest approach corresponding to P and E.

The energy $\epsilon(P, E)$ was summed along the ion trajectory.

Of course, this does not mean that the coefficient of ion reflection from the single crystal face strives to reach zero value. The model study used is of a limited nature and has been intended to illustrate ion scattering on an atomic chain in the presence of a defect. The complete coefficient of reflection must be comparable with calculations on a block containing a big number of atomic layers.

Figures 2 and 3 illustrates isoenergetic curves which cross all loops and contain trajectories with inelastic energy losses within the range (in eV) specified for each curve. For instance in Figure 2 the curve which crosses the regions with the maximum angles of

It was assumed that during the interval between collisions the ion has enough time to relax. But this is not
obvious and there is the possibility that during the time
which elapses between two collisions $(d\sqrt{m_1/2E}) \sim 10^{-15}$
sec) the electron shell of the ion cannot relax. This
means that the inelastic energy losses turn out to be
non-additive and can differ greatly from the calculated
values. Experimental testing of this factor is of doubt-
less interest and the given calculations could serve this
purpose. They must be taken into account when
interpreting the experiments on charge state of ions
scattered in close packed directions.[7,8]

REFERENCES

1. V. M. Kivilis, E. S. Parilis and N. Yu. Turaev, *Soviet Phys.-Dokl.* **12**, 328 (1967).
2. V. M. Kivilis, E. S. Parilis and N. Yu. Turaev, *Soviet Phys.-Dokl.* **15**, 587 (1970).
3. D. S. Karpuzov and V. E. Yurasova, *Phys. Status Solidi (b)* **47**, 41 (1971).
4. V. E. Yurasova, V. I. Shulga and D. S. Karpuzov, *Can. J. Phys.* **46**, 759 (1968).
5. S. D. Marchenko, E. S. Parilis and N. Y. Turaev, *X Int. Conf. Phenom. Ioniz. Gases, 1971*, (Oxford), p. 79, *DAN SSSR* **206**, 313 (1972).
6. O. B. Firsov, *Soviet Phys.-JETPh* **9**, 1075 (1959).
7. C. Snoek *et al. Physica* **35**, 1 (1967).
8. W. F. van der Weg *et al. Physica* **44**, 177 (1969).

THE PECULIARITIES OF ION SCATTERING BY CRYSTALS DUE TO SURFACE HALFCHANNELS, FORMED BY CLOSE PACKED ATOMIC ROWS

Yu. G. SKRIPKA

Physic Department, Donetsk State University, Donetsk, 340055, USSR

The ion scattering by the surface of crystals has been studied when there are halfchannels on the crystal surface, which are formed by close packed atomic rows of the first and the second layers of crystals. The obtained results are compared with one of ion scattering by directions, which have no surface halfchannels.

During the past few years greater interest has been shown in studying the angular and energy distributions of ion scattering by crystals. The study of these characters has led to the discovery of channeling[1-3] and shadows.[4,5] On bombarding of the single crystals by heavy ions, when the scattering is determined by the surface layers of the crystal, the analogous effects have been discovered, i.e. the input blocking effect and output effects.[6-8] It was found that it is possible to obtain information about the channeling and shadows (the scattering by single close packed atomic rows and the scattering by the surface halfchannels).

This article deals with ion scattering by surface halfchannels and the comparison of these results with those, which are obtained for the ion scattering by the structure, which have no halfchannels.

In Ref. 9 the experimental method is described. The copper crystals (100) and (110) were bombarded by 30 keV Ar$^+$ ions.

The energy and angular distributions of ion scattering by crystals have been studied. On Figure 1 is shown the scheme of scattering.

On Figure 2 are shown the energy distributions of ions scattered by crystals for two cases of target orientation: (1) the face (100) when the incidence and scattering planes are coinciding at $\beta = 0°$ with the atomic direction $\langle 100 \rangle$; (2) the face (110) when the incidence and scattering planes are corresponding at $\beta = 0°$ with the atomic direction $\langle 100 \rangle$.

In the first case with increasing β the intensity of the "double" peak is decreasing relative to the "single" reflection peak and in the second case with increasing β the intensity of the "double" peak is increasing at first and then it is decreasing. For $\beta = 6°$ the intensity

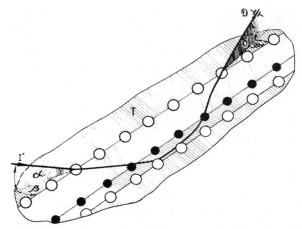

FIGURE 1   The scheme of scattering. The surface half-channels formed by atomic rows of the first and the second crystal layers. α—the sliding angle, β—the azimuthal angle of target rotations, δ—the outlet angle, $\varphi = \varphi' + \beta$—the azimuthal scattering angle.

of the "single" peak and the intensity of the "double" peak are equal. It is necessary to note that in the first case the atomic rows of the second layer are placed under the atomic rows of the first layer. At the same time in the second case the atomic rows of the second layer are placed between atomic rows of the first layer forming the surface halfchannels (see Figure 1).

The energy distributions for various azimuthal scattering angles for $\beta = 0°$ have been obtained. For these energy distributions the angular distributions of scattered ion intensity were taken. On Figure 3 the curves $I(\varphi)$ for the same cases of target orientations

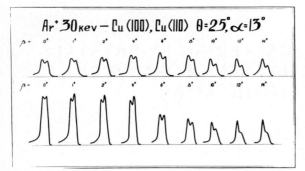

FIGURE 2   Ar$^+$ 30 keV → Cu (100) and Cu (110). θ = 25°, α = 13°. The energy distributions of scattered ions. In the top —Cu (100). β = 0° corresponds to the atomic direction (100). From the left to the right: β = 0°, 1°, 2°, 4°, 6°, 8°, 10°, 12°, 14°. In the bottom—Cu (110). β = 0° corresponds to the atomic direction (100). The β are the same as above.

are shown. It is seen that there is the minimum for φ = 0° for the small outlet angles $\delta(\delta = \theta - \alpha)$ in the first case of target orientation. While increasing output angle the minimum is decreasing and for $\delta \sim 7°$ it disappears. In the second case for $\delta \sim 1°$ the minimum is observed also, but when δ is increased the small maximum in the minimum appears, which increases when δ increases. For $\delta \sim 7°$ the minimum disappears. It is convenient to choose the halfwidth of the spatial distribution as one of the characteristic features of the spatial distributions.

On the top of Figure 4 the halfwidth of the spatial distribution Γ as a function of the polar scattering angle θ for the above mentioned target orientations is shown. The curve 1 corresponds to ions scattered by the face (100) along the atomic direction ⟨100⟩ and the curve 2 corresponds to ions scattered by the face (110) along the atomic direction ⟨100⟩. On the right of Figure 4 we see the projections of the atomic rows of the crystals for the above mentioned target orientations. On the bottom of Figure 4 is shown the intensity of the scattered ions I in dependence of the polar scattering angle θ where the azimuthal scattering angle φ = 0° and the scattering plane is the same as the incidence one.

As it is clear from Figure 4 $\Gamma(\theta)$ are not monotonous. These curves have minima. The minimum of $\Gamma(\theta)$ in the first case is observed at the range of $\theta \sim 23°$ and in the second case for $\theta \sim 21°$. For the small outlet angles the curve for the face (100) lays above the curve for the face (100). In the first case the maximum of $I(\theta)$ is observed for the range $\theta \sim 21°$ and for the range $\theta \sim 19°$ at the second case. It is seen that the minimum of $\Gamma(\theta)$ and the maximum $I(\theta)$ for the same faces do not coincide.

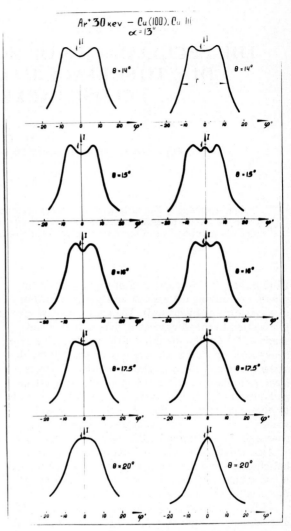

FIGURE 3   Ar$^+$ 30 keV—Cu (100) and Cu (110). α = 13°. The spatial distribution of ions scattered by the crystal. The left part—the face (100), θ = 14°, 15°, 16°, 17°, 5°, 20°, the right part—the face (110), θ = 14°, 15°, 16°, 17°, 5°, 20°. Γ—the halfwidth of the spatial distribution.

The obtained experimental data prove the following:

1) The scattering by the same atomic directions, which lay on different faces of crystals are sharply different. This difference can be caused by different structure of surface atomic rows of the crystal.

2) The monotonous decrease of the intensity of the "double" peak in the case of absence of the half-channels is related to the decrease of the probability of plural scattering outside close packed atomic rows.

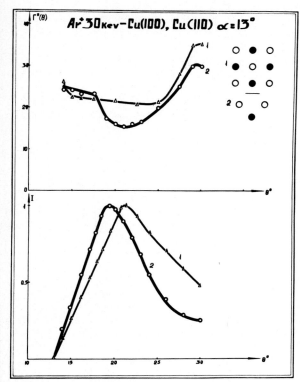

FIGURE 4 Ar^+ 30 keV—Cu (100) and Cu (110) · α = 13°
On the top there are the dependences of the halfwidth of the
spatial distributions Γ upon the polar scattering angle θ
($\theta = \alpha + \delta$). On the bottom there are the dependences of the
scattered ion intensity I upon polar scattering angle θ. The
curves 1 are for the face (100) and atomic direction ⟨100⟩.
The curves 2 are for the face (110) and atomic direction
⟨100⟩.

And the lumps of this peak intensity in the case of
halfchannels are related to the decrease of energy losses
of the scattered ions caused by the surface halfchannels.
This conclusion is founded on the obtained general
displacement of the centre of gravity of energy distributions to high energies in the second case of target
orientation. The analogous displacement of the centre
of gravity of the energy distributions is observed in the
channeling.[3]

3) In the presence of surface halfchannels it is
assumed that the ions are focused into a narrow beam.
It is well seen on Figure 3 (the maximum into the

minimum $I(\varphi)$). The maximum is observed on the
background of the shadow due to the surface close
packed atomic rows.[6]

4) If we suppose, that for single collisions the cross
section is

$$d\sigma \sim \frac{1}{\theta^3}\, d\Omega$$

where θ is the polar scattering angle, for n collisions
we shall have:

$$d\sigma \sim \frac{1}{\theta_1^3}\cdot\frac{1}{\theta_2^3}\cdots\frac{1}{\theta_n^3}\, d\Omega$$

It is clear that for the increase of the multiplicity
of the scattering each angle θ_i is decreased and $d\sigma$ is
increased. As we observe the maxima $I(\theta)$ for $\theta \sim 20°$,
we can conclude that for this θ the multiplicity of
the scattering is higher than for $\theta < 20°$ and $\theta > 20°$.

Thus the experimental data show the qualitative
and quantitative difference of ion scattering in the
case of the presence of halfchannels on the target
surface and in the case when the atomic rows of the
second layer are under atomic rows of the first layer.

ACKNOWLEDGEMENT

The author thanks Dr. V. A. Molchanov and Dr. E. S.
Mashkova for their interest in the work and for many
valuable discussions.

REFERENCES

1. R. S. Nelson and H. W. Thompson, *Phil. Hag.* **8**, 1677
    (1963).
2. I. H. Fluit, J. Kistemaker and C. Snoek, *Physica* **30**, 870
    (1964).
3. C. Erginsoy and V. F. Wegner, *Phys. Rev. Lett.* **13**, 530
    (1964).
4. A. F. Tulinov, V. S. Killiskauskas and H. H. Malov, *Phys.
    Lett.* **18**, N.3 (1965).
5. A. F. Tulinov, *Doclades of Academy of Sciences, USSR*
    162, 546 (1965).
6. E. S. Mashkova, V. A. Molchanov and Yu. G. Skripka,
    Phys. Lett. **29A**, 645 (1969).
7. M. Aissa, E. S. M. Mashkova, V. A. Molchanov and Yu. G.
    Skripka, *Soviet Physics of Solids,* **12**, 2070 (1970).
8. E. S. Mashkova, V. A. Molchanov and Yu. G. Skripka,
    Phys. Lett. **33A**, 373 (1970).
9. E. S. Mashkova, V. A. Molchanov and Yu. G. Skripka,
    Doclades of Academy of Sciences, USSR, **190**, 73 (1970).

ENERGY DISTRIBUTION OF PROTONS WITH PRIMARY ENERGY OF 15 keV BACKSCATTERED FROM A Ni SINGLE CRYSTAL

W. ECKSTEIN, H. G. SCHÄFFLER and H. VERBEEK

Max-Planck-Institut für Plasmaphysik, EURATOM Association, D-8046 Garching, Germany

A nickel single crystal was bombarded with 15 keV protons incident in the vicinity of the ⟨110⟩ and ⟨100⟩ directions. The backscattered protons were energy-analysed with a spherical electrostatic analyser which could be moved around the target. The energy resolution was $E/\Delta E = 140$. The surface peak and a very low scattering intensity from the bulk was observed in a channelling experiment. The latter intensity is zero within the experimental error in double alignment. In this case it can be deduced from the width of the surface peak that less than two atomic layers contribute to it. When the angle between a close-packed direction and the incident beam is increased, at first only the scattering intensity from the bulk increases. When a critical angle is exceeded, the surface peak increases as well. At the same time it broadens and shifts to lower energies. In addition, at energies below the surface peak new peaks in the backscattering spectra appear. Computer simulations indicate that backscattering in this energy range is governed to a great extent by multiple collisions. It is thus impossible to attribute to the observed peaks specific depths from which the backscattering occurs. The enhancement and shift of the surface peak is simulated by the calculations. They cannot, however, explain the observed structure in the spectra.

INTRODUCTION

The backscattering of light ions in the energy range of 5–20 keV from metal surfaces is of interest in connection with the "wall problem" in plasma physics. There are several measurements on polycrystalline material.[1,2,3,4] The observed energy distributions are usually explained in terms of the "single deflection" model.[1,2,3,5] It was, however, pointed out[1,7] that this model is insufficient in this energy range since multiple collisions are likely. Our measurements on polycrystalline material[4] showed that the interference of crystal effects is hard to avoid. This is the reason why investigations on single crystals appeared interesting. Backscattering from single crystals was mostly investigated at higher energies (for example see Ref. 6 and 8a, 8b). To our knowledge there are only a few measurements on the KeV energy range.[3,9,10] The results of different workers are not consistent with one another and the observed structure in the energy distribution is not properly understood. In transmission measurements through thin single crystal films[11] agreement with theoretical conceptions was found.

EXPERIMENTAL SETUP

The apparatus was the same as used in our previous work.[4,12] Its principle is shown in Figure 1. A mass-analysed proton beam is collimated to within 0.8° by two diaphragms. The target is mounted on a goniometer head,[13] by which it can be rotated around three axes perpendicular to each other. The backscattered protons are energy-analysed by a spherical electrostatic analyser which can be swung around the target. Its energy resolution is $E/\Delta E = 140$. The acceptance angle of the spectrometer is 3.8°, which corresponds to a solid angle of 3.5×10^{-3}.

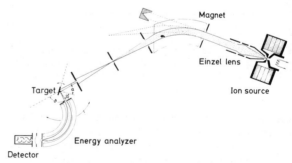

FIGURE 1 Experimental setup (schematic).

All measurements reported here were made with a Ni single crystal with a {100} surface. The crystal was first polished mechanically, then electrolytically. Inside the vacuum it was bombarded by Ne ions for surface cleaning and subsequently heated to ~1200°C by electron bombardment for annealing.

The vacuum in the target chamber was below 10^{-8} torr during the measurements.

EXPERIMENTAL RESULTS

The proton primary energy was 15.36 keV in all measurements reported here. Some examples of back-scattering spectra are shown in Figure 2. The curve at

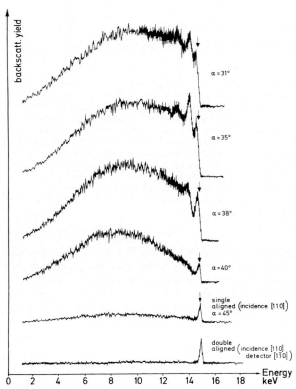

FIGURE 2   Backscattering spectra of 15.36 keV protons entering a Ni crystal in the vicinity of a ⟨110⟩ direction. (The ordinate scale is arbitrary, but the same for all spectra.)

the bottom is a spectrum in double alignment, with the primary beam in the [110] and the spectrometer in the [1$\bar{1}$0] direction. This spectrum shows only the surface peak without any backscattering from deeper layers above the detector background. From the width of the surface peak as compared with the width of the calibrating curve of the spectrometer it is estimated that less than two atomic layers contribute to it. For this estimation an electronic energy loss of 10 eV/Å[14,15] was assumed.

For the next curve the primary beam was still in the [110] direction. By inclining the scattering plane by 11.3° to the (001) plane, the detector was moved out of the close-packed [1$\bar{1}$0] direction. It is, however, still in the (110) low-index plane. With respect to the

double aligned spectrum in this curve only the back-scattering intensity from the bulk is slightly enhanced.

For the upper curves the angle of the impinging protons to the [110] direction was successively increased, i.e. α as defined in Figure 1 was changed from 45°. The direction of the spectrometer with respect to the crystal was kept constant ($\beta = 45°$). At first only a further enhancement of the backscattering intensity from the bulk is observed. In moving away from [110] at angles $\alpha < 40°$, additional peaks occur in the spectra. At the same time the surface peak increases and is shifted to lower energies. The arrows in Figure 2 mark the energy for binary collisions with the surface atoms. This is the position of the surface peak if one assumes that backscattering at the surface takes place in one single collision. The energy difference of the second peak to the surface peak decreases with increasing angle of the primary beam to the [110] in this case. In some cases a third peak occurs.

For angles of the primary beam to the [110] direction up to approximately 5.5° no additional peaks occur. The error in determining this angle is rather large, since the first appearance of the maximum above the background is not easy to observe. According to the calculations (see later) up to this angle which is the blocking angle no direct collisions with atoms of the [110] chain can occur.

In Figure 3 the minimum of the backscattering with 10 keV is depicted when the angle of incidence is varied across the [110] direction. The unsymmetrical shape of the curve is due to the variation of the scatter cross section with the scattering angle ϑ from 75° to

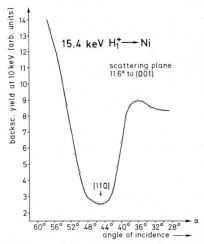

FIGURE 3   ⟨110⟩ channelling minimum for protons with primary energy 15.36 keV backscattered with 10 keV from a Ni crystal.

105°. From the half-width of the curve the channelling angle can be estimated to be 5.5 ± 0.5°. This roughly agrees with Lindhard's[16] estimate, which yields $\psi_2 = 6.5°$. This was also confirmed by the measurements of Andreen.[11]

In Figures 4a, b, backscattering spectra of protons impinging at various small angles to the [100] direction are shown. Only the interesting parts of the spectra at high energies are depicted. Again the direction of the detector with respect to the crystal was constant ($\beta = 55°$). For Figure 4a the scattering plane was the (001) plane; for Figure 4b it was inclined 10° to the (001) plane. When the detector is in a random direction as in Figure 4b, the maxima are more significant than in Figure 4a.

No alterations of the observed spectra during the measurements were observed, i.e. there was no influence of radiation damage, as was observable when the target was bombarded with Ne^+ ions.

COMPUTER SIMULATIONS

In order to understand the observed behaviour, we did some computer simulations. In a planar lattice the trajectory of a particle with given angle and position of incidence was calculated in the following manner:

From the impact parameter relative to the nearest lattice atom the deflection angle and the elastic energy loss are deduced. A continuous electronic energy loss of 10 eV/Å[14,15] was taken into account. This and the geometrical arrangement of the lattice atoms determines the impact parameter and the energy for the next impact. To calculate the deflection angle from the impact parameter, Lindhard's magic formula[17] was used and a Molière potential[18] with the screening radius according to Ref. 17 was assumed.

Figure 5 shows a computer plot as one example of the calculations. The drawing plane is the (100) plane. The lattice atoms are positioned at the small circles. As

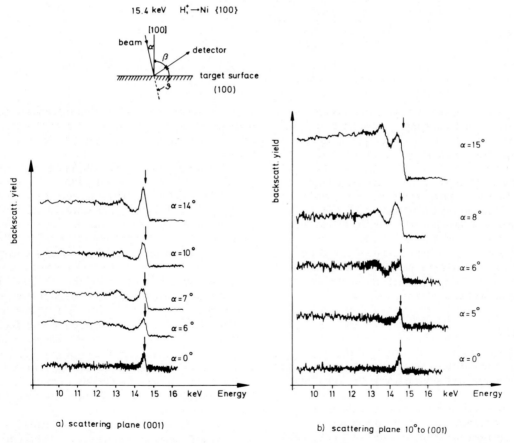

FIGURE 4   Backscattering spectra of 15.36 keV protons entering the crystal in the vicinity of a ⟨100⟩ direction. (a) detector in the (001) plane, (b) detector random.

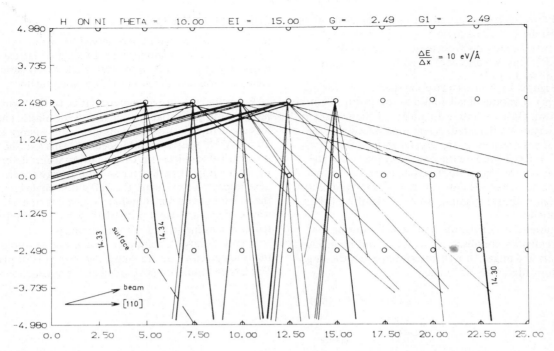

FIGURE 5   Computer plot of proton backscattering from Ni. (Energies indicated at several trajectories are final energies when the particles are outside the crystal.)

the coordinate scales given in Angström units are chosen different, the lattice appears distorted. For the given angle of incidence of 10° to the [110] direction (as for the spectrum with $\alpha = 35°$ in Figure 2) the position at incidence was altered in steps of 0.001 Å, and the individual ion paths were calculated. Only those trajectories are shown which leave the crystal with a total deflection angle of 100° ± 10°. (The angular spread was chosen larger than the actual angle of acceptance of the spectrometer for statistical reasons.) The majority of the calculated trajectories are not shown. Most of them penetrate much more deeply into the crystal and give rise to the broad distribution in the backscattering spectra.

DISCUSSION

At some trajectories the final energies at the surface of the crystal are indicated. Particles with very different path lengths in the crystal have nearly the same final energy.

More detailed calculations show that the majority of particles backscattered from the upper ten atomic layers have energies between 14.3 and 14.9 keV. This is where the peak with the highest energy in the experimental spectrum is situated. This peak is no

longer a true surface peak. Within this energy range are particles which undergo only one large deflection and others which undergo more than one deflection larger than $\approx 20°$.

There are some particles with final energies of (14.1 ± 0.05) keV and the appearance of a second maximum is hinted at. These particles are deflected twice by more than 100°. The statistics are, however, poor as this process is rather unlikely. In elastic collisions these lose a large amount of energy ($\sim$1200 eV) which is much larger than their electronic energy loss of 100 eV. On the other hand, there are particles with final energies of 14.9 keV which lose $\sim$200 eV to the electrons on rather long paths through matter, while their elastic energy loss is only $\sim$260 eV. Also if one assumes that the electronic energy loss may be 2 or 3 times larger[8b] the picture does not change qualitatively.

Calculation of the trajectories in a two-dimensional model is a great simplification as long as more than one collision contributes considerably to the total deflection. Three-dimensional calculations are desirable though very complicated.

In spite of their insufficiency the calculations demonstrate that it is impossible to attribute to the observed peaks in the backscattering spectra different depths from which the backscattering occurs. This is in contrast to Barret's[8a] theoretical results and the experiments by Bøgh[8b] at higher energies.

In the 15 keV energy range, backscattering by several large angle deflections plays an important role. This means that the "single-deflection model" is not suitable for explaining the peaks in the spectra. The depth resolution that would be possible according to the energy resolution of the spectrometer (~100 eV) is largely diminished by the multiple collisions. Nevertheless, there is still a connection between depth and energy within the uncertainties due to the large differences in the elastic collisions. Indeed, according to Lindhard et al.[19] down to one keV the electronic stopping power is larger than the nuclear stopping power. This, however, is no longer true when the very few particles with large-angle deflections and thus large elastic (nuclear) energy losses, which are backscattered from the upper layers are selected by the experiment.

The intensity of the second peak is not explained by the calculations. As the trajectories are no longer in one plane when multiple collisions occur, these would be much more frequent in a three-dimensional calculation. This would enhance the second peak with respect to the first one.

From the calculations it may be assumed that blocking plays a dominant role in the formation of the second peak. This is supported by comparison of Figures 4a and 4b. For the spectra in Figure 4a the detector was blocked by the low-index plane (110) and the second peak is much less pronounced than in Figure 4b, where the detector was in a random position.

CONCLUSIONS

The computer simulations demonstrate that the single deflection model is not suitable for explaining the structure at the high-energy end of the backscattering spectra. Due to multiple collisions there is no unique relation between the energy of the observed peaks and the depth from which the backscattering occurs. The calculations reproduce fairly well the observed increase

and shift to lower energy of the surface peak. A second peak is hinted at by the calculations. They cannot, however, explain the intensity of the second peak with respect to the first one.

ACKNOWLEDGEMENTS

The authors are indebted to Drs. R. Behrisch and B. M. U. Scherzer for valuable discussions and R. Hippele and S. Schrapel for technical assistance.

REFERENCES

1. G. M. McCracken and N. J. Freeman, *J. Phys. B. Ser. 2,* **2,** 661 (1969).
2. K. Morita, H. Akimune and T. Suita, *Jap. Journ. Appl. Phys.* **7,** 916 (1968).
3. E. R. Cawthron, D. L. Cotterell and M. Oliphant, *Proc. Roy. Soc. Lond. A* **319,** 435 (1970).
4. W. Eckstein and H. Verbeek, *J. Vac. Sci. Techn.* **9,** 612 (1972).
5. T. Ishitani and R. Shimizu, *Jap. Journ. Appl. Phys.* **10,** 821 (1971).
6. (a) R. Behrisch, Dissertation TU München, IPP Report 2/68. (b) R. Behrisch, *Canad. J. Phys.* **46,** 527 (1968).
7. D. G. Armour and G. Carter, *Ned. Tijdschr. Vakuumtechnik* **8,** 184 (1970).
8. (a) J. H. Barrett, *Phys. Rev. B* **3,** 1527 (1971). (b) E. Bøgh, *Rad. Effects* **12,** 13–19 (1972).
9. D. P. Smith, *Surf. Sci.* **25,** 171 (1971).
10. E. S. Mashkova and V. A. Molchanov, *Rad. Effects* **13,** 131 (1972).
11. C. J. Andreen and R. L. Hines, *Phys. Rev.* **159,** 285 (1967).
12. W. Eckstein and H. Verbeek, IPP Report 9/7, June 1972, Vacuum sub. for publication.
13. R. Behrisch, G. Mühlbauer and B. M. U. Scherzer, *J. Sci. Instr. Ser. 2,* **2,** 381 (1969).
14. G. F. Bogdanov, V. P. Kabaev, F. V. Lebedev and G. M. Norikov, *Soviet Atomic Energy* **22,** 133 (1967).
15. E. P. Arkhipov and Yu. V. Gott, *Sov. Phys. JETP* **29,** 615 (1969).
16. J. Lindhard, *Mat. Fys. Medd. Dan. Vid. Selsk.* **34,** No. 14 (1965).
17. J. Lindhard, V. Nielsen and M. Scharff, *Mat. Fys. Medd. Dan. Vid. Selsk.* **36,** No. 10 (1968).
18. G. Molière, *Z. f. Naturforschg.* **2a,** 135 (1967).
19. J. Lindhard, M. Scharff and H. E. Schiøtt, *Mat. Fys. Medd. Dan. Vid. Selsk.* **33,** No. 14 (1963).

INFLUENCE OF ION BOMBARDMENT INDUCED RADIATION DAMAGES IN THE CRYSTAL LATTICE ON THE ANGULAR REGULARITIES IN ION SCATTERING

U. A. ARIFOV and A. A. ALIYEV

Institute for Electronics, Uzbek Academy of Sciences, Tashkent, USSR.

The paper examines the angular, spatial and energy distribution of ions scattered by various faces of silicon and germanium single crystals. It is shown that, due to the orderly arrangement of atoms in the crystal lattice, the orientation effects in the angular, spatial and energy distributions are observed at temperatures exceeding the annealing temperature of ion bombardment induced radiation damage. At high temperatures of the sample a reduction of the anisotropy in the angular, spatial and energy distributions of scattered ions was observed which is due to the changing transparency of the crystal resulting from the thermal vibrations of atoms in the lattice.

Results of investigations testify to the possibility of using the orientation dependences of phenomena occurring during the interaction of ions with crystals for determining the annealing temperature of ion bombardment damage and for studying the kinetics of the process of ion bombardment. It is shown that further studies along the same lines as conducted by the authors of this paper could develop a technique for the quantitative estimation of ion bombardment damage in crystals and also in thin epitaxial films.

The peculiarities discovered in the angular, spatial and energy distribution of scattered ions as a result of the orderly arrangement of atoms in the crystal lattice of the target and the thermal vibrations of its atoms can (as mentioned in previous works[1-4]) be explained qualitatively by a theoretical model based on scattering by a series of binary collisions. However, good agreement of theory and experiment is surprising since, on the one hand, theoretical studies are based on the concept of the ideal and defectless crystal[9,10] whereas, on the other hand, it is known[11] that as a result of bombardment of solids by fast atomic particles a large number of defects appear which damage the structure of the crystal. This presence of extensive damage which does not affect the peculiarities in angular, spatial and energy distributions of scattered ions due to the orderly arrangement of atoms of the crystal demands an explanation. The only argument, as mentioned in[9] was the fact that in our previous works,[1-4] aimed at studying the influence of the crystal structure and thermal vibrations of atoms in the lattice on the process of ion scattering, the experimental conditions ensured effective annealing of bombardment damage. In this connection it seemed very interesting to undertake studies similar to those in[9,10] but using for a target single crystal samples of silicon and germanium whose annealing temperature is considerably higher than that of metals.[11]

Studies of the angular dependence of energy distributions of scattered ions was conducted on an installation described in Ref. 2 where an electrostatic condenser with a resolving power $\Delta E/E_0 \approx 0.7\%$ was used for analysing secondary ions. The unit also had facilities for the ion bombardment of the target at various angles of incidence ϕ by using an adjustable ion source. Analysis of the energy of secondary particles emitted at various escape angles θ and scattering angles β was done by a special device which changed the orientation of the target. The angular distribution was recorded by a rotatable beam catcher.

Results of investigations of the angular, spatial and energy distributions of secondary ions during the bombardment of silicon and germanium single crystal faces by alkali metal ions within the range of low energies (under 5 keV) are reported below.

Figure 1 illustrates the curves of dependence of the ion scattering coefficient $K_p(\phi)$, which were obtained during the bombardment of face (001) of a germanium single crystal by Na^+ ions at $E_0 = 1200$ eV. The angle of incidence was changed by rotating the target around axis $\langle 010 \rangle$. Curves 1–4 have been recorded at various temperatures of the target. It is seen that at sufficiently high temperatures the curves pass through a series of minima and maxima, i.e. they display typically anisotropic dependences. At comparatively low temperatures (provided the sample had not been subjected to thermal treatment before the measurements) there occurs, as in the case of a polycrystal, a smooth increase of K_p with the increase in angle ϕ.

It has been observed that the curves of K_p recorded during the bombardment of face (001) of Ge by Na^+

114   U. A. ARIFOV AND A. A. ALIYEV

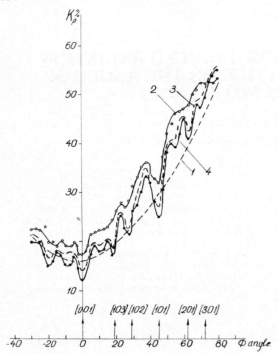

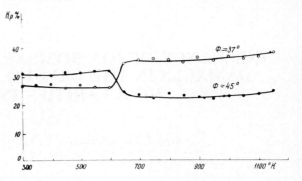

FIGURE 2   Temperature dependence of K_p which corresponds to the maximum and minimum of curve $K_p(\phi)$ in Figure 1.

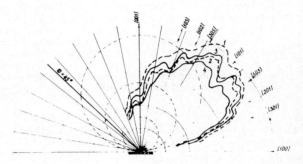

FIGURE 1   Dependence of ion scattering coefficient K_p on the angle of incidence for bombardment of an (001) face of Ge by Na^+ ions at $E_0 = 1200$ eV. Here curves 1–4 are at temperatures: (1) 300°K, (2) 600°K, (3) 900°K, (4) 1400°K.

FIGURE 3   Angular distribution of secondary ions recorded for Na^+ ions on (001) Ge at $E_0 = 1200$ eV. Curves 1–4 correspond to various temperatures of the sample: (1) 300°K, (2) 600°K, (3) 900°K, (4) 1400°K.

ions immediately after thermal processing of the sample and after a lengthy (~10 hours) bombardment at room temperature sharply differ. In the second case the curve of $K_p(\phi)$ is similar to that recorded during the bombardment of a Ge polycrystal. An investigation of the behaviour of curve $K_p(\phi)$ depending on the duration of ion bombardment shows that this change is mainly due to the accumulation of bombardment damage in the Ge sample. The persistence of the anisotropy in curve $K_p(\phi)$ which was recorded for target temperatures $T > 600°$K indicates that the annealing temperature of Ge lies within the range of $T > 600°$K.

The dependences of K_p on the temperature of the sample (Figure 2) were measured at two fixed values of the angles of incidence (corresponding to the maximum and minimum in curve $K_p(\phi)$) with the aim of clarifying the transition of one type of $K_p(\phi)$ dependence into another. Their examination shows that, within a certain narrow temperature range which delineates regions characteristic of arranged and non-arranged structures, there occurs a sharp jump of the K_p coefficients. In the case of curve $K_p(T)$ recorded at angle ϕ which corresponds to the minimum of curve $K_p(\phi)$, this jump is

in the direction of reducing K_p while at angle ϕ which corresponds to the maximum of curve $K_p(\phi)$ the jump is in the direction of increasing K_p. As in Ref. 9 it has been discovered that the temperature transition of curve $K_p(T)$ depends on the type of bombarding ion, i.e. with the increase of the mass of bombarding ion the transition temperature of curve $K_p(T)$ is shifted towards higher temperatures. It turned out that the annealing temperature for bombardment damage in Si and Ge estimated by the above-mentioned method coincides within the energy range of 1–5 keV with the annealing temperature obtained by other methods.[11]

Figure 3 illustrates polar diagrams which characterize the angular distribution of secondary ions during bombardment of face (001) of Ge by Na^+ ions at $E_0 = 1200$ eV. Here the distribution of secondary ions was studied within the plane of scattering which passes through axis (100) of Ge and diagrams 1–4 were derived at various annealing temperatures.

It appears that, at relatively low temperatures of the sample, the fine structure of angular distribution (resulting from the secondary ion yield blocking, i.e. the shadow effect) is not observed which is due to the bombardment damage introduced by prolonged ion bombardment. In diagrams 2 and 3 (where the temperature of the target $T_2 = 600°K$, $T_3 = 900°K$ is much higher than the annealing temperature for germanium) the fine structure of angular distribution is most pronounced. At high temperatures of the target $(T_4 = 1400°K)$ there is some smoothing in the fine structure, which is apparently due to the changing transparency of the crystal due to thermal vibrations of the atoms of the lattice. Indeed the thermal vibrations of the lattice atoms result in a reduction in the anisotropy of angular, spatial and energy distributions of ions scattered by single crystals.[3,4]

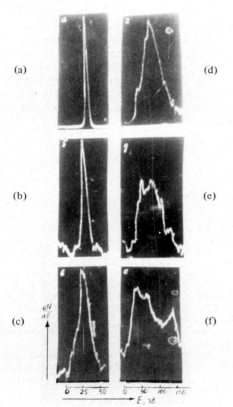

(a)

(b)

(c)

(d)

(e)

(f)

$\frac{dN}{dE}$

0   25   50   0   50   100   150

E, эв

FIGURE 4   Oscillograms of the energy spectrum of secondary ions recorded under bombardment of face (111) of Si by Na$^+$ ions at various energies. Energy of primary Na$^+$ ions are: (a) $E_0 = 50$ eV, (b) 100 eV, (c) 150 eV, (d) 200 eV, (e) 300 eV, (f) 400 eV.

Figure 4 illustrates a series of oscillograms for the energy distribution of secondary ions produced during the bombardment of face (111) of Si by Na$^+$ ions with various primary energies. It is seen that with the increase of E_0 the high-energy region of the peak (right-hand section) for slow ions broadens rapidly and, beginning

with the value of primary energy $E_0 = 100$ eV, there occur individual peaks of single and double scattering. Simultaneously the half-widths of peaks broaden and additional peaks appear. Further studies of energy distributions depending on the temperature and duration of ion bombardment indicate that changes in the peculiarities of angular, spatial and energy distributions are the result of radiation damage and annealing; for example, the absence of a fine structure in the energy spectrum of ions scattered by Si and Ge single crystals at temperatures below that of radiation damage annealing provides convincing proof of the correctness of such a point of view.

Figure 5 illustrates the dependence of R (here R is the ratio of double peak energy spectrum intensity to single peak intensity) on ϕ obtained for Na$^+$ on (001) Si at $E_0 = 1200$ eV.

It is seen that for a crystalline state of the target surface (i.e. at a temperature higher than the annealing temperature of radiation damage the dependence of R on ϕ, as in the case of metal,[3,4] has a fine structure. It has been shown that for a target made amorphous by ion bombardment, the energy spectrum has only one peak whose intensity does not depend on the azimuthal angle of target rotation.

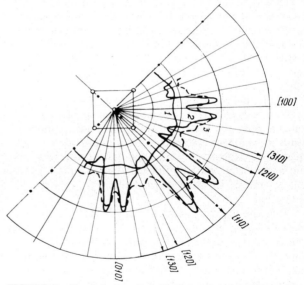

[100]

[310]

[210]

[110]

[120]

[130]

[010]

FIGURE 5   Dependence of relative intensity of double peak in ion spectrum on azimuthal angle of target rotation, ϕ. Na$^+$ on (001) face of Si at $E_0 = 1200$ eV. (1) 300°K, (2) 1000°K, (3) 600°K.

It must also be noted that not only ion bombardment, but also excessive repetition of thermal processing of the target (heating–cooling) at a vacuum of $\sim 5 . 10^{-8}$ torr gradually leads to the formation of oxide layers on the surface of Si and Ge samples and subsequently to the formation of amorphous layers which shield the single crystal structure of the target.

Thus the results of research show that the orientation effects which are displayed in the angular, spatial and energy distributions due to the orderly arrangement of atoms in the crystal lattice, can be registered only at temperatures which exceed the annealing temperature of radiation damage induced by ion bombardment. These results also confirm the possibility of using the orientation dependence of phenomena occurring during the interaction of ions with crystals for determining the radiation damage annealing temperature and observing the kinetics of radiation damage in the course of ion bombardment. With further investigations along the same lines as conducted by the authors of this paper a technique could be developed for the quantitative estimation of ion bombardment damage in crystals and also in thin epitaxial films. Furthermore, the form of the angular, spatial and energy distributions of scattered ions and alterations in these due to radiation damage, orientation and temperature of the target can be explained if we take into account the fact that defects in the crystal lattice produced by ion bombardment play a leading part in determining the structure of the energy, spatial and angular distributions. The sharp changes in the behaviour of energy, angular and spatial distribution of scattered ions during the transition from the annealing temperature of the sample ($\sim 600°$K) also testify to the role played by the damage and the restoration of the ordered crystalline structure at high temperatures.

Taking the above-mentioned into account, the satisfactory agreement of experimental results with calculations[5-8] based on a plain atomic chain model study is no longer unexpected and can provide a qualitative explanation for all the basic effects observed during the study of the influence of crystalline structure defects and lattice atom thermal vibrations on ion scattering.

Thus the summary results of previous (for instance Refs. 1–4) and present studies show that the differences caused by the atomic factors in the crystal lattice (orderly arrangement of atoms, bonding ties between atoms, thermal vibrations of the lattice atoms, etc.) result in the appearance of a number of peculiarities which differ in the interaction of ions with solids from interaction with gas targets. As illustrated in Refs. 1–4 the most characteristic of these differences in this respect are the influences of the crystalline structure and the temperature of the target. It is known that in the case of gas the temperature of the target, provided it is not too high, has no noticeable influence on the interaction of ions with it. In the case of a solid the thermal vibrations of the atoms in the lattice alter the transparency of the crystal lattice and correspondingly the probability of ion collision with the atoms of the sub-surface layers of the target and also the depth of penetration by the ion into the crystal lattice. In addition, the target temperatures, i.e. the thermal vibrations of the atoms of the lattice, result in the reduction of the effective energy of atomic bonds and in the distortion of atomic chains which participate in the transmission of successively focused collisions. In the case of radiation defects, the heating of the crystal to a high temperature results in the annealing out of these defects and, correspondingly, to radical changes in the nature of the interaction.

REFERENCES

1. U. A. Arifov, A. A. Aliyev, *Proc. 9 Internat. Conf. on Phenomena in Ionized Gases,* Bucharest 1969 p. 104.
2. U. A. Arifov and A. A. Aliyev, *JETP* **57**, 1877 (1969).
3. U. A. Arifov and A. A. Aliyev, *DAN SSSR* **192**, 1244 (1970).
4. U. A. Arifov and A. A. Aliyev. *Proc. 10 Internat. Conf. on Phenomena in Ionized Gases,* London (1971) p. 77.
5. E. S. Parilis, V. M. Kivilis and N. Y. Turayev, *Proc. 9 Internat. Conf. on Phenomena in Ionized Gases,* Bucharest (1969).
6. V. M. Kivilis, E. S. Parilis and N. Y. Turayev, *DAN SSSR* **192**, p. 1259 (1970).
7. V. E. Yurasova, D. S. Karpuzov and V. I. Shulga, Proc. 9 Internat. Conf. on Ionized Gases, Bucharest (1969) p. 103.
8. V. E. Yurasova, D. S. Karpuzov and V. I. Shulga, *Izv. AN SSSR. Ser. Phys.* **33**, p. 819 (1969).
9. E. S. Mashkova and V. A. Molchanov, *Izv. AN SSSR. Ser. Phys.* **33**, 757 (1969).
10. U. A. Arifov, A. A. Aliyev and E. Turmashev, *Radio Engineering and Electronics* **17**, 359 (1972).
11. G. J. Dienes and G. H. Vineyard, *Radiation Effects in Solids* (Interscience Publishers, New York, 1957).

BOMBARDMENT INDUCED SURFACE DAMAGE IN A NICKEL SINGLE CRYSTAL OBSERVED BY ION SCATTERING AND LEED

W. HEILAND and E. TAGLAUER

Max-Planck-Institut für Plasmaphysik, EURATOM Association, D-8046 Garching, Germany

In an experiment which combines ion scattering at low energies and low energy electron diffraction (LEED) the influence of Ar^+ ion bombardment on the structure of a Ni (110) single crystal surface is investigated. At an energy of 1 keV Ar^+ ion scattering from an undisturbed single crystal face is governed by multiple scattering which can be described by sequences of binary collisions. Computer simulations of the scattering process show that by introducing surface defects changes in the backscattering spectra occur. The surface damage from the ion beam gives indeed rise to increasing binary scattering, thereby decreasing multiple scattering. These changes are observable above 10^{15} ions/cm². The results suggest the production of point defects and extended defect structure. This is in qualitative agreement with the measurement of the widths of LEED spots, where the widths increase and become asymmetric with increasing dose. The production of point defects reaches saturation at 10^{16} ions/cm², at the dose rate applied.

1 INTRODUCTION

Backscattering of low energy noble gas ions has been used for the analysis of surface composition and surface structure. At energies below a few thousand electron volts only ions backscattered from surface atoms can escape neutralization in an appreciable number.[1,2] Since the interaction is characterized by two-body collisions the energy spectra of the backscattered ions show peaks which correspond to the masses of the surface atoms. The theory of elastic collisions provides the positions of the maxima in the energy spectra in many experimental situations whereas the widths and absolute intensities of the peaks are not yet well understood. This is mainly due to the lack of a quantitative theory of the neutralization probability. The arrangement of the atoms of a surface can be studied by analysing multiple scattering[3,4] or shadowing[5,6] effects. These phenomena are especially pronounced for grazing angles of incidence of the ion beam and for scattering with large impact parameters, i.e. low energies and high masses of the incident ions.

An ion beam in the energy range under consideration inevitably produces surface damage in the target. This should affect the ion scattering spectra from a previously annealed crystalline surface in the course of ion bombardment, particularly if structure dependent multiple scattering is observed. In the present work backscattering of 1000 eV Ar^+ ions from a Ni (110) face was investigated and the influence of the ion bombardment induced surface damage on the scattering spectra was studied. These measurements are compared to observations of LEED patterns taken from the same surface. It has been shown by several authors that

LEED is very sensitive to surface damage. The degradation of peak heights in intensity profiles was used to calculate a mean damage area per incident ion.[7,8] From distributions of sharp and diffuse diffraction spots surface steps can be characterized.[9,10] Also the peak widths in angular profiles can be related to surface defects.[11] This method was used here because it seems to be appropriate for our purpose and is easily available in our experiment.

2 EXPERIMENT

The experimental setup (Figure 1) allows the bombardment of metal targets with noble gas ions at energies

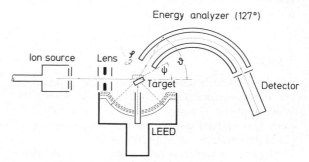

FIGURE 1   Scheme of the experimental setup.

between 50 eV and 2000 eV at a background pressure of 2×10^{-10} torr. The current densities can be controlled between 10^{-8} and 10^{-5} A/cm². The cross section of the ion beam at the target surface was measured by introducing a secondary electron image converter at the position of the target. This gives an

approximate measure of the intensity distribution in
the ion beam. The given dose values are calculated for
an assumed rectangular beam profile of equal height
and a width corresponding to the half width of the
distribution. Ion current was measured with a positive
bias of 25 V applied to the target to avoid errors due
to secondary electrons.

The LEED system is a standard Varian 4-grid
arrangement. By introducing the mentioned imaging
device it was checked that the electron beam samples a
spot of the target surface which is close to the centre of
the ion beam focus. Intensity distributions were
recorded from the photographed LEED patterns using
a microdensitometer. The slit used corresponds to an
area of 200 x 50 μm on the fluorescent screen. The
method proves to be highly reproducible and keeps the
actual LEED measuring time close to the exposure time
(1 min). This seems to be important since the ion spectra
are influenced by the intermediate electron bombardment. Since the surface structure is temperature and
time dependent this might be due to annealing during
the exposure time enhanced by the thermal energy
provided by the electron beam.

The target is a Ni single crystal with a surface
parallel to the (110) crystal plane. For cleaning the
crystal was prepared according to the procedure
described by Germer and MacRae.[12] The surface
quality is checked by LEED and He$^+$ ion backscattering. The energy of the backscattered ions can be
measured in a continuous range of scattering angles
between 0° and 120°. The target manipulator controls
the angles of incidence relative to the surface and to
the crystal orientation with an accuracy of ±1°. It is
also equipped with an electron bombardment heating
system and a thermocouple to measure the crystal
temperature.

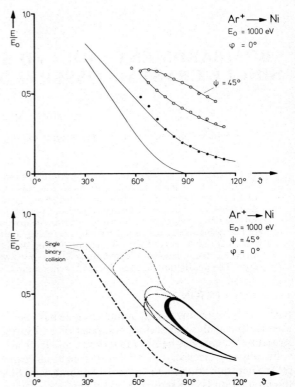

FIGURE 2   Dependence of the Ar ion backscattering energy
on the scattering angle. Upper part: experiment (• binary
scattering; ○ multiple scattering). The binary scattering is in
good agreement with the theory (solid line). The lower solid
line is the theoretical value for the binary recoil, which is not
observed under the given conditions. Lower part: computer
simulation (linear chain model, Born–Mayer potential). Thick
solid line: perfect chain and perfect parts of chain with
defects. All other events occur only in case of defects. Dotted
line: point defect in chain; dashed line: step; dash-dotted line:
single adatom on chain; thick dash-dotted line: binary recoil
from all imperfect chains.

3 RESULTS

For the ion bombardment and scattering experiment
the following conditions were chosen: Ar$^+$ ions with
1000 eV primary energy; angle of incidence 45°;
laboratory scattering angle 85°; scattering along a
[110] direction in the Ni (110) surface.

The backscattering spectra under these conditions
are governed by multiple scattering events.[13] No
definite peak is detected at the binary scattering energy
$E_B = 0.22 E_0$ as long as the surface damage is kept low
(first spectrum in Figure 3). A numerical treatment of
the Ar$^+$ backscattering from an idealized surface, i.e.
undisturbed atom chains, shows that no scattering at
this energy should be observed (Figure 2). The main

ion intensity occurs at energies higher than E_B as is
typical for multiple scattering. But the exact origin of
these peaks is still unknown since the simple linear
chain model does not work very well in this case.
Figure 2 shows the angular dependence of the energy
of ions backscattered from a surface with defects as it
results from the experiments and from numerical calculations. The experimental energies from single binary
collisions are in fairly good agreement with the theory.
The data points for multiple scattering lie on typical
loops. The calculations of scattering from a one
dimensional chain (for details see Ref. 13) with different
types of disorder show similar loops. However, not a
single one gives quantitative agreement with the
experiment.

Figure 3 shows a set of Ar^+ backscattering spectra at different stages of surface damage. The intensity of the peak due to single binary scattering at $E_B = 0.22 E_0$ increases, whereas the multiple scattering intensity (between 0.3 and 0.7 E_0) decreases with increasing dose.

Figure 4a shows the corresponding LEED patterns. It demonstrates that in the angular profiles the peak widths and the asymmetry increases with increasing dose (Figure 4b). This result suggests that the induced

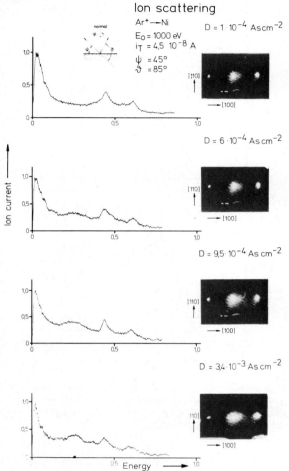

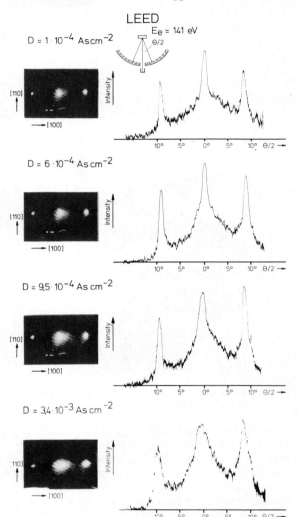

FIGURE 3a   Dose dependence of Ar^+ ion backscattering at constant energy and constant angles. The scattering plane is perpendicular to the (110) surface and parallel to the [110] direction in the surface.

FIGURE 4a   Dose dependence of LEED peaks under the same ion bombardment conditions as in Figure 3a. Electron energy 141 eV.

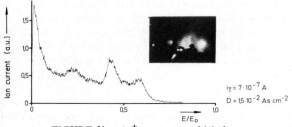

FIGURE 3b Ar^+ spectrum at high dose.

surface damage consists of atomic defects and also of extended defect structures like steps or facets. For example, the development of (111) facets during ion bombardment is assumed in Figure 5. This gives rise to a structure in the [100] direction with chain distances of 7.04 and 10.56 Å (2 and 3 times the lattice constant), which causes the observed splitting of the (00) beam.

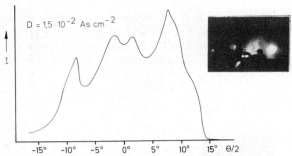

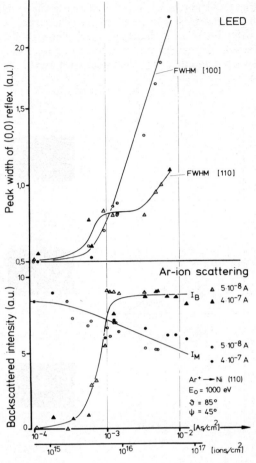

FIGURE 4b   LEED pattern under condition of Figure 3b.

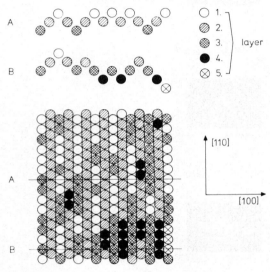

FIGURE 5   A model for the surface structure produced during ion bombardment.

FIGURE 6   Dose dependence of the peak widths of LEED spots and the intensity of the backscattered Ar^+ ions. Binary scattering from defects, I_B and multiple scattering from ordered parts of the surface, I_M.

In Figure 6 the Ar^+ scattering and the LEED data is plotted as a function of ion dose. For the scattering data the integral intensities for binary scattering I_B and multiple scattering I_M were taken in the energy regions 0.2 to 0.25 E_0 and 0.3 to 0.7 E_0 respectively. The experimental threshold above which a dose dependence can be observed is rather high. The binary scattering from more or less isolated atoms has an exponential increase until it reaches saturation at about 10^{16} ions/cm^2. The multiple scattering intensity decreases slowly. Comparable behaviour is found for the peak widths of the LEED spots, which are taken before the actual peak splitting is observed. The width of the [110] peak also reaches a saturation value at about the same dose as I_B. The width of the [100] peak increases monotonically with dose in the considered region. This behaviour is to be compared to the multiple scattering peak and suggests that both quantities are not related to point defects only.

The saturation behaviour is only clearly observed at relatively low primary current densities. There seems to be a dose rate dependence which is rather complex and far from fully covered by our experiments. Pre-treatment and target history may have some influence, too. Nevertheless an obvious interpretation of the saturation behaviour seems to be that equilibrium between damage production and damage annealing is reached. This is supported by the observation of the temperature dependence of the backscattering spectra (Figure 7). After moderate bombardment the binary intensity disappears and the multiple intensity increases by keeping the crystal at 90°C. Parallel LEED observations also show recovery with the peak widths narrowing down to the initial values.

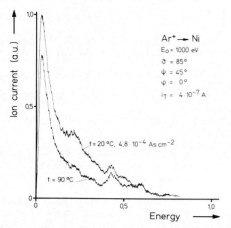

FIGURE 7   Backscattering spectra after ion bombardment and after annealing at 90°C for 5 minutes.

4 DISCUSSION

The analysis of ion scattering and LEED data from an Ar^+ bombarded Ni surface shows qualitatively comparable results. The thresholds for observation beyond the experimental error are very high, at about 3×10^{15} ions/cm^2, which is of the order of one atomic layer removed by sputtering. The dependence of the observed effects on ion dose is about exponential as one would expect from a simple theory. For instance, the adoption of the analysis of Jacobson and Wehner for our case would mean peak width-dose relation like

$$\Delta\theta_{hk} \sim \frac{\lambda}{d_{hk}N_0} e^{\bar{a}D}$$

λ = electron wavelength, d_{hk} = lattice parameter, N_0 = number of surface atoms that contribute to the reflex at θ_{hk} from the annealed crystal, D = Dose, $\bar{a}$ = mean damage area per ion. The peak width induced by the apparatus is neglected, but actually it is in the order of 0.5°, so the angular profiles are a convolution of the apparative width and the peak width due to the damage. Consequently also the induced damage area per ion in the present experimental region can only be roughly estimated to about 0.4 Å^2 or less. From this rather small value it has to be concluded that recovery during ion bombardment plays a significant role. The observation of surface damage in Ge by Jacobson and Wehner and in Mo and W by Bellina and Farnsworth[8] showed an exponential dependence only for lower ion doses.

The results show that the surface structure is not completely disturbed by the ion bombardment in the sense of getting amorphous. Even at the highest doses of about 100 ions per surface atom LEED patterns can still be observed and the multiple scattering intensity decreases only slightly with dose. Thus it must be concluded that a damaged surface structure is created with e.g. steps and facets (although no direct evidence for facets was established by observing the angle-voltage relation for different orders). Also the multiple scattering results mean that at least chains of a few atoms still exist. If it is possible to give a more quantitative analysis of the different peaks in multiple scattering of an annealed surface perhaps a more sensitive damage analysis is possible by that means. Also it might be useful to operate with different ions and geometrical parameters for analysis and for bombardment. For instance, experiments at very low angles of incidence[3] are probably more sensitive to structure defects.

The question to what extent lattice distortions due to embedded ions are important could be investigated if a complete multiple scattering analysis were available, since multiple scattering is sensitive to atomic separations. The present Ar spectra show no evidence for embedded Ar, but the sensitivity for Ar–Ar scattering is not very high, since there is a high background in the low energy region. The relatively strong asymmetry compared to the small change in multiple scattering intensity is possibly due not only to damage but also to a distortion effect caused by implantation into deeper layers.

It can be seen from Figure 6 that the sum of the intensities $I_B + I_M$ is not constant as might be expected from a straightforward consideration: I_M is proportional to the density of atoms in their ideal position and I_B should be proportional to the number of displaced atoms. The overall intensity increases instead and reaches a flat maximum at about 10^{16} ions/cm^2. The neutralization probability of an ion is a function of the distance of the ion from the surface.[15] Therefore it can be expected that ion backscattering, e.g. from single atoms on a lattice, is enhanced compared to the scattering from other configurations.

SUMMARY

In a backscattering experiment the energy spectra of 1 keV Ar ions incident at 45° on a Ni(110) surface show peaks due to multiple scattering. The surface damage from the beam gives rise to increasing binary scattering, thereby decreasing the multiple scattering intensity. These changes are observable above 10^{15} ions/cm^2. The results suggest the production of point defects and extended defect structures. This is in qualitative agreement with the measurement of the width of LEED spots, where the widths increase and become asymmetric with increasing dose. The production of point defects reaches a saturation at 10^{16} ions/cm^2.

ACKNOWLEDGEMENTS

We thank our colleagues Drs. W. Eckstein, H. Verbeek and H. Vernickel for many helpful discussions. We also wish to thank H. G. Schäffler for performing the numerical calculations. Thanks are due to Franz Schuster for his valuable technical assistance.

Appreciation is also due to Dr. Schlesinger from TU München for the use of the microdensitometer.

REFERENCES

1. D. P. Smith, *Surf. Sci.* **25**, 171 (1971).
2. D. J. Ball, T. M. Buck, D. MacNair and G. H. Wheatley, *Surf. Sci.* **30**, 69 (1972).
3. S. H. A. Begemann and A. L. Boers, *Surf. Sci.* **30**, 134 (1972).
4. E. Taglauer and W. Heiland, *Surf. Sci.* **32**, 67 (1972).
5. W. Heiland and E. Taglauer, *J. Vac. Sci. Technol.* **9**, 620 (1972).
6. H. H. Brongersma and P. M. Mul, *Chem. Phys. Lett.* **14**, 380 (1972).
7. R. L. Jacobson and G. K. Wehner, *J. Appl. Phys.* **36**, 2674 (1965).
8. J. J. Bellina, Jr. and H. E. Farnsworth, *J. Vac. Sci. Tech.* **9**, 616 (1972).
9. M. Henzler, *Surf. Sci.* **22**, 12 (1970).
10. J. E. Houston and R. L. Park, *Surf. Sci.* **26**, 269 (1971).
11. D. Wolf, Thesis, Universität München (1972).
12. L. H. Germer and A. U. MacRae, *J. Appl. Phys.* **33**, 2923 (1962).
13. W. Heiland, H. G. Schäffler and E. Taglauer, paper presented at the Surface-Vacuum Interface Conference, Twente (Holland) 1972; *Surf. Sci.* **35**, 381 (1973).
14. R. L. Park, *J. Appl. Phys.* **37**, 295 (1966).
15. H. D. Hagstrum, *Phys. Rev.* **336**, 96 (1954).

ACCOMMODATION COEFFICIENTS FOR LOW-ENERGY IONS ON METAL SURFACES

U. A. ARIFOV, D. D. GRUICH, G. E. ERMAKOV,
E. S. PARILIS, N. Y. TURAYEV and F. F. UMAROV

Institute for Electronics, Uzbek Academy of Sciences, Tashkent, USSR.

The interaction of low-energy ions (E = 3 to 100 eV) with the surface of a solid cannot be treated in terms of gas dynamics. The scattering of particles at an energy of $E_0 > 20$ eV may be explained in terms of binary collisions with the atoms of the target. The validity of the single collision model with free surface atoms for medium energies from 10 to 100 eV and even down to 1 eV was confirmed on the basis of both experimental and computational data. This paper describes experimental studies of the secondary ion emission from the (100) and (110) faces of a Mo single crystal and both experimental and theoretical studies of alkaline ion accommodation coefficient on polycrystalline Mo within the energy range E = 3 to 50 eV with a varying direction of the primary ion beam from normal to the glancing angle of incidence (φ = 0 to 75°). On the basis of the retarding potential method using a spherical condenser whose central electrode was the target, we measured the energy distribution of secondary ions. Calculations have been performed for the energy of scattered ions and the high energy portion of accommodation coefficient on the basis of single and double binary collisions using the Born–Mayer potential and taking into consideration the influence of adsorption forces at the surface of the target.

Experimental and theoretical studies of the accommodation coefficients of gas atoms with thermal velocities on metal surfaces have been performed by various researchers.

The interaction of low-energy ions (E = 3 to 100 eV) with the surface of a solid cannot be treated in terms of gas dynamics. The scattering of particles at an energy of $E_0 > 20$ eV may be explained in terms of binary collisions with the atoms of the target. The validity of the single collision model with free surface atoms for medium energies from 10 to 100 eV and even down to 1 eV was confirmed on the basis of both experimental and computational data.[1-5] It has been shown experimentally and theoretically[6,7,8] that at $E_0 = 0.15 \div 90$ keV the energy spectra of ions display peaks of single and double scattering on free atoms of the target. The exchange of energy between the colliding particles and the value of the accommodation coefficient depend on the mass ratio of these particles. Nevertheless it is not possible to describe the ion-surface interaction on the basis of binary collisions alone because within the range of up to 50 eV the secondary ion emission is limited by sorption forces at the surface of the target[9,10]

This paper describes experimental studies of the secondary ion emission from the (100) and (110) faces of a Mo single crystal and both experimental and theoretical studies of alkaline ion accommodation coefficient on polycrystalline Mo within the energy range E_0 = 3 to 50 eV with a varying direction of the primary ion beam from normal to the glancing angle of incidence (φ = 0 to 75°). On the basis of the retarding potential method using a spherical condenser whose central electrode was the target, we measured the energy distribution of secondary ions. Calculations have been performed for the energy of scattered ions and the high energy portion of accommodation coefficient on the basis of single and double binary collisions using the Born–Mayer potential and taking into consideration the influence of adsorption forces at the surface of the target.

The targets used in our experiments were de-gased in vacuum $\approx 3 \times 10^{-8}$ torr for two days by electron bombardment at a temperature of 2400°K. Prior to each measurement the target was cleaned of its film of adsorbed gases and atoms from the primary ion beam by a short high-temperature flash.

To determine the threshold of secondary ion emission we used an electrometer with a sensitivity of $1 \times 10^{-14}_A$ and for obtaining the retarding potential curves we applied the double modulation method.[11] The highest incident ion current density on the target surface was $j \approx 1 \times 10^{-8}$ A/cm^2. Thus a mono-layer of residual gas and primary ion beam atoms is formed on an initially clean molybdenum surface in about two minutes. All the measurements were performed within a short time interval of 3 to 4 seconds after flashing the target at high temperature. Under these circumstances the target surface coverage by impurities was not greater than 3% of the mono-layer.

Alkaline ions were selected for the experiment because within the energy range under study ($E_0 = 3$ to 50 eV) the neutral component of the secondary ion emission from metals is not greater than 3%.[12] The experimental instrument and the electrical circuit used in research have been described in detail in Ref. 8.

Figure 1a and Figure 1b illustrate the series of curves

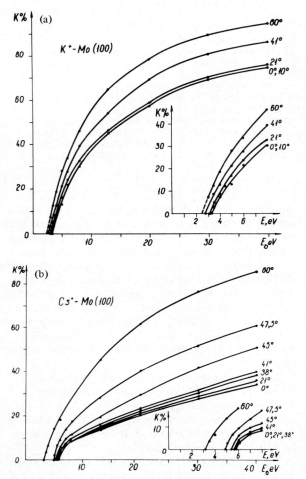

FIGURE 1a, b   Ion emission coefficient $K(E_0, \varphi)$ vs. primary energy E_0 and various angles of incidence φ for K^+ ions (a) and Cs^+ ions (b) on face (100) of clean cold Mo.

for the secondary ion emission coefficients depending on the energy and the angle of incidence φ of alkaline metal ions on the (100) face of a molybdenum single crystal. In the lower right hand corner of Figure 1a and Figure 1b there are large-scale illustrations of those

sections of the curves where secondary ion emission stops.

As mentioned earlier in Refs 9 and 10 at the interaction of low-energy alkaline ions ($E_0 = 20$ to 500 eV) with a clean metal surface the secondary ion emission consists of elastically scattered ions and a group of low-energy ions. At $E_0 \leqslant 50$ eV most of the emission consists of scattered ions. Reduction of secondary ion emission coefficient $K(E_0, \varphi)$ along with the reduction of E_0 from 40 to 3 eV as illustrated in Figure 1, results from the influence of the sorption barrier on the surface of the target. Apparently secondary emission stops when the energy of the secondary ion is equal or less than the height of the potential barriers on the target surface.

With increased φ the scattering of ions at smaller angles becomes possible and as a result they retain the greater part of their initial energy[13] and the threshold of secondary ion emission is thus shifted towards lower values of E_0. It must be noted that the threshold of secondary ion emission is determined mainly by double scattering of ions since their energy can be greater than that of single scattered ions.[9]

Figure 1c illustrates the dependence of the threshold

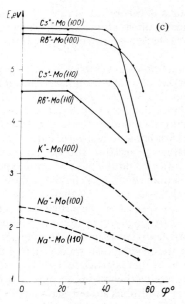

FIGURE 1c   Threshold energy of secondary ion emission versus angle of incidence for alkaline ions on (100) and (110) faces of Mo.

values of incident ion energy on the angle of incidence of alkaline metal ions on the (110) and (100) faces of a molybdenum single crystal. It seems that the threshold values of energy at which the secondary ion emission stops depends on the mass of the incident ions which is in agreement with the conservation laws for binary elastic collisions.

A series of curves for accommodation coefficients was constructed on the basis of the experimental energy distributions

$$\alpha(\varphi, E_0) = 1 - \frac{\int_0^{E_{\lim}} K(E_0, E)\, dE}{E_0} \qquad (1)$$

These coefficients are for slow alkaline ions on pure polycrystalline Mo. The energy removed by secondary ions is determined by the area under the retarding potential curve in Figure 2a and 2b; E_0—the primary energy.

Figure 3 illustrates a series of experimental curves of $\alpha(\varphi, E_0)$ for K^+ ions on polycrystalline Mo. It is seen from the experimental curves that α increases with the reduction of φ and has a poor dependence on energy within the range of 5 to 50 eV. In the case of normal incidence of the primary beam on the target ($\varphi = 0$) at $E_0 \leqslant 15$ eV the curve displays an increase and after α reaches the value of 0.91 there occurs saturation. This results from the existence of a secondary ion emission threshold due to the sorption barrier on the target surface. At high values of φ, when the energy exchange between colliding particles becomes smaller the experimental curves display no saturation.

Let us assume that the adsorption forces form on the surface a flat potential barrier of λ height for single charge positive ions (Fig. 4). Scattering occurs on individual atoms on the crystal surface and the energy of the ions at the moment of collision is $E_0 + \lambda$, whereas after the collision it is $E_1 = E_{a,b} - \lambda$, where $E_{a,b}$ is the energy of the ion correspondingly after a single or double scattering. The theoretical analysis in this paper is limited to examining single and double collisions as a source of the high-energy part of the scattered beam.

The trajectory of the ion which has been scattered in a direction which forms angle β' with the normal is refracted after crossing the potential barrier and forms angle β'' with the normal. Thus

$$\frac{\sin \beta''}{\sin \beta'} = \left[\frac{E_{a,b}(\beta)}{E_{a,b}(\beta) - \lambda} \right]^{1/2} \qquad (2)$$

At $\beta = \beta_{\lim}$ and $\beta'' = \pi/2$ there occurs complete internal reflection and emission of particles scattered at the given angle stops.

In the case of a normal incidence scattering in the given direction is possible in two ways: as a result of single scattering (a), as a result of double scattering (b). There are two angles β'_a and β'_b which on the basis

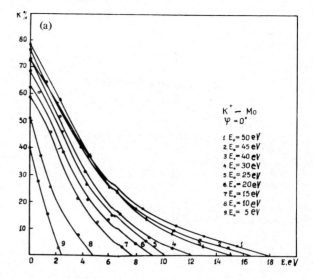

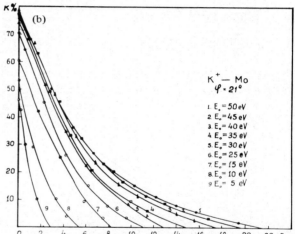

FIGURE 2   Retarding potential curves $K(E_0, E)$ at various primary energy E_0, normal incidence $\varphi = 0$ (2a) and $\varphi = 21°$ (2b) for K^+ ions on clean cold polycrystalline Mo.

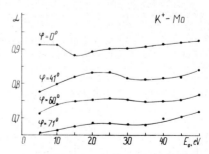

FIGURE 3   Energy accommodation coefficient α vs. primary energy E_0 at various angles of incidence φ for clean cold polycrystalline Mo.

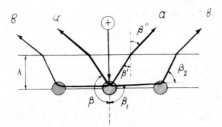

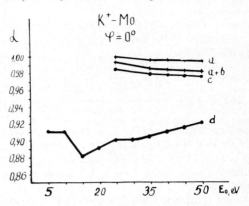

FIGURE 4   Diagram of single (a) and double (b) scattering. β = scattering angle for single collision, β_1 = the first and β_2 = the second scattering angle for double collision. β' and β''–angles of incidence and refraction on potential barrier. λ = its height.

of Eq. (2) correspond to the same angle β'', since functions $E_a(\beta)$, $E_b(\beta)$ differ:

$$E_a = \frac{E_0 + \lambda}{(\mu + 1)^2} (\cos \beta + \sqrt{\mu^2 - \sin^2 \beta})^2$$

$$E_b = \frac{E_a(\beta_1)}{(\mu + 1)^2} (\cos \beta_2 + \sqrt{\mu^2 - \sin^2 \beta_2})^2$$

$$\mu = \frac{m_1}{m_2}, \beta'_a = \pi - \beta, \beta'_b = \pi - (\beta_1 + \beta_2) \qquad (3)$$

Correspondingly the ions scattered in the given direction have two values of energy (Figure 5).

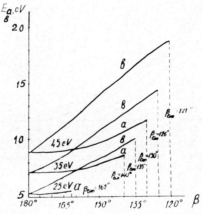

FIGURE 5   The energy of single (a) and double (b) scattered ions versus scattering angle β. The angles β_{lim} correspond to full internal refraction.

Branches of double scattering (b) always yield greater energy than that of single scattering.

Calculation of the high-energy portion of the accommodation coefficients was performed on the basis of the binary single and double collisions. When taking into consideration only single collisions the values of the accommodation coefficient proved to be

higher than the experimental ones (solid curve "a" in Figure 6). At $E_0 = 25$ eV the single scattered ions (a)

FIGURE 6   Experimental and calculated energy accommodation coefficients at normal incidence ($\varphi = 0$) for K^+ ions $a-$ calculation for single scattering only; $a + b$–calculation for the sum of single and double scattering; $c-$experimental accommodation coefficient for the high-energy portion of the secondary ion emission due to single and double scattering $d-$the full experimental accommodation coefficient.

cannot overcome the barrier of adsorption forces and the accommodation coefficient calculated on the basis of single collisions becomes equal to 1.

In order to explain the experimental curve it is necessary to take into consideration also the double collision.

After double scattering the energy of ions exceeds the energy of single scattered ions and its dependence on the angle of scattering is such that double scattered ions can overcome the barrier at $E_0 < 25$ eV. The calculated α for the single and double scattering (curve $a + b$) is in better agreement with the experimental high-energy portion of the accommodation coefficient due to single and double scattered ions (c) Figure 6.

The curve (c) is constructed from the high-energy portion of experimental distribution (Figure 2a) by substituting in (1) the area $\int_{E_a(180°)}^{E\lim} K(E_0, E)\, dE$

instead of the full area. Here $E_a(180°)$ is the energy of single backscattering–the minimal energy for single and double collisions. The curve (d) is the full experimental accommodation coefficient including the low-energy portion of secondary ion emission. A comparison of the calculated ($a + b$) and experimental (c) curves shows that the high-energy part of secondary ion emission can be explained by binary single and double collisions.

REFERENCES

1. C. A. Visser, J. Wolleswinkel and J. Los. "High and intermediate energy molecular beams." *Proc. of the Intern. Symposium*. Entropie, No. 30, pp. 61–64, (1969).
2. J. Falcovitz, L. Trilling, H. Y. Wachman and J. L. Keck. "Rarefied gas dynamics." *Proc. of the Intern. Symposium VII*, 1972 (Academic Press, N.Y. and London).
3. R. O. Barantsev, *Progress in Aerospace Science* Vol. 13, p. 19, edited by D. Kuchemann (Pergamon Press, Oxford and N.Y., 1972).
4. R. A. Oman, "Rarefied gas dynamics." *Proc. of the Intern. Symposium VI*, pp. 1331–44 (1969).
5. A. J. Erofeyev, *Applied Mech. and Tech. Physics* No. 2, p. 135, 1967. *CAGI Trans.* **1**, No. 4, p. 52 (in Russian) (1970).
6. E. S. Parilis and N. Y. Turayev, *DAN SSSR* **161**, 84 (1965).
7. E. S. Mashkova, V. A. Molchanov, E. S. Parilis and N. Y. Turayev, *Phys. Lett.* **18**, 7 (1965).
8. D. D. Gruich, G. E. Ermakov, Z. Khalmirzayev and U. A. Arifov, *Atomic Collision on Solid Surface* kap. II, pp. 28–55. Edited "FAN" Uz SSR (in Russian), Tashkent (1972).
9. U. A. Arifov, D. D. Gruich and L. Y. Chastukhina, *Izv. AN SSSR, Ser. Phys.* **28**, 1402 (1964).
10. U. A. Arifov and D. D. Gruich, *DAN Uz SSR* **11**, 20 (1964).
11. U. A. Arifov, *Interaction of Atomic Particles with a Solid Surface* (Edited Consultants Bureau, New York, 1969).
12. D. D. Gruich, N. Rakhimbayeva, G. Ikramov and T. Arifov, *Izv. AN Uz SSR, Ser. Phys.* **1**, 53–60 (1964).
13. U. A. Arifov, D. D. Gruich, G. E. Ermakov, E. S. Parilis, N. Turayev and Z. Khalmirzayev, *Proc. 10th Intern. Conf. on Phenomena in Ionized Gases* (Oxford, 1971) p. 78.

MEASUREMENT OF THE BACK-SCATTERING OF FISSION FRAGMENTS FROM THIN METAL FILMS

A. VETTER, G. FIEDLER, K. GÜTTNER and H. SCHMIDT

II. Physikalisches Institut der Justus-Liebig-Universität, 63 Giessen, Germany

The back scattering of ^{235}U fission fragments from thin polycrystalline gold and lead films has been studied by transmission electron microscopy. This method gives information about the total reflection rate, the distribution of the reflection angles and the penetration depth for low angles of incidence. Satisfactory agreement was obtained between Monte-Carlo calculations and the experimental results.

1 INTRODUCTION

Few investigations have been carried out for the back-scattering of fission products from the surface of solids, even though scattering from the walls of an experimental container can seriously influence a measurement. The first quantitative results were obtained by Engelkemeir[1,2] and by Albrecht[3] in which it was shown that a high percentage of fission fragments were backscattered from gold and platinum surfaces. For grazing angles of incidence, reflection rates greater than 50% were measured. These values were in agreement with Monte-Carlo calculations.[4]

In the following, the backscattering of ^{235}U fission products from gold and lead films will be examined. Besides the reflection rate and the angular distribution of the scattered particles, results are obtained concerning the penetration depths. These results are compared with Monte-Carlo calculations.

2 EXPERIMENTS AND CALCULATIONS

2.1 Experimental Methods

The scattering experiments were carried out in a two meter long tube which was placed in the thermal flux of a reactor core. At one end, there was an ^{235}U film which served as a fission fragment source. A nearly parallel beam of particles was created with an assembly of apertures. The targets were placed at the other end of the tube and were irradiated at grazing incidence. The targets contained both the scatterer and the detector, and the geometry of the targets can be seen in Figure 1. The detector is a sandwich consisting of layers of 200 Å collodion, 15 Å Pt and 300 Å formvar. The scattering foil of gold or lead is located underneath this assembly. After irradiation the detectors were shadowed with Pt–Ir to increase the contrast of

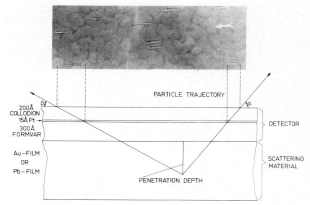

FIGURE 1   Upper part: Electron micrograph of the detector. Lower part: The target construction showing a fission particle trajectory having single scattering.

the tracks in the upper collodion film and to enhance the image. A transmission electron microscope with a magnification of 40,000 times was used for observation. For preparation the detector was removed from the scatterer with dilute hydrochloric acid. After cleaning with distilled water the detector was placed on an electron microscope grid.

Even in single collodion films tracks can be recognized. The entry and exit position of a fission particle passing through this layer can clearly be seen in Figure 2a.[5] Furthermore, fission products produce characteristic light tracks in thin polycrystalline Pt or Pd targets (see Figure 2b). With a sandwich foil as in Figure 1 it is possible to distinguish between the entry and the exit position of a reflected particle. Because of the different shapes of the tracks in the collodion and Pt foils the path of the particle can be followed. The reflection rate may be determined by counting the number of tracks of incident and reflected particles.

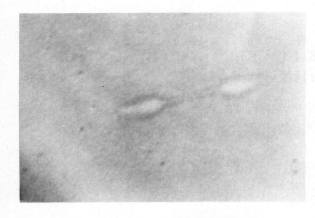

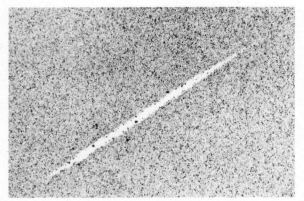

FIGURE 2 Fission particle tracks (top) in collodion (bottom) in platinum.

Knowing the projection along the surface for both the incident and reflected particle tracks, it is possible to calculate the entrance and reflection angles and the penetration depths geometrically under the assumption that the particles have been reflected by a single scattering event.

The scattering materials were vacuum deposited polycrystalline films of gold and lead with thicknesses ranging from 630 Å to 1200 Å and 1100 Å to 1620 Å, respectively. The angles of incidence α could be varied between $1.7°$ and $9.7°$ (see Figure 1). The fission fragments had the well known mass and energy distribution, and the calculations were done for particles having median nuclear charge, mass number and energy: $M = 117, Z = 46, E = 85$ MeV.

2.2 Monte-Carlo Calculations

These calculations are based on a model that has been explained elsewhere in detail.[4]

The fundamental method consists of following particle histories in the metal films. For the single

interactions simple expressions are derived employing a potential of the form

$$V(r) = \frac{Z_1 \cdot Z_2 \cdot e_0{}^2}{r} \exp\left(-r/a\right)$$

Z_1 and Z_2 are the nuclear charges of the incident particles and of the film atoms, e_0 is the unit charge and a is the screening parameter. For this constant we used

$$a = a_0/\sqrt{Z_1{}^{2/3} + Z_2{}^{2/3}}$$

where $a_0 = 0.529 \cdot 10^{-8}$ cm. On the upper potential a screened Coulomb potential

$$\overline{V}(r) = E_c\left(\frac{a_c}{r} - 1\right)$$

is fitted. E_c and a_c are energy dependent constants. By this we got a selection of the scattering angle dependent on the energy of the particle. The energy losses consist of two parts. The energy loss due to electron excitations has been taken into account by the Bethe formula. The energy loss due to the nuclear collisions can be calculated following classical rules for elastic collisions. The energy losses were fitted to experimental values.

3 RESULTS AND DISCUSSION

The percentage of particles reflected is dependent on the angle of incidence and this relationship is displayed in Figure 3. This figure also contains curves calculated for the maximum and minimum values of foil thickness used. For gold and lead the curves show an exponential decay with a slope which falls off with increasing foil thickness. The irregular shape of the curves for lead may be caused by the roughness of the film surface. As the film thickness increases, the gold films maintain their fine polycrystalline structure, and thus, a plane surface, but the lead films become rougher from coagulation into domains. The discrepancy between the calculation and the measurement may be attributed to the wide mass and energy distribution of the fission particles. A further reason could be the low energy cut-off ($E_{\min} = 5$ MeV) used in the calculation.

The distribution of the angle of reflection was measured for the thinnest and thickest films. Only those tracks were measured and calculated having trajectories near the plane of incidence. This simplification is justified because only a few percent of the particles are scattered with angles greater than $10°$ out of the plane of incidence. The distributions for Au are shown in Figure 4. For small angles of incidence α the

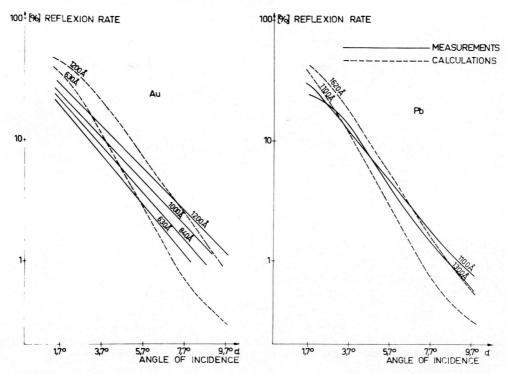

FIGURE 3   Percentage distribution of the reflected particles as a function of the angle of incidence.

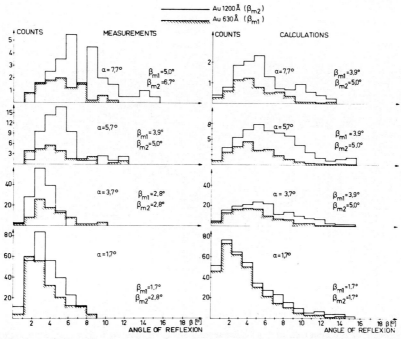

FIGURE 4   Distribution of the reflection angles for several angles of incidence for gold. β_{m1} and β_{m2} are the reflection angles of maximum yield for film thickness of 630 A and 1200 A, respectively.

132   A. VETTER, G. FIEDLER, K. GÜTTNER AND H. SCHMIDT

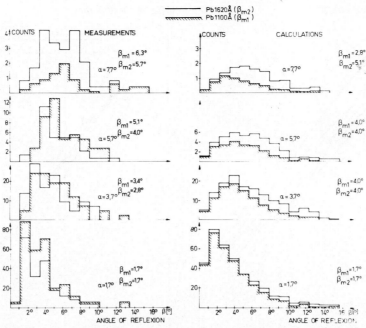

FIGURE 5   Distribution of the reflection angles for several angles of incidence for lead. β_{m1} and β_{m2} are the reflection angles of maximum yield for film thicknesses of 1100 Å and 1620 Å, respectively.

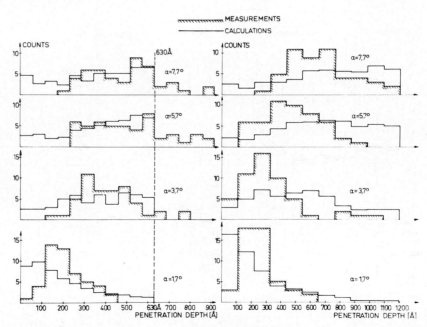

FIGURE 6   Distribution of the penetration depths for several angles of incidence. The graphs on the left hand side were made for a gold film of 630 Å thickness, and those on the right hand side were made of gold film thickness of 1200 Å.

distributions increase steeply and have a maximum yield at β_m and a less steep decay as the angles of reflection become larger. With increasing α, the slope of the curves becomes flatter, and the maximum of the distributions of the reflected particles becomes lower in intensity and less distinct. For small angles of incidence the maximum yield of the reflected particles occur at an angle β_m approximately equal to α. Comparing the distributions for the two thicknesses it can be seen that the reflection angle for the maximum number of counts becomes larger with increasing foil thickness. The corresponding results of lead can be seen in Figure 5. The results are in satisfactory agreement with our calculations.

The distribution of the penetration depths for the gold films is shown in Figure 6 and it may be seen that as the angle of incidence increases, the penetration depths shift to larger values. The penetration depths in the experiment were inferred under the simplifying assumption of single particle scattering, which is not completely valid. Some measured penetration depths

appearing in Figure 6 are longer than the scattering film thickness. This and the differences between the theoretical and experimental results might be attributed to multiple scattering.

ACKNOWLEDGEMENTS

The authors are thankful to Prof. Ewald for his continuous interest. We are indebted to Dr. Gunther and Dr. Siegert and the staff of the Reactor of TU Munich for the help in performing the irradiations and Dr. Siegert for valuable discussions. The financial support of the Bundesministerium für Bildung und Wissenschaft is gratefully acknowledged.

REFERENCES

1. D. Engelkemeir a. G. N. Walton, United Kingdom Atom Energy Author. Rep. AERE-R 4716 (1964).
2. D. Engelkemeir, *Phys. Rev.* **146**, 304 (1966).
3. J. Albrecht a. H. Ewald, *Z. Naturforsch.* **26a**, 1296 (1971).
4. K. Güttner, *Z. Naturforsch.* **26a**, 1290 (1971). K. Güttner, H. Ewald and H. Schmidt, *Rad. Effects* **13**, 111 (1972).
5. G. Fiedler a. W. Kalkbrenner, *VI. International Conference on Corpuscular Photography* (Florence, 1966), p. 449.

SURFACE STUDY BY APPLICATION OF NOBLE GAS ION BACKSCATTERING TECHNIQUE

K. AKAISHI

Institute of Plasma Physics, Nagoya University, Nagoya, Japan

A simply designed noble gas ion scattering system enclosed in a Pyrex glass container was constructed. Using this system, we studied sorption characteristics of the residual gases (H_2O; CO, . . . etc.) in the glass system or CO gas on the bare or Pd coated, Ni surface. In the scattering experiments, the contaminated states by the active gases of these surfaces could be known from observation of the scattered He ions peak by oxygen atoms. In the case of gold coated Ni surface the active gas adsorption was depressed and only Au peak was detected.

1 INTRODUCTION

It has been the object of many researchers to reduce plasma contaminants, especially in case of plasma heating experiments. Low mass gas atoms or molecules (C, O, CO, H_2O) are the dominant contaminants. It is well known that preheating of the vacuum chamber reduces the surface impurities. Plasma firing or gas discharge cleaning method has often been used in laboratory plasma apparatus, since the baking technique is limited by the magnetic coils surrounding the system.

The effect of gas discharge cleaning method on the stainless steel chamber has been experimentally studied by Govier and McCracken.[1] They concluded that this technique is suitable to reduce the surface impurities (H_2O, CO . . .) except for hydrogen.

On the other hand, if the quantity of the gases adsorbed on the metal wall could be controlled by coating the surface, then the gas discharge cleaning method may exhibit more effectiveness. From this point of view, Komiya, *et al.*[2] showed that a gold film on the stainless steel sheet has reduced CO adsorption as compared with the bare case.

In our case, we examined the surface impurities in the case of bare and palladium or gold coated nickel sheet, using the low energy helium ions scattering technique,[3] where we learned that a thin film of gold is very convenient to depress gas adsorption.

2 EXPERIMENTAL

The low energy ions scattering system is depicted in Figure 1. It simply consists of an ion source, a target with two filaments (that at the front is for the evapor-

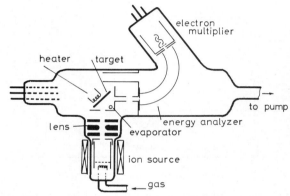

FIGURE 1   Schematic of the ion scattering apparatus.

ator while the other for outgassing), a 127° electrostatic energy analyzer (the mean radius is 5 cm, the slit width is 1 mm) and an electron multiplier. All these parts are installed in a Pyrex glass container.

To observe the surface impurity on the Ni surface, the system was baked out at 250°C for 14 hrs, and pumped down to 3×10^{-8} torr by the Hickma pump with a liquid nitrogen trap. A current of 10^{-7} A helium ions (current density to the target is about 100 $\mu A/cm^2$) coming out from the ion source where the pressure was increased up to 10^{-5} torr, was scattered perpendicularly on the target surface entering the analyzer slit, then flown down through the analyzer electrode to reach the electron multiplier, which in turn is connected to a DC electrometer, hence could be detected.

Applying the preceding steps, the Ni target was tested. In case of preheating, the main surface impurities were O, Na, Cl and Si, and when it was heated up to 800°C for 2 hrs, it was noticed that these peaks were decreased while that of Ni was increased sharply. Figure 2 shows a typical energy distribution of He ions in the case of the clean Ni target. Continuing the same experiment, the Ni peak height, after the pretreatment, began to decrease gradually as the clean Ni surface re-adsorbed the residual gases[4] (H_2O or CO) in the Pyrex glass system, while the O peak started to increase. The variation rates of the peak heights of Ni and O as a function of time after the pretreatment are shown in Figure 3. The result of Figure 3 is compared with the similar observations of CO adsorption on Mo surface published by D. P. Smith.

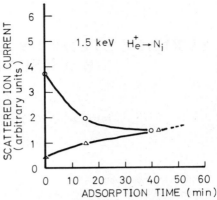

FIGURE 3   Plots of O and Ni peak heights vs. adsorption time. Where the circles and the triangles represent the peak heights of Ni and O, respectively. The current density was about 100 μA/cm².

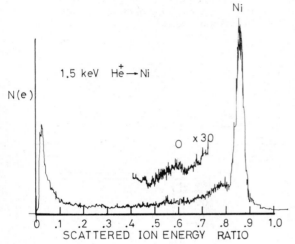

FIGURE 2   Energy distribution of 1.5 keV He ions scattered from a clean Ni target. The scattering angle was 90° and the angle of incidence was 45°. The primary ion beam intensity was about 100 μA/cm² and the spectrum was taken 5 min after the heat treatment 800 °C. 2 hrs.

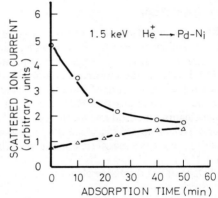

FIGURE 4   Plots of O and Pd peak heights vs. adsorption time. Where the circles and the triangles represent the peak height of Pd and O, respectively.

As a next step, the palladium coated Ni target was observed. A few ten of layers of Pd atoms were evaporated on the previously cleaned Ni surface. In this case, also, the fresh Pd film started gradually to be contaminated with the residual gases. This process is shown in Figure 4.

Finally, we studied the gas adsorption process when the Ni target was coated by a few ten of layers of gold. The result of the experiment showed that the Au peak could only be detected, when the analyzer output spectrum was scanned on the DC amplifier (see Figure 5). We also examined the time dependence of Au peak by CO gas adsorption and residual gases adsorption. However it was difficult to detect an O peak beside the Au peak.

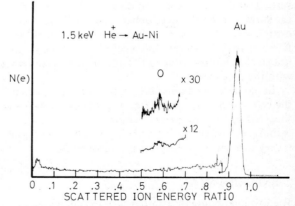

FIGURE 5   Energy distribution of 1.5 keV He ions scattered from the Au coated Ni target, after the target was exposed to the CO gas pressure of 1.5 x 10⁻⁵ torr.

3 DISCUSSION AND CONCLUSION

It has been shown that the He ions scattering technique can be used to observe the active gas adsorption state on the bare, or Pd coated, Ni surface. In our experiments, we concluded that when previously mentioned surfaces were contaminated with the active gases, the O peak, corresponding to the gas coverage, was predominant. This O peak could be decreased when the target surface was coated by the gold. This showed us that the gold film weakly reacted with the active gases and only a very small amount of the active gases stayed on the film. This result agreed with the results of Komiya et al.[2] Hence the gold coating process might be considered as a suitable auxiliary process to keep the surface impurity emission negligibly small.

REFERENCES

1. R. P. Govier and G. M. McCracken, *J. Vac. Sci. and Tech.* **7,** 552 (1970).
2. S. Komiya, T. Narusawa and C. Hayashi, *J. Vac. Sci. and Tech.* **9,** 302 (1972).
3. D. P. Smith, *Surface Sci.* **25,** 171 (1971).
4. T. W. Hickmott, *J. Appl. Phys.* **31,** 128 (1960).

THEORETICAL MODELS IN SECONDARY IONIC EMISSION

P. JOYES

Laboratoire de Physique des Solides associé au C.N.R.S.
Bât. 510, Université Paris-Sud 91 Orsay (France)

We examine various theoretical models of secondary ionic emission and try to show that now many connections clearly appear between this phenomenon and questions of interest in various fields of Physics. We discuss first the kinetic model which has been applied to pure targets (1969) and which can be extended to compounds by using the new ideas about correlation diagrams in asymmetric systems. We also discuss some "surface effects" models. The purpose of these calculations is to evaluate the probability that a conduction electron, strongly delocalized in the target, follows an emitted atom. A new model which uses the Anderson's description of the screening of a charged particle inside a metal is presented. A comparison between theoretical and experimental results is made, mainly for the kinetic model.

1 INTRODUCTION

When the surface of a solid target is bombarded with a beam of primary ions of about 10 keV, a secondary emission from the target occurs. The secondary products are of different kinds, there are electrons, photons, monoatomic neutrals and ions (simply or multiply charged), polyatomic neutrals and ions. Here, we are concerned only with the secondary ionic emission. General discussions and reviews of works on this subject were done by Kaminsky in 1965,[1] Fogel in 1967[2] and Carter and Colligon in 1968.[3]

The secondary ionic emission has first been used in analysis techniques (Ion Microanalyser 1962[4,5]), then to allow a better understanding of the phenomenon various apparatus were built with which particular physical aspects were examined.[6-10] Simultaneously, theoretical models were studied (Kinetic model,[11-15] Surface-effect models[16,17]). By comparing theoretical and experimental results we shall discuss the main features of these models. One of the conclusions is that, now, we are arriving at a stage where the connection between secondary emission and many interesting physical questions becomes obvious: let us quote, for instance, the dependence of secondary emission on target properties (surface conditions, work function, inelastic collisions in solids) or on secondary product properties (shake-off probability, stability of clusters).

We shall exclude here "chemical emission" which mainly proceeds via breaking of foreign molecules fixed on the surface of the target, like for instance oxygenized molecules. Indications on this subject are given in various papers.[18-22]

2 THE KINETIC MODEL

2.1 Description of the Kinetic Model and Application to Pure Targets[11-15]

The primary ions entering the target displace particles of the target which also collide with other ones. In these collision cascades a part of the particles may be displaced violently enough to cause one of their bound electrons to be ejected. Among these excited particles some will reach the surface without losing their excitation (mainly by Auger effect) and therefore some will escape the surface in a neutral state but still carrying their bound level excitation. These will give outside

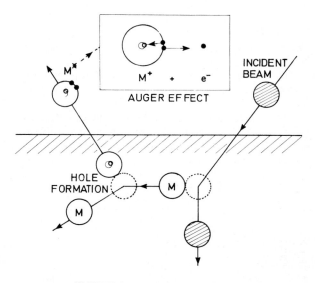

FIGURE 1   The kinetic mechanism.

the target a secondary ion and an Auger electron (Figure 1). If the electronic excitation has enough energy, the Auger-deexcitation may give a doubly charged ion and two electrons or even a triply charged ion and three electrons according to the various shake-off probabilities.[15]

To check this model a first experiment was to look for the secondary Auger electrons whose existence is predicted by this model. This was done by Hennequin[23,10] who in fact discovered them. Other experimental evidence came afterwards and we shall give them presently.

Let us describe now, how the calculations, based on this model, were undertaken.

The first element needed was a description of the spatial, angular and energy distribution of the collisions in the target. This was done by using the Boltzmann transport equation[24] a method which is often used in this kind of problem especially for sputtering or radiation effects.[25-27]

The next step is to calculate the probability $p_t(E)$ for a bound level excitation to be created when a collision at energy E occurs. To solve this problem for inelastic collision inside solids we extended the Fano and Lichten model[28,29] worked out for inelastic collisions in gases. This part of our calculation will be developed in the next paragraph where we show how recent ideas about asymmetric collision can well explain new experimental data on secondary emission of compounds.

Another point to be dealt with was the problem of the Auger life-time of a bound level hole in a metal when the two electrons, involved in the Auger effect, have their initial state in the conduction band. Our calculation was done in the one-electron approximation[12] and its results were checked by comparison with x-ray line width for some metals.[30]

The problem of the way out of excited particles[15] was attacked by again using the Boltzmann transport equation.

We shall not go into the details of the calculations which are already reported in various papers; Table I

TABLE I

Results of the kinetic model for Magnesium. N_I is the total number, per incident particle, of secondary ions whose energies are larger than 30 eV. $\bar{E}$ is their average energy. Incident energy: 8 keV, normal incidence.

	Exp (9)	Theo.
$N_I(Mg^+)$	2.8 10^{-3}	1.6 10^{-3}
$N_I(Mg^{++})$	3.7 10^{-4}	1.8 10^{-4}
$\bar{E}(Mg^+)$ (eV)	200	220
$\bar{E}(Mg^{++})$ (eV)	300	220

gives an example of the results obtained for a pure magnesium target.

2.2 Hole Creation During the Collision of Two Atoms of a Solid

As we already stated, the problem of the excitation of a bound electron during the collision of two particles in a solid or in a gas can be treated in similar ways.

Let us first recall the Fano and Lichten promotion model used for gases.[28] As the distance R between the two nuclei becomes smaller and smaller, the electronic levels become molecular levels whose energies vary with R. When two levels cross, exchanges between them occur, and higher levels, empty before the collision, can be partly filled.

In solids we are concerned with the same problem, however not only levels but also bands intervene. Let us take the example of Al → Al collision in aluminium.

The initial levels are $2p$ levels completely filled which, when R decreases, give rise to 4 molecular levels (Figure 2). The $4f\sigma$ level is promoted quickly and its energy crosses the energies of the conduction band. This situation can be described by saying that the $4f\sigma$ level becomes a virtual bound state:[31,32] it is no longer well defined but is broadened by its interaction with the continuum. When the upper energies of this level interact with free states above the Fermi energy,

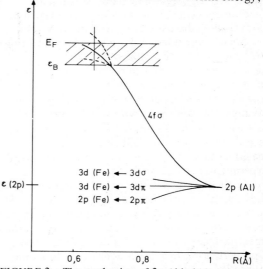

FIGURE 2   The mechanism of $2p$ Al hole creation in pure Al targets.

electrons can escape from the virtual bound state.

In reference[14] we show that, by introducing a life-time of the electrons in the virtual bound state, one can describe the electron exchange between virtual bound state and continuum.

An important element of this model is the knowledge of the $4f\sigma$ energy evolution. As long as this level
has not interacted with the continuum it is reasonable
to consider that its evolution is the same for the
Al → Al collision in the metal as in the gas phase. Therefore the knowledge of the behaviour of the $4f\sigma$ energy
level in the gas phase will be of good help for our
problem. This point will be developed in the next paragraph where we consider the asymmetric collisions
occurring in the bombardment of metallic compounds.

3 Kinetic Emission from Metallic Compounds Al–O and Al–Cu. Comparison with Experiments

3.1 The correlation diagrams

According to the
above-mentioned idea, the study of Al → X collision in
Al–X compounds (X can be any element) necessitates
the study of the Al → X collision in the gas phase.

Firstly, we draw the so-called "correlation diagram"
in which the atomic levels ($R = \infty$) are joined, by straight
lines, to the levels of the united atom ($R = 0$). These
diagrams do not give the exact evolution of the levels,
they only indicate whether the levels have a large or a
small total promotion. Nevertheless, conclusions can
be drawn in as far as a large total promotion energy is
necessary for a level to be highly excited.

The rules which allow to draw asymmetric correlation
diagrams have recently been given by Barat and
Lichten.[†][33,34]

3.2 Al–O compounds

In the case of gaseous phase
Al → O collision we see (Figure 3a) that the level which
originates at the $2p$ Al level and which has the largest
promotion is of $3d\sigma$ type (instead of $4f\sigma$ in Al → Al
Figure 3b)). This level joins at $R = 0$ an energy situated
at about -9 eV, whereas in the Al → Al case the $2p$ Al
level gives rise to a $4f\sigma$ level which joins the $4f$ iron
level which one can locate at nearly: $-1/n^2 \sim -0.85$ eV.

Now let us come back to Al → O collisions in Al–O
compounds with small O concentration, where the
band structure is nearly that of pure aluminium. Roughly
speaking one can say that, in the framework of this
model, the creation of $2p$ Al-hole is forbidden. To show
this, let us suppose that the average energy of the $3d\sigma$
level embedded in the conduction band is the same as
the $3d\sigma$ energy in the gaseous state. Then we see that
if we broaden this level in the same way as in a previous
paper,[14] even at $R = 0$, the upper energy of the
broadened level, located at $\epsilon + \frac{2}{3}(\epsilon - \epsilon_B)$ (the meaning

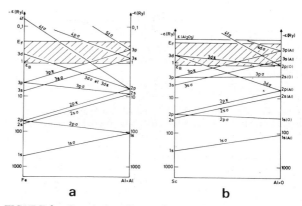

FIGURE 3   Correlation diagram in gaseous Al → Al (a), and
Al → O (b) collisions. The energies ϵ_B, E_F, $E(Al_2O_3)$ refer
respectively to the bottom of the conduction band, the Fermi
level and the first unoccupied level in Al_2O_3.

of the symbols is given in the figures), does not reach
the free states above E_F. In this discussion we assume
that the levels whose energy is larger than the oxygen
$2s$ energy ($3s\sigma$, $3d\pi$, $3p\pi$, $4f\sigma$) remain embedded in the
aluminium conduction band. We also notice that the
crossing of $3d\sigma$ and $3p\sigma$ which is complete does not
create any excitation on $3d\sigma$. Thus, according to our
model, the probability of $2p$ Al-hole creation by Al → O
collision is almost zero; $2p$ Al-holes will be mainly
created by Al → Al collisions.

To check this conclusion we shall follow the variation
of the kinetic emission (Al kinetic ions and Auger
electrons) with respect to the Al concentration C(Al).
One can say, in a qualitative manner, that the number
of Al → Al collisions in Al–O compounds will depend on
$C(Al)^2$, so one can expect to observe a rapid variation
of the kinetic emissions with C(Al).

If, on the other hand, we consider larger O-
concentrations we have to take into account a change
in the band structure since the first empty level in
Al_2O_3 is higher than in pure aluminium. According to
this it becomes more and more difficult to create $2p$
Al holes, even in Al → Al collisions, and the $2p$ Al hole
production must therefore decrease faster than $C(Al)^2$.

Let us now examine experimental results. Brochard
et al.[35,36] give a curve showing the dependence of the
secondary Al^{++} current on the partial O_2 pressure in
the collision chamber. One sees that the variation is
very fast, much faster than a linear one, which is in
agreement with our conclusions. Experiments about
secondary aluminium Auger electrons emitted by Al–O
compounds have also been carried out.[37,38] The
dependence of this secondary electron current on O_2
partial pressure shows also a rapid variation. The results

---

† The separated atom levels ($R = \infty$) are joined to the
united atom levels ($R = 0$) as follows: we join in each side the
levels which have the same number of zeros in the radial wave
functions by using, each time, the lowest available level.

of the preceding experiments for four partial pressures are given in Table II we note that, as predicted by the kinetic model, both secondary currents (Al^{++} and aluminium Auger electrons) present the same behaviour.

TABLE II

Currents of secondary Al^{++} [35,36] and Al Auger electrons[37,38] emitted by pure Al targets for various O_2 partial pressures. A normalization is done by dividing the currents by the values obtained at $p(O_2) = 10^{-6}$ torr.

$p(O_2)$ (torr)	Al^{++} current	Al Auger electrons current	Qualitative theoretical analysis
10^{-6}	1	1	Rapid
$4 \cdot 10^{-6}$	1	1	decrease
$2 \cdot 10^{-5}$	0.5	0.85	and same
$6 \cdot 10^{-5}$	0.25	0.3	behaviour

2.3.3 Al–Cu compounds   In the case of Al–Cu compounds the interpretation is more difficult. The correlation diagram (Figure 4) shows that, as in the Al → Al case, the level proceeding from the $2p$ Al level which has the largest promotion is of $4f\sigma$ type, it joins the $4f$ level of molybdenum.

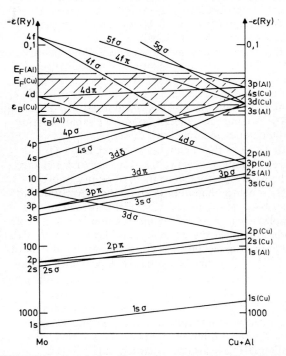

FIGURE 4   Correlation diagram in gaseous Al → Cu collisions. The energies ϵ_B, E_F refer respectively to the bottom of the conduction band and the Fermi level.

To compare this situation with Al → Al, we have to place the $4f$ Mo level with respect to the $4f$ Fe level. Because of the larger atomic number one can think that the first is lower. Therefore, in small Cu concentrations (where the Al structure can still be used) the $2p$ Al hole production must decrease. However, this argument is not very strong because of the difficulty of situating the $4f$ levels.

It is interesting to look at experimental data in gaseous collisions. Saris *et al.*[39] have bombarded Z atomic number gaseous targets with Ar^+ ions and have registered the variations of $2p$ Ar holes production. Passing from Ar → Ar to Ar → Cu reduces the hole production of an amount of 50%. This fact might be explained by the decrease of the $4f$ level energy of the united atom in the way quoted before or by the decrease in the $2p$ wave function radius of the target[39] which makes more difficult the formation of molecular levels derived from $2p$ atomic levels.

Whatever the explanation may be, the same effect is likely to occur in our problem, which would give a strong reduction of the $2p$ Al hole production on passing from Al → Al to Al → Cu.

This conclusion is the same as the one already reached by Brochard *et al.*[35,36] to explain the fact that Al^{++} intensities in CuAl alloys vary roughly with C(Al)

Thus the kinetic model appears to be a useful tool to understand various experimental features; it also suggests other experiments on compounds or alloys.

3 SURFACE EFFECT MODELS

3.1 Introduction

The kinetic model contributes to explain the secondary emissions of ions and electrons; obviously it does not cover the whole phenomenon.

Veksler *et al.*[40] noticed that, when the energy of primary ions decreases, kinetic emission ceases before ion emission. We noticed also[15] that in the case of aluminium targets, whereas Al^{++} and Al^{+++} consistently have a strong kinetic character, only high energy Al^+ have it.

Various authors constructed models where the electronic excitation is directly created in the outer shells and we shall give a brief description of some of these works.

Let us note that, for a particle leaving the metal, the probability to carry a bound level hole (kinetic effect) to be ionized by a "surface effect" are both small so it not necessary to modify our kinetic model assumptions the probability that the two mechanisms affect the same particle is very small, the two secondary currents will simply add to each other. One can also notice that, in the range of secondary energies studied ($\leqslant 1$ keV), only the kinetic mechanism can be responsible for the creation of multicharged particles.

3.2 Probability that a Delocalized Conduction Electron Follows an Atom Leaving a Target

In this paper we shall speak mainly about two models (Schroeer, Joyes-Toulouse). Blaise† will develop his method in another paper of this conference.[41,42]

The problem of the ionization of a metallic atom in its crossing of the metal surface is very intricate and simplifying assumptions must be introduced. One of them consists in calculating the population of an atomic level centred on a target particle which initially is strongly coupled to the continuum and is continuously decoupled from it as the particle escapes the sample. This situation is described by means of the following hamiltonian (Figure 5)

$$H = \Sigma_k \epsilon_k a_k^+ a_k + E_d a_d^+ a_d$$
$$+ \Sigma_k (V_{dk} a_d^+ a_k + V_{kd} a_k^+ a_d) \qquad (1)$$

where Anderson's[43] notations and definitions are used: k and d refer respectively to conduction and atomic state, E_d and V are specified time-varying functions which give the average energy of the atomic level and the strength of its coupling to the continuum, respectively. In (1) Coulomb electron interactions are neglected.

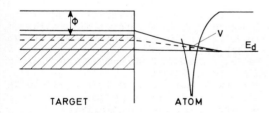

FIGURE 5 Evolution of broadening and displacement of a level centred on an atomic particle leaving a metallic target. E_d, ϕ and V give respectively the energy of the bound level, the work function of the metal and the strength of the coupling between the bound level and the continuum.

Many difficulties still remain in this problem and various attempts to solve them have been proposed.

† The ionization probability is obtained by comparing the band structure of the metal with the excited level structure of the emitted atom. The emitted particle needs a minimal excitation energy E^x to be ionized, this excitation is given by two or three conduction electrons which have a total energy larger than E^x and which make together a resonance transition from the metal to the atom; the larger $E^x - \phi$ the larger the probability to get a positive ion (ϕ: work function).

3.2.1 Calculation by Schroeer

Schroeer[16] uses the so-called adiabatic approximation. The probability that an excited state with wave function $\psi_f(r, t)$ be occupied during the H variation is given by:

$$R = \left| \int_{-\infty}^{+\infty} \frac{\langle \psi_f^*(r, t) | \partial H(r, t)/\partial t | \psi_i(r, t) \rangle}{\Delta E(t)} \right.$$
$$\left. \times \exp \frac{i \Delta E(t) t}{\hbar} \, dt \right|^2$$

where $\psi_i(r, t)$ is the ground state of $H(r, t)$ and $\Delta E(t)$ the energy difference between the two states. Unfortunately the matrix elements $\langle \psi_f | \partial H/\partial t | \psi_i \rangle$ cannot easily be obtained and this author assumes various simple expressions for it. He also assumes that ΔE is constant and equal to $E_d - \phi$ (E_d is the ionization potential, ϕ the target work-function). He finally finds that R varies as a power of ($E_d - \phi$):

$$R \alpha (E_d - \phi)^n$$

with n negative, a result which is well supported by experimental data.[44] This inverse power dependence has also been found by Jurela.[45]

3.2.2 Calculation by Joyes and Toulouse[17]

It is possible to give a solution of the problem in a somewhat different manner which allows taking into account the reaction of the conduction electrons near the Fermi level. These electrons are easily excited since they do not need a large energy and so, after any time-varying perturbation, many of them are excited, and the Fermi sea needs a long time to become calm again.[46–51] This is a non-adiabatic effect which leads to an excitation of the emitted particle.

The population of the d level at time τ is given by the limit of a Green function $G_{dd}(\tau, \tau')$, when τ' tends to τ:

$$G_{dd}(\tau, \tau') = -i\{T\langle a_d(\tau) a_d^+(\tau') \rangle\}$$

which satisfies an integral equation deduced from 1.[52] The details of the calculation will appear in a forthcoming paper. The final result has two parts:

— an adiabatic term which would give the population of the level if the relaxations were instantaneous

— a non-adiabatic term which depends on the relative position, in the initial state, of E_d and ϕ, and more precisely on the delocalized charge which has to be squeezed into the atomic orbital of the emitted particle. Physically this means that when the electrons are strongly delocalized in the target, they cannot easily follow the atom.

Such results might be used to analyse experimental

data especially about ion emission from dilute alloys of first row transition elements in normal metals.[53] These alloys are interesting because they present resistivity effects (the residual resistivity, at $\sim 900°K$, increases from AlTi to AlCr and then decreases from AlCr to AlCu) which are due to the situation of energy E_d with respect to E_F and which have been explained by using the Anderson's hamiltonian.

Hence, one can say that the main interest of this proposed model is to provide means for studying the secondary emission phenomena using controlled approximations checked in other fields of Solid State Physics. One can also note that by including for instance in (1) a correlation term

$$U n_d \uparrow n_d \downarrow$$

will allow studying magnetic systems.

Other authors have introduced excitation models based on similar ideas. Let us quote for instance the discussion of the spectral analysis of light emitted by secondary particles in which Terzic et al.[54] examine among other processes, the effect of broadening of energy levels near a surface. We shall finally note that the problem of the electronic transitions in the atom surface interaction appears in other works on secondary ionic emission[55] and in other fields of physics as potential emission of electrons.[56–58]

4 CONCLUSION

In this paper we do not intend to give a complete presentation of theoretical works on secondary ionic emission; let us quote however the thermal models. C. A. Andersen has proposed a thermodynamical model,[22,59] different from all those described before in which no temperature effects were included. Two parameters are adjusted in this calculation and so the model has only been applied to many elements compounds (when two concentrations are known). An application to simpler systems, where other models have been applied, would also be interesting.

Among the various aspects of secondary ionic emission clearly connected to important questions in physics, one should also quote the secondary emission of polyatomic clusters. One idea seems to be well established now: the secondary intensities of a given species depends mainly on its stability.[60–62] So the secondary ionic emission provides a tool to study these small metallic particles and to check calculations. A better knowledge of the constitution of the metallic bond should result from these works.

One can say that now, besides its application in analysis techniques, secondary ionic emission can also be considered as a field of growing interest in Physics.

REFERENCES

1. M. Kaminsky, *Atomic and Ionic Impact Phenomena on Metal Surfaces* (Springer-Verlag, Berlin Heidelberg, New York, 1965).
2. Ya. M. Fogel, *Soviet Phys. Uspekhi* **10**, 17 (1967).
3. G. Carter and J. S. Colligon, *Ion Bombardment of Solids* (Heinemann Educational Books Ltd. London, 1968).
4. R. Castaing and G. Slodzian, *J. de Microscopie* **1**, 395, (1962).
5. G. Slodzian, *Ann. Phys.* **9**, 591 (1964).
6. Y. Le Tron, Thèse de 3° Cycle, Orsay (1963).
7. R. Castaing and J.-F. Hennequin, *C.R. Acad. Sc. Série B* **262**, 1008 (1966).
8. J.-F. Hennequin, *Rev. Phys. Appl.* **1**, 273 (1966).
9. J.-F. Hennequin, *J. Phys.* **29**, 655 (1968).
10. J.-F. Hennequin, *J. Phys.* **29**, 1053 (1968).
11. P. Joyes and R. Castaing, *C.R. Acad. Sc. Série B* **263**, 384, (1966).
12. P. Joyes and J.-F. Hennequin, *J. Phys.* **29**, 483, (1968).
13. P. Joyes, *J. Phys.* **29**, 774 (1968).
14. P. Joyes, *J. Phys.* **30**, 224 (1969).
15. P. Joyes, *J. Phys.* **30**, 365 (1969).
16. J. M. Schroeer, The Second International Nevac Symposium (1972).
    J. M. Schroeer, T. N. Rhodin and R. C. Bradley, *Surf. Science,* **34**, 571 (1973).
17. P. Joyes and G. Toulouse, *Phys. Let.* **39A**, 267 (1972).
18. D. Guenot, Diplôme d'Etudes Supérieures, Orsay, 1966.
19. G. Slodzian and J.-F. Hennequin, *C.R. Acad. Sc. Série B* **263**, 1246 (1966).
20. A. Benninghoven, *Z. Naturf.* **22a**, 841 (1967).
21. A. Benninghoven, *Z. Phys.* **220**, 159, (1969).
22. C. A. Andersen, *J. Mass. Spectro. and Ion Phys.* **2**, 61 (1969).
23. J.-F. Hennequin, P. Joyes and R. Castaing, *C.R. Acad. Sc. Série B* **265**, 312 (1967).
24. P. Sigmund, *Phys. Rev.* **184**, 383 (1969).
25. J. E. Westmoreland, *Rad. Effects* **6**, 187 (1970).
26. K. B. Winterbon, P. Sigmund and J. B. Sanders, *Det Kongelige Danske Videnskabernes Selskab. Mat. Fys. Med.* **87**, 14 (1970).
27. J. Bottiger, J. A. Davies, P. Sigmund and K. B. Winterbon, *Rad. Effects* **11**, 69 (1971).
28. U. Fano and W. Lichten, *Phys. Rev. Let.* **14**, 627 (1965).
29. W. Lichten, *Phys. Rev.* **164**, 131 (1967).
30. C. Bonnelle, Thèse de Doctorat, Paris, 1964.
31. J. Friedel, *Cand. J. Phys.* **34**, 1190 (1956).
32. J. Friedel, *Nuovo Cim.* **52**, 287 (1958).
33. M. Barat, Congrès des Collisions Atomiques, Rennes, France (1971).
34. M. Barat and W. Lichten, *Phys. Rev.* **A6**, 211 (1972).
35. D. Brochard, Thèse de 3° Cycle, Orsay (1971).
36. D. Brochard and G. Slodzian, *J. Phys.* **32**, 185 (1971).
37. J.-F. Hennequin and P. Viaris de Lesegno, *C.R. Acad. Sc. Série B* **272**, 1259 (1971).
38. P. Viaris de Lesegno, Thèse de 3° Cycle, Orsay (1972).
39. F. W. Saris, *Prag. Rep. VII ICPEA C* (North-Holland, Amsterdam, 1971).
40. V. I. Veksler and B. A. Tsipinyuk, *Sov. Phys. J.E.T.P.* **33**, 753 (1971).
41. G. Blaise and G. Slodzian, *C.R. Acad. Sc. Série B* **266**, 1525 (1968).
42. G. Blaise and G. Slodzian, *J. Physique* **31**, 93 (1970).
43. P. W. Anderson, *Phys. Rev.* **124**, 41 (1961).

44. H. Beske, *Z. Naturforsch* **22a** 459 (1967).
45. Z. Jurela, *Atomic Collision Phenomena in Solids* (North-Holland, Amsterdam, 1970).
46. J. Friedel, *Phil. Mag.* **43**, 153 (1952).
47. J. Friedel, *Phil. Mag.* **43**, 1115 (1952).
48. B. Roulet, J. Gavoret and P. Nozieres, *Phys. Rev.* **178**, 1072 (1969).
49. P. Nozieres, J. Gavoret and B. Roulet, *Phys. Rev.* **178**, 1084 (1969).
50. P. Nozieres and C. T. de Dominicis, *Phys. Rev.* **178**, 1097 (1969).
51. M. Natta and P. Joyes, *J. Chem. Phys. Sol.* **31**, 447 (1970).
52. D. R. Hamann, *Phys. Rev. B* **2**, 1373 (1970).
53. G. Blaise, Thèse, Orsay (1972).
54. I. Terzic and N. Perovic, *Surf. Science* **21**, 86 (1970).
55. W. F. Van Der Weg and P. K. Rol, *Nuclear Instruments and Methods* **38**, 274 (1965).
56. H. D. Hagstrum, *Phys. Rev.* **150**, 495 (1966).
57. H. D. Hagstrum and G. E. Becker, *Phys. Rev.* **159**, 572 (1967).
58. H. D. Hagstrum and G. E. Becker, *Phys. Rev. Let.* **22**, 1054 (1969).
59. C. A. Andersen, *Science,* **175**, 853 (1972).
60. P. Joyes, *J. Phys. Chem. Sol.* **32**, 1269 (1971).
61. G. Staudenmaier, *Rad. Effects* **13**, 87 (1972).
62. M. Leleyter and P. Joyes, this Conference, Munich (1972).

SECONDARY ION EMISSION OF ALLOYS IN RELATION WITH THEIR ELECTRONIC STRUCTURE

GUY BLAISE

Laboratoire de Physique des Solides, associé au C.N.R.S., Université Paris-Sud, 91405 Orsay, France

When an atom crosses a metal-vacuum surface, the perturbation acting upon outer electronic shells produces excited states the energy of which exceeds, in some cases, the ionization energy of the atom; then, one deals with auto-ionizing states. We propose a secondary positive ion emission process from metals and alloys based upon the formation of auto-ionizing states. In such a process the ion yield K^+ of an element is proportional to the occupation probability of auto-ionizing states. This probability is closely connected to the density of states around the element inside the metal or alloy, in a small bandwidth energy near the Fermi level: the greater the fraction of valence electrons of the atom in this bandwidth, the higher the occupation probability will be. This connection explains enhancement observed experimentally in the ion emission of transition elements (from Ti to Cu) in dilute alloys. This effect is particularly noticeable when d electrons of transition elements are distributed in a virtual bound state which crosses the Fermi level as in the case of chromium in nickel matrix or transition elements in aluminium or copper matrix.

1 INTRODUCTION

When a metal M is bombarded by inert gas ions with a few keV energy, a small part of the sputtered particles is ionized as positive ions M^+. These ions arise from atoms displaced by collisions inside the metal and ejected with energies which do not exceed, for the most part, a few hundred eV. As a consequence, the velocity of the atoms is much smaller than the velocity of the conduction electrons. Then we can consider, in some degree, that the conduction electrons will follow the atoms in their displacements inside the metal and will maintain them as neutral, outside.

Therefore, to account for a positive ion emission, it is necessary to consider that the ionization process occurs in vacuo. Such a process is possible if the atoms leave the metal as neutral but in highly excited electronic states the energy of which E^* exceeds the ionization energy I; one deals with auto-ionizing states.[1] In that case, the atoms will be ionized spontaneously through an auto-ionization process which will take place in vacuo.

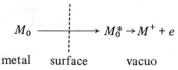

The perturbation acting upon the outer electronic shells of the atom, while it crosses the metal-vacuum surface, is responsible for the formation of these auto-ionizing states.

2 THE FORMATION OF AUTO-IONIZING STATES

An auto-ionizing state results from a multi-electronic excitation of the atom.[1] The typical example is the state $3d^9\ 4s\ 5s(^4D)$ of copper the energy of which is $1/10$ eV above the ionization energy. This state is represented, Figure 1, in a one electron model, beside the electronic structure of copper metal.

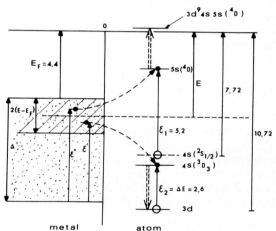

FIGURE 1 Electronic structures of copper atom (auto-ionizing state $3d^9\ 4s\ 5s(^4D)$) and copper metal.

We notice that the mean level E of the levels $4s$ and $5s$ of the atom lies below the Fermi level E_F of the metal. Consequently, this auto-ionizing state is equivalent in energy to a state occupied by two electrons of the conduction band of the metal.

We have laid down as an hypothesis that the formation of auto-ionizing states results from an electronic
transition in which the mean energy level E of the
electrons is maintained.[3]

In a more general way, if we consider the excited
states $3d^9\, ml\, n'l'$ of copper, represented in a one
electron model, the formation of auto-ionizing states is
possible, according to the process described previously,
since the mean level E_m corresponding to the first
auto-ionizing state (that is the state that contains just
the ionization energy according to the considered
process) is below E_F.

In other words, the condition required for the
formation of auto-ionizing states is:

$$E_m - E_F > 0 \qquad (1)$$

First, this process was applied to the ion emission
from transition metals of the $3d$ series (Ti to Cu).[3] In
that case, the condition (1) implies different configurations of the auto-ionizing states, leading to
different ionization processes. For instance, if $\nu + 1$ is
the number of $d + s$ electrons of the atom, the following processes have been considered:

$$3d^{\nu-1}\, nl\, n'l' \rightarrow (3d^\nu)^+ + e \qquad (2)$$

$$3d^{\nu-2}\, nl\, n'l'\, n''l'' \rightarrow (3d^{\nu-1}\, 4s)^+ + e \qquad (3)$$

Typically, process (2), involving 2 excited electrons,
applies to copper ion emission whereas process (3),
involving 3 excited electrons, applies to nickel ion
emission.

The ionization probability P^+ of an atom leaving a
metal must be proportional to the occupation probability $P(E_m - E_F)$ of its auto-ionization states; that is
states with a mean level E between E_m and E_F.

In fact, we must also take into account electronic
exchange due to the interaction between the atom and
the metal, while the atom crosses the metal-vacuum
surface. Such processes have been studied by Hagstrum.[4]
In the present case, we consider processes leading to a
destruction of the auto-ionizing states. So, we include
a survival probability $P_0(v)$ of the auto-ionizing states,
depending on the velocity v of the particle. Finally:

$$P^+ \propto P(E_m - E_F)P_0(v) \qquad (4)$$

We have shown[5] P_0 depends on the configuration of the
auto-ionizing states: for instance, the auto-ionizing
states $3d^7\, nl\, n'l'\, n''l''$ of a nickel atom (process 3) have
a greater survival probability than the auto-ionizing
states $3d^9\, nl\, n'l'$ of a copper atom, ejected with the
same velocity.

This interpretation corresponds to the fact that the
energy distributions of Cu^+ and Ni^+ ions are not
similar in the low energy range (up to about 30 eV)
(Figure 2),[5] since the energy distributions of neutral

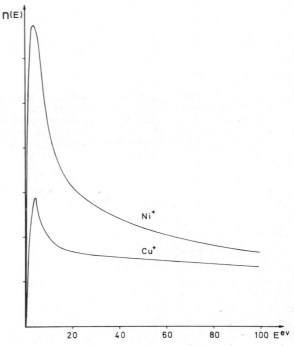

FIGURE 2   Energy distributions of Cu^+ and Ni^+ ions.

atoms are similar.[6] However $P_0(v)$ approaches unity
when the velocity of the particles increases. In the
same time, the energy distributions of ions become
similar. Consequently, the relative ion emissions of
the series of transition elements, corresponding to ions
with high energy, are directly proportional to the
occupation probability $P(E_m - E_F)$ of auto-ionizing
states.

The probabilities $P(E_m - E_F)$ have been calculated
with the following assumptions[3]:

—The initial state of a transition, leading to auto-
ionizing states, is given by the distribution of the
electrons of the metal, with a density of states $n(\epsilon)$
near the Fermi level.

—The final state corresponds to a free atomic state
with density $g(E)$.

—We suppose $g(E)$ is constant for auto-ionizing
states.[7]

—The matrix element of the transitions, H_{fi}, does
not depend on the initial energy ϵ of the electrons.[4]

Then, we obtain:

$$P(E_m - E_F) = \frac{\int_{E_F}^{E_m} C(E)\,dE}{\frac{1}{2}N^2}$$

or

$$= \frac{\int_{E_F}^{E_m} C(E)\,dE}{\frac{1}{3}N^3} \qquad (5)$$

according as ionization processes (2) or (3) apply. N is the total number of $d + s$ electrons of the atom.

$C(E)$ represents the number of states, deduced from the density $n(\epsilon)$ of the initial state, with a mean energy level E.

A good agreement was obtained between experiments and calculations for pure metals of the transition series.[3]

3 ION EMISSION OF ALLOYS

The previous description of the formation of auto-ionizing states can also be applied to ion emission of alloys. Here, we present the main results on the emission of transition elements of the $3d$ series in diluted solutions in nickel, iron, copper and aluminium matrices.

Let us consider an alloy $M_{1-C}T_C$ in which the concentration C is low enough so that solute atoms do not interact each other and do not modify the ion emission process of the matrix element M. Typically this concentration C is less than 1%.

Experimentally we define a constant k as the ratio of the ion yields $K(T^+)$, $K(M^+)$ proportional to the respective concentrations:

$$\frac{K(T^+)}{K(M^+)} = k\,\frac{C}{1 - C}$$

Measurements of k have been carried out with Slodzian's ion micro-analyser.[8] The k values reported here are deduced from experimental ion yields measured with 100 eV energy range of the ions[5]; that is in the similar part of the energy distributions of ions.

According to experimental analysis of the energy distributions, below 20 eV, and calculations $E_m - E_F$,[5] we concluded that process (2) applies to a solute copper atom, whereas process (3) applies to the other solute elements, that is Ti, V, Cr, Mn, Fe, Co, Ni.

Theoretically $k = P(T^+)/P(M^+)$ is the ratio between the occupation probability of auto-ionizing states

$P(T^+)$ of a solute atom T and the probability $P(M^+)$ of a matrix atom M. $P(M^+)$ is the same as for pure metal M. On the other hand, $P(T^+)$ must be calculated again, according to the electronic structure of the solute atom T into the alloy. In many cases, this electronic structure is quite different from the electronic structure of a pure metal T,[9] so it is with the ion yields.[10–12]

This connection between the electronic structure of an atom inside a solid and its ionization probability allows us to explain enhancement observed experimentally in the ion emission of solute atoms in alloys.[5–12]

As a matter of fact, it appears from Eq. (5) that the greater the fraction of electrons in the small bandwidth energy $E_m - E_F$, the higher the occupation probability of auto-ionizing states will be.

We discuss now, qualitatively, the electronic structure of alloys and consequences relative to the ion emission (quantitative calculations will be reported (Ref. 5)).

When a solute atom is introduced into a metal, the excess nuclear charge displaces locally the mobile electrons until the displaced charge screens out the new nuclear charge exactly. We may say that the excess electron charges introduced with the solute atom remains around its nucleus at short distances.[9]

The d electrons of transition elements play the principal part in the screening of the atom because they are numerous and localized in narrow bands with high density near the Fermi level. For the same reasons, they play also the principal part in the formation of auto-ionizing states. So, we shall neglect, subsequently, the contribution of $4s$ electrons which are spread in wide bands with a small density near E_F.

Nickel Based Alloys

We have represented Figure 3 the electronic structures of alloys: dashed circles represent the electronic structure of nickel matrix atoms and full lines the electronic structure of solute atoms. Exchange interaction within a d shell of nickel tend to split it into 2 halves of opposite spin directions.[9]

If the solute atom is on the right of the matrix (that is for Cu), the potential it provides is more attractive than the potential of the matrix atom. So, the d states of the solute atom tend to decrease in energy with respect to the d states of the matrix. On the contrary, if the solute atom is on the left but close to the matrix (Co, Fe), the d states are repelled toward the Fermi level because the potential is less attractive than the potential of the matrix atom. In that case, there is only a small shift of the density of states toward the Fermi level with respect to the density of states of

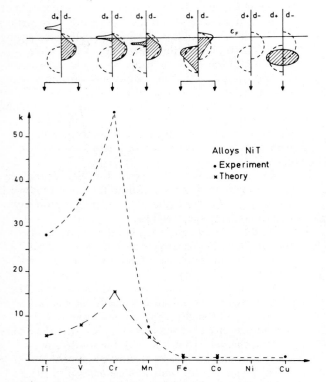

FIGURE 3   Electronic structures of nickel based alloys (upper part). Experimental (points) and theoretical (crosses) k values.

the matrix. At last, when the solute atom is on the left and far from the matrix (Mn, Cr, V, Ti) the repulsive potential is strong enough to extract the d states from the d states of the matrix. Now these states lead to a Friedel's virtual bound state the width of which is very narrow (less than 0.5 eV for Ni based alloys) and the density very high.[13] Such a virtual bound state results from a resonance effect between the d states of the solute atom and the states of the conduction band $4s$ of the matrix.[9] For chromium the virtual bound state crosses the Fermi level whereas for V and Ti it is above, so it is empty.

It appears from this discussion that the fraction of d electrons localized in the small bandwidth $E_m - E_F$ (the order of magnitude is about 0.5 eV) tend to increase from copper to titanium in the transition series with a noticeable case for chromium when the virtual bound state crosses the Fermi level. This fact is related with k values (Figure 3) which increase first slowly, from Cu to Fe, in relation with a small shift in the d states; then very quickly when a virtual bound state appears and crosses the Fermi level (Cr) and, at

last, k decreases when the virtual bound state is completely empty (V, Ti).

The agreement between experimental and theoretical k values is not very good from Cr to Ti. In fact, the ionization process is very sensitive to the density of states near E_F and this density is not known with a great accuracy.

Iron Based Alloys

In that case (Figure 4) a virtual bound state does not appear. The d states of the solute atom are progress-

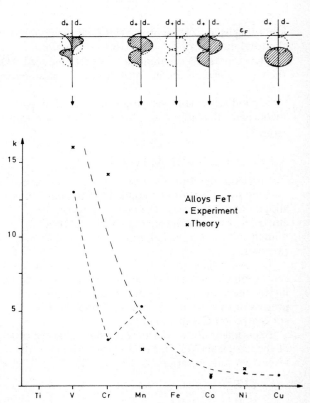

FIGURE 4   Electronic structures of iron based alloys (upper part). Experimental (points) and theoretical (crosses) k values.

ively shifted toward the Fermi level from Cu to Ti; this is connected with increasing constants k.

Copper Based Alloys (Figure 5)

The d electrons are localized in a virtual bound state with 1 eV width about. This state is split for elements in the middle of the series. The virtual bound state goes up through the Fermi level from Ni to Ti and k values increase in the same time.

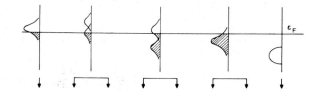

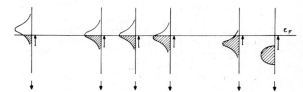

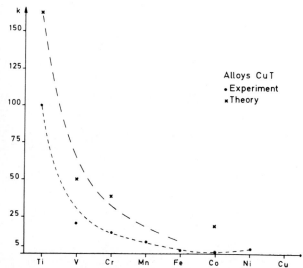

FIGURE 5   Electronic structures of copper based alloys (upper part). Experimental (points) and theoretical (crosses) k values.

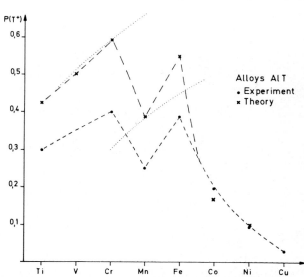

FIGURE 6   Electronic structures of aluminium based alloys (upper part). Experimental ion yields $K(T^+)$ of solute transition elements (points) and theoretical probabilities $P(T^+)$ of auto-ionizing states (crosses).

Aluminium Based Alloys (Figure 6)

The electronic structure is approximately the same as for copper based alloys, that is d electrons localized in a virtual bound state (width ~ 2 eV), but without splitting. The virtual bound state goes up through the Fermi level from Cu to Ti. First we observe increasing probabilities $P(T^+)$ and ion yields from Cu to Fe (Figure 6), after which the ion yields oscillate from Fe to Ti. We can show this oscillation is closely connected to the number N_d of d electrons in the virtual bound state, as it has been measured by different techniques[14] (Table I).

From Fe to Ti, we may consider all the d electrons participate to the formation of auto-ionizing states because the order of magnitude of the energy $E_m - E_F \gtrsim 1$ eV[5] is the same as the width of the occupied part of the virtual bound state. Equation (5) becomes:

$$P(T^+) \simeq \frac{\frac{1}{3}N_d^{\,3}}{\frac{1}{3}N^3}$$

As a consequence, $P(T^+)$ values, mentioned in Table I, show a dip for manganese and titanium as experimental ion yields (Figure 6).

TABLE I

	Ti	V	Cr	Mn	Fe	Co	Ni	Cu
$d+s$ electrons: N	4	5	6	7	8	9	10	11
d electrons N_d	3	4	5	5	6.6	7.25	8	10
$P(T^+)$	0.42	0.5	0.6	0.39	0.55	0.18	0.1	

4 CONCLUSION

Although the calculations are rough enough, in some cases, we may consider the most remarkable features of the ion emission M^+ of alloys have been interpreted from auto-ionization processes. The correlation established between the ionization probability of an atom and the electronic structure of this atom inside the metal, leads us to believe that the ion emission might be used for investigations on the electronic properties of solids.

REFERENCES

1. E. U. Condon and G. H. Shortley, *The theory of atomic spectra* (Cambridge, 1957).

2. H. S. M. Massey and E. H. S. Burhop, *Electronic and ionic impact phenomena*, Second Edition Vol. 1.
3. G. Blaise and G. Slodzian, *J. Phys.* **31,** 93 (1970).
4. H. D. Hagstrum, *Phys. Rev.* **96,** 2, 336 (1954).
5. G. Blaise and G. Slodzian, *J. Phys.* to be published (1973).
6. H. Oechsner, *Z. Physik.* **238,** 433 (1970).
7. W. C. Blais and J. B. Mann, *J. Chem. Phys.* **33,** 1 (1960).
8. G. Slodzian, *Ann. Phys.* **9,** 591 (1964).

9. J. Friedel, *Nuovo Cimento* **7,** 287 (1958).
10. H. E. Beske, *Z. Naturf.* **22a,** 459 (1967).
11. J. Schelten, *Z. Naturf.* **23a,** 109 (1968).
12. G. Blaise and G. Slodzian, *C.R. Acad. Sc.* **273,** série B, 357 (1971).
13. J. Friedel, Rendiconti della Scuola Internazionale di Fisica "Enrico Fermi" XXXVII, Corso, 1966.
14. R. Aoki and T. Ohtsuka, *J. Phys. Soc. Jap.* **26,** 3, 651 (1969).

INFLUENCE OF TEMPERATURE ON THE SECONDARY ION EMISSION OF A MONOCRYSTALLINE ALUMINIUM TARGET

R. LAURENT and G. SLODZIAN

Laboratoire de Physique des Solides associé au C.N.R.S. Bât 510,
Université Paris-Sud, 91405 Orsay, France

We have studied with a Slodzian's microanalyser in which the target is horizontal, the secondary ion emission of solid and liquid aluminium versus temperature from 300°K to 1000°K.

We found an increase with temperature of the secondary emission in channeling directions and a decrease in the other directions. The behaviour of the clusters was also studied. We found no great difference between solid and liquid state for the clusters intensities. A slight difference is found in the energy distribution of secondary ions when the crystal is bombarded in different directions.

1 INTRODUCTION

The dependence of the sputtering yield on the target temperature was studied by several authors.[1,7,8,11] They found that the temperature has little effect except near the melting point.[24]

The most important effects occur with single crystals. The anisotropy of the crystal becomes smaller when temperature rises. The channeling theory of Lindhard[12] was used by Elich, Roosendaal and Onderdelinden[1] to explain the variation of the sputtering when the primary beam is incident in a direction of channelling. It was found with copper that in directions of channelling the sputtering increases with temperature. In a maximum opacity direction, a decrease of the sputtering yield was observed.[1,6,7]

Not only the sputtering yield is changed when the crystal is rotated and the temperature increased but the secondary electron and ion emission are modified in the same experimental conditions.[4–7,10,13,17–23] In this work we are looking for the effects of temperature on the secondary ion emission of monocrystalline aluminium.

2 APPARATUS

We have used a Slodzian's microanalyser[14,15] in which the target is horizontal. So we can study both solid or liquid metal. The target is heated by electron bombardment. The temperature is measured with a chromel-alumel thermocouple placed in direct contact with the target. However, the temperature measured is not exactly the true temperature of the specimen since we measured the melting point of aluminium at 645°C instead of 660°C. This is probably due to thermal leakage.

Several difficulties appeared in these experiments:

a) Mechanical deformation of the sample holder at high temperature.

b) Non-flat surface of liquid aluminium because the crucible is not well wetted.

We can rotate the target around an axis normal to the surface while the incident angle of the primary beam is kept constant; its value is about 60°.

We use a 6 keV argon primary beam. The density is higher than 600 μA/cm^2. The vacuum is better than $8 \cdot 10^{-7}$ torr even at high temperature. In these experimental conditions we could make sure that the chemical effects are avoided. Indeed, it is very important to prevent contamination of the sample by residual gases and the oxidation of the surface at high temperature.

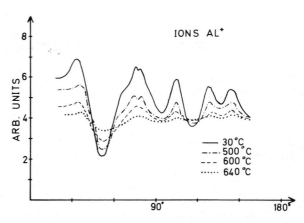

FIGURE 1   Secondary emission obtained at different temperatures with a single crystal rotated around an axis normal to its surface.

3 RESULTS

3.1 Al⁺ Ions
3.1 Al^+ Ions

When we rotate the target around an axis normal to the surface we obtain (Figure 1) a curve which shows the crystal anisotropy in which channelling plays the principal part. When the temperature rises, the anisotropy of the crystal becomes smaller and, particularly, the openings of the channels are reduced. At two given orientations we have followed the evolution of Al^+ ion emission with temperature. We have chosen a direction which gives a maximum of emission and a direction where the emission is minimum. The Figure 2 shows that as the temperature increases the intensities in the two directions come closer to each other. On each side near the melting point the intensities are the same. In addition as the temperature rises we can see that there is a slow increase when the crystal is in a position where the primary beam is channelled, whereas, when the crystal is in a position which gives a maximum of the ion emission, the intensity decreases.

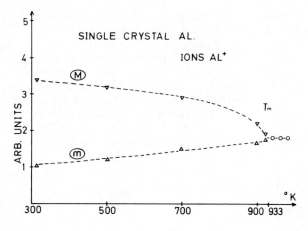

FIGURE 2   Evolution of the intensities in a position giving a maximum and in a position giving a minimum.

3.2 Al^{++} and Clusters Ions

We have found similar curves as those in Figure 1 at room temperature with secondary ions such as Al^{++}, Al_2^+, Al_3^+, Al_4^+ and Al_5^+. The anisotropy is the same as for Al^+. But the contrast (ratio of the intensities at a maximum and a minimum) can be different for various types of ions as shown in Table I. These numbers are the ratio between the intensities of a given maximum and a given minimum.

TABLE I

	Al^{++}	Al^+	Al_2^+	Al_3^+	Al_4^+	A
$\dfrac{I_{max}}{I_{min}}$	3.4	2.5	2	1.6	2.5	2

When the temperature rises the contrast decreases for all these ions and we have curves similar to those of Figure 1. As it is shown on the Figures 1 and 2 there is a decrease of the average intensity of ions Al^+. The behaviour of Al_4^+, Al_5^+ and Al^{++} is the same as Al^+ (Figure 3). But, the decrease is stronger for Al_2^+ and especially for Al_3^+.

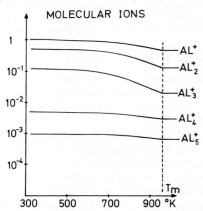

FIGURE 3   Evolution of the normalized intensities of the different species when averaged on a 360° revolution of the target.

3.3 Energy Spectra

We have looked for the behaviour of the energy spectra of Al^+. It should be recalled that in our instrument, the collection efficiency of the secondary ions depends on their initial energy and emission angle. For initial energies lower than 1 eV, the collection efficiency is 100%; as the energy becomes higher than 1 eV there is a discrimination which leads to a collection which is more and more restricted to ions emitted with direction near the normal at the surface target. For instance, for 100 eV particles only those particles emitted in a half angle of 10^{-2} rd are collected. To take into account this discrimination would require a precise knowledge of the angular distribution. If one wishes a crude estimate, one may assume a Lambert's law for the angular distribution. Several authors had not paid enough attention to the fact that they may lose most of the low energy ions when they are using a field free space around the target, since the exit plays the part of a lens which could defocus strongly these ions. On the other hand, the energy distribution, in the low range energy, is obtained owing to the retarding field provided by the electrostatic mirror which is included in the microanalyser; the precision is good enough to show differences in work function of different metals.[15]

At room temperature we have found differences between the energy spectra when the sample is in a position of minimum as compared to a position which gives a maximum. If the contrast is high enough we have more low energy ions in a position of maximum or, in other words, more fast ions (100 eV) in a channelling position (Figure 4). When the metal is liquid there is a marked peak at low energy.

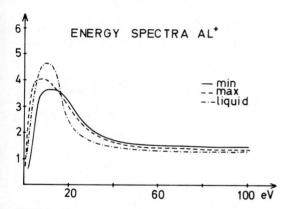

FIGURE 4   Energy spectra at room temperature and above the melting point. The normalization is such that the integral value between 0 and 100 eV is the same for all the three curves.

4 DISCUSSION

The results concerning the temperature dependence of secondary ion emission when the sample is rotated are very similar to those obtained in sputtering experiments:[1,2,7] as the temperature increases there is both a lowering in contrast and a decrease of the average intensity. Two effects can explain the reduction of contrast:

a) When the primary beam is incident in a channelling direction, the temperature increase produces an increase of the crystal opacity and therefore leads to a higher sputtering rate,[1] higher secondary electron emission rate[6] and higher secondary ion emission rate (this work).

b) When the primary beam is incident in an opaque direction, the sputtering[7] and the opacity[6] are reduced by the increase of temperature.

In addition, an increase of temperature reduces the

average length of focused collisions and accordingly the sputtering rate (averaged on all directions).

The energy distribution of secondary ions is a little more difficult to understand because we need the energy distributions of neutral atoms. In general, these distributions are more difficult to establish[25] and in our experimental conditions (angle of incidence, temperature) there is a lack of information. The relative enhancement of the emission of "slow" ions when a position of a minimum is compared to a position of maximum, probably rests on the fact that the focused collisions and the channelling of displaced particles are responsible for carrying high energy toward the surface. When the primary beam is channelled, the sputtering rate is lowered because the collision cascades are initiated deeper. At first sight, it seems that, in these conditions, the mean energy of the sputtered particles should be decreased but this does not take into account that focused collisions and channelling effects are more efficient to carry higher energies toward the surface. This idea is supported by the fact that when the sample is melted the proportion of fast ions is decreased.

It should be noted that a part of the emission of Al^+ ions is due to particles bearing a hole on their $2p$ shell[16] which is created during a collision inside the crystal. It is clear that these holes have a very short life time (10^{-14} to 10^{-15} sec) and that the survival of these excited particles depends on their kinetic energy: fast particles have a better chance to reach the surface and to carry outside their electronic excitation and become ionized afterwards. This effect could enhance the relative intensity of fast particles when the sample is in a channelling direction.

The emission processes of polyatomic ions are not well known. The dependence of this emission with the orientation of the crystal under the primary beam shows that the contrast is lower for Al_2^+ and Al_3^+ than for Al^+ but it is approximately the same for Al_4^+ and Al_5^+. To explain this, we can only speculate on the probability of ejection of a polyatomic ion. A simple idea is that in order to sputter a polyatomic cluster, several atoms sitting at the surface of the target have to receive an impulse directed in the outward direction and compatible with the binding energy of the cluster[26] at the same time. If the primary beam is in a channelling direction, the collision cascades are initiated deeper, the collisions tend to be "thermalized" and there is a better chance to have correlated collisions at the surface. As a consequence, there might be a higher probability for a simultaneous ejection of neighbouring atoms with the adequate energy; this could be the case for Al_2^+ and Al_3^+. Of course, it is difficult to make these assertions quantitative and it is difficult to understand exactly why Al_4^+ and Al_5^+ ions do not exhibit the same behaviour as Al_2^+ and Al_3^+.

It is worthwhile to note that the intensities of polyatomic ions are the same before and after the melting point. This implies that the roughness of the surface does not play a very important part in the formation of clusters.

5 CONCLUSION

From our experiments, it is clear that it is necessary to study the temperature dependence with monocrystalline targets. Secondary emission from polycrystalline samples[3,9,11] provides us an average on many grains which does not give a direct insight on the processes involved. According to the grain sizes and the grain growth with the temperature, the emission may be averaged on a small number of grains and lead to fluctuating results.

At first sight, secondary emission and sputtering are following similar temperature and orientation behaviour. The details of the energy distribution of secondary ions could be better understood if we knew in each case the energy distribution of neutral particles. However, we have to keep in mind that the characteristic properties of ion emission could also intervene to modify the shape of the distribution.

In general, it seems that the known processes of ion emission and sputtering can explain all the features observed; the liquid state itself does not give results very different than those obtained in a solid state.

The interpretation of the results on polyatomic species are more speculative but it seems quite certain that the surface defects do not play a significant part since there is no great difference between the emission of the solid and the liquid states.

REFERENCES

1. J. J. Ph. Elich, H. E. Roosendaal and D. Onderdelinden, *Rad. Effects* 10, 175–184 (1971).
2. J. J. Ph. Elich, H. E. Roosendaal and D. Onderdelinden, *Rad. Effects* 14, 93–100 (1972).
3. R. C. Bradley, *Jour. Appl. Phys.* 1–8 (1959).
4. V. M. Agranovitch and D. D. Odintsov, *Sov. Phys. Dokl.* 10, no. 6, Dec. 1965 (Trans. from *Dokl. Akad. Nauk. SSS* 162, 778–780, June 1965).
5. O. I. Kapusta, S. Ya. Lebedev and N. M. Omel'yanouskay, *Bull. Acad. Sci. USSR Phys. Ser. V* 35, 239–244 (1971).
6. I. N. Evdokimov and V. A. Molchanov, D. D. Odintsov an V. M. Chicherov, *Sov. Phys. Sol. State* 8, 2348 (1967) (Tr from *Fizika Tverdogo Tela* 8, 2939–2944 (1966).
7. I. N. Evdokimov, V. A. Molchanov, D. D. Odintsov and V. M. Chicherov, *Sov. Phys. Dokl.* 12, 1050 (1968) Trans *Dokl. Akad. Nauk. SSSR* 177, 550–553 (1967).
8. C. E. Carlston, G. D. Magnuson, A. Comeaux and P. Mahadevan, *Phys. Rev.* 138, 3 (1965).
9. G. E. Chapman, B. W. Farmery, N. W. Thompson and I. H. Wilson, *Rad. Effects* 13, 121–129 (1972).
10. Yo. V. Martynenko, *Sov. Phys. Solid. States* 8, 515 (196 Trans. from *Fizika Tverdogo Tela* 8, 637–642 (1966).
11. R. C. Krutenat and C. Panzera, *Journal of Appl. Phys.* 41 4953 (1970).
12. J. Lindhard, *Mat. Fys. Mett. Dan. Vid. Selsk.* 34, no. 14
13. M. Bernheim, Ion Surface Interaction Conference, Garching, 1972.
14. R. Castaing, J.-F. Hennequin, L. Henry, G. Slodzian, *Focusing of Charged Particles,* Vol. 2, Ed. by Septier (Academic Press, New York, 1967).
15. G. Blaise and G. Slodzian, *C.R. Acad. Sci. Paris* 271, 1216–1219 (1970).
16. P. Joyes, *J. Phys. Paris* 29, 483–487 (1968).
17. O. I. Kapusta, S. Ya. Lebedev, N. M. Omel'yanovskaya *Sov. Phys. Sol. State* 12, 707 (1970).
18. M. V. Bukhanov, V. E. Yurasova, A. A. Sysoev, G. V. Sar and B. I. Nikolaev, *Sov. Phys. Sol. State* 12, 313 (1970).
19. I. A. Abroyan, V. P. Lavrov and A. I. Titov, *Sov. Phys. S State* 7, 2557 (1966).
20. E. Dennis and R. J. Macdonald, *Rad. Effects* 13, 243 (19
21. Yo. V. Martynenko, *Phys. Stat. Sol.* 15, 767 (1966).
22. S. Ya Lebedev *et al., Sov. Phys. J. Tech. Phys.* 34, 1101 (1964).
23. A. A. Adylov *et al., Sov. Phys. Sol. State* 11, 1441 (1970
24. R. S. Nelson, *Phil. Mag.* 11, 291 (1965).
25. R. V. Stuart, G. K. Wehner and G. S. Anderson, *J. of Ap Phys.* 40, 803 (1969).
26. G. Staudenmaier, *Rad. Effects* 13, 87 (1972).

INFLUENCE OF CHANNELLING ON SECONDARY ION EMISSION YIELDS

M. BERNHEIM

Laboratoire de Physique des Solides, associé au CNRS, Bât. 510, Université Paris-Sud, 91405 Orsay, France

The sputtering yield of a single crystal changes very sharply with the relative orientation of the primary beam and the lattice. This is related to the channelling effect of the primary ion through the lattice.

The secondary ion emission yield varies when a monocrystalline target is rotated around an axis normal to the surface and the incidence angle of the primary beam is kept constant at about 60 degrees.

The intensities of different types of ions, monoatomic ions of 0 to 15 eV and fast ions 100 eV, 200 eV, 500 eV, multi-charged and polyatomic species, were recorded as a function of the target orientation.

We investigate different ways of reducing the variation of penetrations of the primary ion into the target to suppress the differences of sputtering in order to be able to analyse thin polycrystalline target by making use of the secondary ion emission.

INTRODUCTION

he sputtering yield S (number of target atoms ejected r primary ion) of a single crystal changes very sharply ith the relative orientation of the primary ion beam ith the lattice.[1-3] This is related to channelling effects the primary beam through the lattice. It is known at the total positive emission (composed mainly by flected ions) is also affected by the relative orientation the primary beam.[4]

In a similar way, we have observed that the emission different secondary ionic species (characteristic of e elements present in the target) vary with the imary beam channelling. This appears particularly the distribution images of polycrystalline targets otained with the Ion Microanalyser[5] (Figure 1). The fferences in contrast on these images comes either om differences in the sputtering yield or from fferences in ionization efficiencies (the ionization ficiency τ being defined as the number of secondary

ions to the number of atoms emitted by the target and the ionization yield K as the number of given kind ions per primary ions).

Instruments like the Secondary Ion Microanalyser are also used for measuring the concentration variations in the depth direction[6]: as the sputtering proceeds, deeper layers are brought to the surface and at the same time the secondary emission is recorded; it is possible to obtain an in depth concentration profile. When this type of analysis is applied to polycrystalline samples, the different sputtering yields reduce strongly the depth resolution and may disturb interface studies. Consequently, the reduction of such sputtering effects is of great importance.

2 EXPERIMENTAL CONDITIONS

These effects depend on the direction of the primary beam with the sample lattice. By turning the target around the axis of the collecting system (which is normal to the sample Figure 2), it is possible to follow this dependence without changing the collection efficiency.

The primary beam hits the target with an energy $e(V - U)$; the incidence angle $(90° - \theta)$ results from the deflection of the beam by the electrostatic field E (uniform field). This field accelerates to 3800 eV the secondary ions; the emission lens focuses these ions on the entrance aperture of the magnetic prism which limits their initial transverse energy to 0.7 eV.[5]

During the target rotation we can record the current of each ionic species. To avoid chemical emission,[7] the primary bombardment is applied on an area larger than

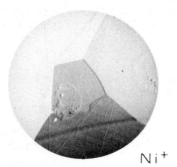

Ni$^+$

GURE 1   Ion image of a nickel polycrystalline sample mbarded by primary oxygen ions.

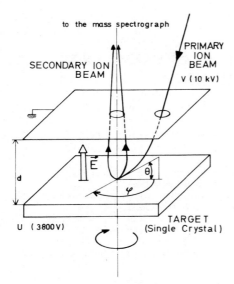

FIGURE 2   Schematic arrangement of the target chamber.

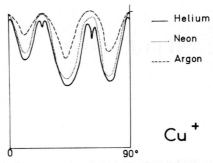

FIGURE 3   Secondary emission of an approximation (100) copper crystal bombarded by different primary ions (for a better comparison the secondary ion currents are normalized at $\varphi = 0°$).

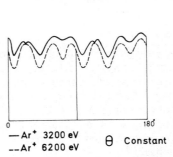

FIGURE 4   Effect of the primary ion energy (the secondary ion currents are normalized).

the geometric field viewed by the emission lens, in a good vacuum ($1 \ 10^{-7}$ torr during the bombardment) and with a high ion density (more than $5 \ \mu A/mm^2$).

We are able to verify the absence of chemical ion emission[7] (when an oxygen jet is directed on the target surface, there is no emission variation until the pressure reaches a given value depending on the ion density and the target nature).

3 EXPERIMENTAL RESULTS

Figure 3 shows the dependence of secondary emission as the target is rotated. We may define the "contrast" as the ratio of a maximum to a minimum. We did not observe any change in the emission contrast of metallic sample as a function of the duration of the bombardment; this means that metallic samples keep their crystalline structure under bombardment. This is not true for (100) and (110) silicon where no changes of the ion yield during the rotation could be observed. Apparently such samples lose their crystalline structure under bombardment. The same conclusion is reached by authors studying secondary electron emission[8,9] and focusons diagrams at room temperature.[10]

a Primary Beam Parameters
It is known that channelling is dependent on the nature of the channelled ion and its energy. So it is not surprising to see that the secondary emission exhibits similar features when the sample is bombarded with various primary ions (Figure 3) and when their energy is changed (Figure 4).

It is also possible to change the angle of the beam with the target surface by adjusting the voltage U and the distance d (Figure 2) but this adjustment modifies also the primary ion energy $e(V - U)$. So, on a copper single crystal (100), we observe (Figure 5) large modifications of the contrast when θ is changed. When the incidence angle $(90° - \theta)$ is close to $45°$ the contrast is increased. Deep and broad minima appear as the primary ion beam direction is near a (110) direction.

b Secondary Emissions
In the usual working conditions of the Ion Microanalyser ($U = 3800$ V, $V = 10.000$ V, $\theta = 28°$, primary argon ions) other experiments have been carried out:

On Figure 6 are drawn emissions of secondary ions leaving with different initial energies, an approximative (100) aluminium crystal. The variations are very similar if the scales are taken into account; the same result is obtained for molecular species (Figure 7).

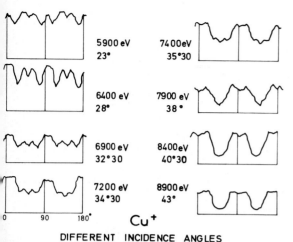

DIFFERENT INCIDENCE ANGLES

FIGURE 5   Effect of angle θ and primary ion energies $(V - U)$ variations.

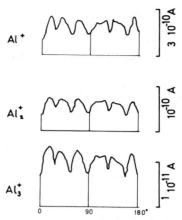

FIGURE 7   Aluminium single crystal molecular ion emission.

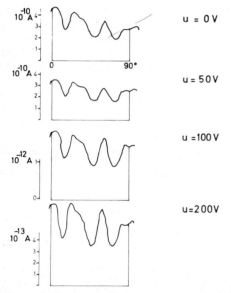

FIGURE 6   Aluminium sample (100 approximatively) high initial energy eu ions.

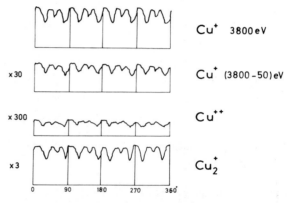

FIGURE 8   Copper single crystal different emissions (the Cu^{++} and Cu^+ (3800-50) eV are produced by excitation transfer).

transfer can be used to monitor the sputtering yield of pure copper, the sputtering and the secondary emission yields seem to vary in the same proportion so that the ionization efficiency τ stays constant. To obtain more accurate results, sputtering yield measurements would be of great interest.

c Thin Film Analysis

As we have already noticed, in depth distributions measurements of one polycrystalline target is difficult because of the sputtering yield differences.

The previous experiments suggest various ways of reducing the primary ion channelling and their consequences on the sputtering yield:

—by reducing the primary ion energy (but the primary ion density drops down at the same time)

For a (100) copper single crystal, we have noticed that Cu^{++} double charged ions follows variations similar to those of Cu^+ ion or Cu_2^+ molecular ions. The Cu^+ ions formed as the Cu^{++} ions by excitation transfer[11] vary also in the same way.

According to these experiments, the channelling of the primary beam seems to be responsible for the secondary ion yield variations. Particularly, if we assume[12] that the emission of copper ions by excitation

—by using very low incidences angles (ion beams tangent to the target surface are difficult to obtain with a uniform density).

—by using large molecular ions big enough to be unable to channel through the target lattice. Diatomic primary ions are not convenient to that purpose as shown on Figure 9.

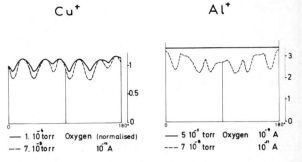

FIGURE 10   Bombardment by argon ions under oxygen pressure.

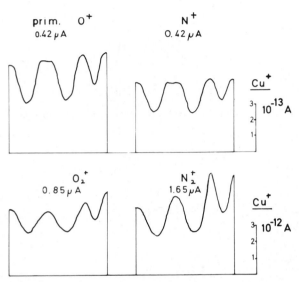

FIGURE 9   Effect of a diatomic primary beam on the copper emission.

Similar effects can be obtained if the chemical bondings on the target surface are modified, for instance, by directing on the target active gases like oxygen during the ion bombardment. In the case of aluminium and zirconium single crystals, all lattice effects disappear (Figure 10). Thus in such conditions, we are able to analyse thin polycrystalline films of these metals. It is worthwhile noticing that if we bombard such samples with oxygen ions O^+ or O_2^+ in a good vacuum (avoiding to direct on the target surface neutral oxygen atoms) the crystal effects are not appreciably wiped away.

But this method does not apply for copper (Figure 10). In this case by reducing the primary ion density, energy or incidence and blowing oxygen on the sample, crystalline effects are always observed.

4 CONCLUSION

We have tried to show some effects on the primary ions channelling and their relations to ion emission or to sputtering. Different methods can be used for reducing the primary ion channelling and their effect on sputtering but till now we are unable to propose a general way.

REFERENCES

1. G. D. Magnusson and C. E. Carlston, *J. Appl. Phys.* **34**, 3267 (1963).
2. J. M. Fluit, P. K. Roll and J. Kistemaker, *J. Appl. Phys.* **34**, 690 (1963).
3. D. Onderlinden, *Appl. Phys. Letters* **8**, 189 (1966).
4. A. Van Wijngardeen, E. Reuther and J. N. Bradford, *Can. J. Phys.* **47**, 411 (1969).
5. G. Slodzian, *Ann. Phys.* **9**, 591 (1964).
6. P. Contamin and G. Slodzian, *C.R. Acad. Sci. Paris* **267**, 805 (1968).
7. D. Guenot, Diplôme d'Etudes Supérieures, Paris 1966.
8. B. Fagot and Ch. Fert, *C.R. Acad. Sci. Paris* **258**, 1180 (1964).
9. J. P. Fontbonne, N. Colombie and B. Fagot, *C.R. Acad. Sci. Paris* **270**, 1573 (1970).
10. G. S. Anderson, G. K. Wehner and H. J. Olin, *J. Appl. Phys.* **34**, 3492 (1963).
11. J.-F. Hennequin, G. Blaise and G. Slodzian, *C.R. Acad. Sci. Paris* **268**, 1507 (1969).
12. M. Bernheim, G. Blaise and G. Slodzian, *J. Ion Phys. Mass. Spectr.* (10) (1972).

ENERGY DISTRIBUTION OF SECONDARY IONS FROM 15 POLYCRYSTALLINE TARGETS

Z. JURELA

Boris Kidrič Institute of Nuclear Sciences, Beograd, Yugoslavia

Fifteen metallic and semiconductor polycrystalline targets of 99.9% purity were bombarded with an Ar^+ ion beam of 40 keV energy. The energy distribution of different kinds of secondary ions of basic target material (atomic and molecular, positively and negatively charged), as well as of negative ions (C^-, C_2^-, O^- and O_2^-) formed by sputtering the surface impurities from bombarded targets was measured. The energy distribution of secondary ions was measured in the range of $E_2 > 1$ eV. The upper measurement limit depended on the yield of observed ion species and it reached $E_2 \leqslant 3000$ eV. The shape of the energy distribution curves could be explained satisfactorily by the law for the non-adiabatic emission process.

1 INTRODUCTION

Secondary ion emission is studied in connection with the determination of yields of different ion species. It is possible to draw conclusions about the bulk and surface purity of the bombarded target, as well as of the composition of residual gas from the spectra of secondary ions. Secondary ion emission under the conditions of atomically clean surface may be used for quantitative and semiquantitative chemical analysis of all kinds of solid targets.

The second approach to the study of the emission of secondary ions involves the measurements of their energy distribution. The results of such measurements are of great help for establishing the mechanism of secondary ion emission. Energy distribution without mass analysis were measured over the whole range[1,2] $0 < E_2 < E_1$, where E_1 is the energy of a primary ion beam. The appearance of peaks in the high-energy part of the spectrum was explained satisfactorily by the theory of scattering in binary collisions.[3,4] Conversely, the low energy peaks of sputtered neutrals[5] and secondary ions[6] could be explained only by the process of local evaporation at impact zones of primary particles. Degrees of ionization of sputtered particles for some conductive, semiconductive and nonconductive targets observed in that aspect[7] gave realistic temperatures of sputtering centres. Somewhat lower values of those temperatures[8] were obtained from a study of the dependence of the sputtering coefficient on the target temperature. Andersen and Hinthorne[9] have recently given the thermodynamic description of the secondary ion emission. They obtained good results by applying it to the quantitative analysis of several samples, supposing that the local temperature is 11,000°K. However, that value is considerably higher than the critical temperature for the majority of their samples. Out of this, it may be concluded that the emission of secondary ions should be studied from both aforementioned aspects. Most of the measurements, in which the bombarded surface was relatively clean[10-12] were over a range of several tens and hundreds electron-volts. Exceptionally clean experimental conditions and a wide range of measurements were seldom accomplished.[13,14] A more sensitive detector system was used in order to extend the range of measurement of the energy distribution of secondary ions up to $E_2 = 3000$ eV.[15,16]

The simplified focused-cascade scheme of sputtering given by Veksler[17] anticipates knees and inflection points in the energy distribution curves. However, if with primary ion energy $E_1 > 10$ keV the energy distribution of secondary ions was measured in the region far above the energy allowed by the focuson theory ($E_2 \leqslant E_f$) the assumptions made by Thompson[18] for sputtering neutrals seem more justifiable. His random cascade model assumes angular isotropy of flux of dislocated atoms in a solid and the absence of any kind of flux dependence on the coordinates. Particle ejection results principally from the generation of random collision cascades by the bombarding ion. It is also assumed that the energy in the cascades results from binary collisions and the mean free path is independent of energy. This leads to the distribution of E_2^{-2} in the range $E_b < E_2 < 10^3$ eV where E_b is the binding energy of atoms in the crystal lattice. A good agreement between theoretical and experimental results[18-20] is obtained. Furthermore, generation of focused collision sequences, both single and assisted by the cascade predicts the spectrum of ejected particles to behave according to the E_2^{-1} law in the range $E_2 \leqslant 10$ eV.

Sputtering and secondary ion emission are non-adiabatic processes.[21,22] All past results of the measurements of energy distribution and the mean energy of sputtered neutrals and secondary ions, as well as the results of this work point to that conclusion. For non-adiabatic processes of secondary ion emission Veksler,[17] starting from the simplified cascade-focuson scheme, obtained for the relation of energy distribution of ions and neutrals the following expression

$$d^2 W_i = \gamma \times E_2^{1/2} \times d^2 W_a, \qquad (1.1)$$

where W_i and W_a are the probabilities for formation of sputtered ions and neutrals by focused cascades, γ is the numerical coefficient, and E_2 the energy of sputtered ions and neutrals.

Measurements extended up to 3000 eV. Targets were chosen in such a way that every group of the periodic system of elements was represented by at least one element. This should permit the observation of the influence of chemical properties of various elements on the energy distribution of secondary ions. The comparison of several elements from the same group of the periodic system serves for the determination of that dependence on their physical properties. It is particularly interesting to compare the results for metals and semiconductors. The energy distribution from the following metals was studied: Mg, Al, Mn, Co, Ni, Cu, Mo, Ag, Ta, W, Pt and Au, and from semiconductors: C, Si and Ge.

2 EXPERIMENTAL CONDITIONS

The apparatus and experimental conditions are the same as in earlier measurements:[15] a primary Ar^+ ion beam of 40 keV energy, an incidence angle of $45°$, a current density of 60 to 100 $\mu A/cm^2$, and a residual pressure of 2×10^{-6} torr. Secondary ions were additionally accelerated to $eU_a = 1000$ eV. The differential measurement method was applied. The resolving power of the magnetic analyser was $R \simeq 400$, and of the retarding potential about 330. In this way and by means of this system the energy analysis of secondary ions with initial energies $E_2 > 1$ eV is possible. The error in measurement decreases with the increase of initial energy and at $E_2 = 1000$ eV it is $\pm0.3\%$ while at $E_2 = 10$ eV it is $\pm15\%$. The system of measurement is simple. The desired retarding potential is easily adjusted and the spectrum of secondary ions whose total energies are higher than the effective height of the potential barrier is then registered. Only a slight mistake in determination of the zero energy level is possible. This is due to rather high retarding potential (>1250 V) and to the fact that the effective height of the potential

barrier approximates 0.796 x retarding potential. The maximal discrepancy of some experimental points in several successive measurements is less than $\pm5\%$. All energy distribution curves have been drawn on the plotter of the computer. Therefore, the points are connected with straight lines. It slightly stresses the structure of the distribution curves.

The extension of the range of measurements to low energies (0.1 eV $< E_2 \leqslant 1$ eV) is possible by using the same system. It is only necessary to decrease the voltage for the additional acceleration of secondary ions. Such an extension of the range of measurement to low energies is possible only for targets with extremely high yield of secondary ions. The other way is to increase the resolving power of the system by decreasing the width of the entrance slit. This is one of the aims of future work.

3 ENERGY DISTRIBUTION OF SECONDARY IONS

Only some sorts of secondary ions of target material and surface impurities are included in the measurements. It is known from previous measurements[15,16] that the spectrum of secondary ions of target material consists of multiply- and singly-charged, positive and negative, mono- and poly-atomic ions. The ions of oxygen, carbon, oxide, carbide are detected in the spectrum of surface impurities. We shall observe only such species which were obtained with most targets and which had a sufficient yield to permit the measurement of the energy distribution over a wide energy range.

3.1 Monatomic Ions of Target Material
Monatomic singly charged ions are dominant in the spectra of positive and negative secondary ions.[15,16] Naturally, it depends a great deal on the purity of a bombarded surface. The yield of positive ions from the majority of elements is much higher than that of negative ions.[23] For the latter the same order of magnitude is found only for targets whose atoms have great electron affinity, such as graphite, silicon, gold etc.[16,23] In contrast, from targets whose atoms have negative electron affinity it is impossible to obtain negative ions of target atoms, as from magnesium and manganese.[16] This results from the instability of such ions.

The energy distribution of positive ions of target materials is shown in Figures 1 and 2, and for negative ions in Figure 3. As the range of measurement extends from 1 eV $< E_2 \leqslant 3000$ eV the diagrams are given in log–log scale. Distribution curves are mainly monotonically decreasing functions with the knee in the range 10 eV $< E_2 < 100$ eV. This points to the fact that the secondary ion emission cannot be explained by a single

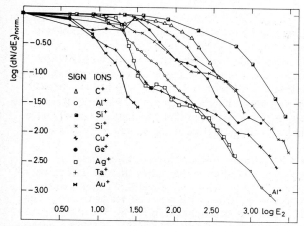

FIGURE 1 Energy distribution curves for positive atomic secondary ions of target material.

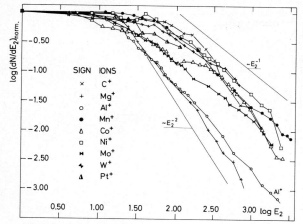

FIGURE 2 Energy distribution curves for positive atomic secondary ions of target material.

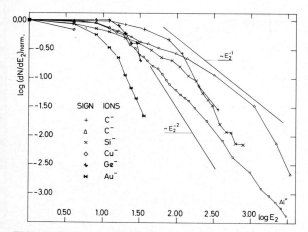

FIGURE 3 Energy distribution curves for negative atomic secondary ions of target material.

mechanism. The shape of the low-energy part of the curves approaches the function $f_1(E_2) \sim \exp(-mE_2)$, where $0.01 < m < 0.10$, and E_2 is the energy of the secondary ions. The shape of the high-energy part of the distribution curves, on the right of the knee, approaches the function $f_2(E_2) \sim E_2^{-n}$, where $1 < n < 2$, and for the Al^+ curve it is $n \simeq 1.4$. So the high-energy part of Al^+ and some other curves shown in Figures 1 and 2, in the wide energy range $20 \text{ eV} < E_2 \leqslant 3000 \text{ eV}$ approaches the law $f_2(E_2) \sim E_2^{-1.4}$.

The energy distribution curves of C^+, Mg^+, Si^+ and Au^+ ions have somewhat sharper drop off. For those four curves, the exponent n has the following values: 1.7, 1.6, 1.75 and 2.0, respectively. The energy distribution curves of ions from other targets have a slower drop off. This is particularly evident for the curves of Mn^+, Ni^+, Ta^+, W^+ and Pt^+ ions. For the first three curves the exponent n has the value 1.1, and for the last two 0.9. (Staudenmaier's result for W^+ ions is $n = 0.5$, Dennis' for Cu^+ ions is $n = 1.35$.) The common characteristic of the other six distribution curves (Si^+, Co^+, Cu^+, Ge^+, Mo^+ and Ag^+) is that beside knees they also have inflection points. Further on, the curves for Co^+, Ge^+ and Ag^+ beside maxima at low energy also show maxima at $\simeq 55$ eV for Ag^+, $\simeq 150$ eV for Co^+ and at $\simeq 1000$ eV for Ge^+.

The energy distribution curve of C^+ ions, shown in Figure 1 was obtained from pyrolytic graphite and that in Figure 2 from common graphite. They both are much alike, only the yield of C^+ ions obtained from common graphite is much higher. This points to the influence of the target structure on the yield of secondary ions. Two energy distribution curves of Si^+ ions are shown in Figure 1. The curve marked x was obtained for the normal conditions of bombardment, i.e. such as performed for other targets. The other curve was obtained as a result of a four to five times longer bombardment of the same target. In this case the conditions were cleaner, but the surface was more damaged. One may conclude that the shape of some energy distribution curves may considerably depend on the state of the surface of a bombarded target. The mentioned effect could be explained by the presence of an oxide layer on the target surface in the first measurement and its removal in the second one.

The energy distribution of negative ions was measured for five targets: graphite, silicon, copper, germanium and gold. The yield of negative ions from other targets was insufficient for the energy distribution measurements. The shape of the curves, shown in Figure 3, is similar to those in Figures 1 and 2. The range of measurement for graphite and gold is the same as for positive ions, and for silicon and germanium somewhat narrower. Other than with positive ions, the yield of negative ions for pyrolytic graphite is higher than for the common one, and the range of measurements is one order of magnitude larger. It is characteristic for all the curves in Figure 3, that their high-energy part has a sharper drop off, than is the case with the corresponding curves for positive ions. In this case the values for the coefficient n are: 1.9 for C^-, Si^- and Ge^-, and 2.7 for Au^-.

3.2 Molecular Ions of Target Material

Taking into account only the peak intensities in the spectrum of secondary ions it may be concluded that the yield of molecular ions from some targets approaches 45% of the yield of monatomic ions.[15] Since the mean energy of diatomic ions is much smaller than for the monatomic ions, the yield of diatomic ions is, therefore, correspondingly smaller. Thus, the yield of diatomic ions for the majority of targets is <1% of the yield of monatomic ions. A higher ratio is found for silicon $\simeq 2\%$, for nickel 4%, for molybdenum and germanium 5% and for cobalt and tungsten 7%. Triatomic molecular ions were detected only from some targets, and their yield was approximately for one order of magnitude smaller than of diatomic ions. Similar results were obtained by some other authors.[24-26] However, the yield of triatomic and higher clusters, for special experimental conditions, could be of the same order of magnitude as the yield of monomers or dimers.[14,27-29] Such conditions for negatively charged clusters are obtained by covering the spot on the bombarded sample with monolayer of cesium atoms,[27,28] which drastically decreases the work function. The increased yield of positively charged clusters[14,29] and clusters of compound molecules[30] are, probably, caused by the high density of dissipated energy in the surface layers, similar as in some laser irradiation.[31]

The energy distribution curves of diatomic ions are shown in Figure 4. As the range of measurements is rather small, $E_2 \leqslant 80$ eV, the diagrams are presented on linear scale. The shape of the energy distribution curves of diatomic ions in the observed region is similar to that for Al^+ ions, but they drop off much sharper. Si_2^+ and Mn_2^+ curves are exceptions and partially also the Mg_2^+ curve. Apart from that, the first two curves have their second maximum at the energy $E_2 > 30$ eV.

3.3 Monatomic Ions of Surface Impurities

We shall only observe the energy distribution curves of C^- and O^- ions, which were measured over a wide energy range. The distribution curves of C^- ions from different targets are shown in Figure 5, and those of O^- ions in Figure 6. Distribution curves of C^- ions decrease

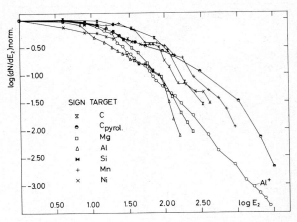

FIGURE 5   Energy distribution curves for secondary C^- ions from different targets.

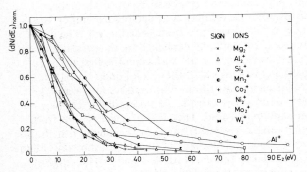

FIGURE 4   Energy distribution curves for positive molecular secondary ions of target material.

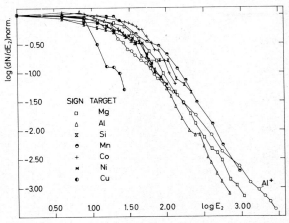

FIGURE 6   Energy distribution curves for secondary O^- ions from different targets.

approximately like the corresponding curves of negative ions and far sharper than the curves of positive monatomic ions of the target material.

Distribution curves of O^- ions, with the exception of the copper target, have a similar course. Namely, the slope coefficient n of all curves has close values: 1.7 for Si, 2.0 for Mg and 2.1 for other targets (Al, Mn, Co, Mg and Cu). The high yield of O^- ions was obtained from the targets with high affinity toward oxygen, such as Mg, Al and Mn. From other targets the yield of O^- ions is considerably smaller, while from the noble metals (Ag and Au) O^- ions were not detected. This fact leads to the assumption that the emission of O^- ions originates from the oxide layer on the target surface, and not from the layer of adsorbed gases. Otherwise, the yield of O^- ions should have been approximately the same from all targets, or even higher from those having higher sputtering coefficient (for example from Au targets). According to the above said, for $C^-, \ldots, C_n^-$ ions it is more probable that they originate from the carbide layer which covers the target surface, than from the layer of adsorbed gases.

3.4 Molecular Ions of Surface Impurities

Out of a large number of species, we shall study only C_2^- and O_2^- ions. A very sharp drop off, far sharper than for any other kind of ions, is characteristic for all the distribution curves of C_2^- ions. This means a small half-width of the distribution and the low mean energy of C_2^- ions, considerably lower than for Al^+ ions. The energy distribution of O_2^- ions was measured only from Mg and Mn targets. From other targets the O_2^- peak was either not detected, or its intensity was insufficient for energy distribution measurements. In that case distribution curves have somewhat slower drop off, than for C_2^- ions. This kind of ions might have been formed by sputtering of complex oxides on the target surface.

4 COMPARISON OF THE OBTAINED RESULTS WITH THEORY

By combining the results of Thompson[18] regarding the shape of energy distribution of sputtered neutrals and those of Veksler[17] regarding the relation of energy distribution of sputtered ions and neutrals, it follows that the energy distribution curves of secondary ions should obey the law $\gamma \times E_2^{-3/2}$ in the energy range $10 \text{ eV} < E_2 < 1000 \text{ eV}$. The coefficient γ represents the ratio of the number of sputtered ions and neutrals with unit energy ($E_2 = 1 \text{ eV}$). The numerical values of γ calculated according to the expression given by Veksler change slightly in dependence on the atomic

number Z of the targets. The greatest discrepancy in the value of the coefficient γ for the observed 15 elements is for magnesium and tungsten. The following ratio was obtained $\gamma(Mg)/\gamma(W) < 5$. This is far below the value known from the earlier measurements[23] for the ratio of the degrees of ionization of sputtered particles, $\alpha_+(Al)/\alpha_+(Au) \simeq 10^4$. As in this case, we are interested only in the shape of the energy distribution curves of secondary ions, and not in their absolute values, we shall only compare the experimental results with the law $E_2^{-1.5}$.

The value of the exponent n obtained theoretically have the high-energy parts of 2 distribution curves (Ge^+ and Ag^+) shown in Figures 1 and 2. Somewhat smaller values were obtained for the Al^+ and Si^+ curves, and higher values for Mg^+, C^+ and for the second curve of Si^+. The greatest discrepancy from the theoretical value of the exponent are found for Mn^+, Cu^+, Mo^+ and Ta^+ curves, and also for W^+, Pt^+, Co^+ and Au^+ curves. A considerable difference in the slope of some curves of the energy distribution may be partially caused by different surface conditions on particular targets (see 2 Si^+ curves), crystallographic structure and the other physical properties of bombarded targets. For every investigated element the energy range of random collision cascades and focused collision sequences should be precisely determined.

The experimental data on the energy distribution of sputtered neutrals are scarce. Only in few papers[18-20] the range of measurements approaches $\geqslant 10^3$ eV (copper and gold). As it was possible to measure the energy distribution of secondary Au^+ ions only in rather a narrow energy range, the comparison was restricted to the distribution curves of Cu neutrals and Cu^+ ions. This comparison is shown in Figure 7. The distribution curves are normalized and presented on log-log scale. Curve 1 represents the energy distribution of sputtered Cu atoms,[19] and curve 2 the energy distribution of Cu^+ secondary ions, obtained in these measurements. Curve 3 represents the theoretical distribution which Cu^+ ions should have according to the experimental curve for Cu atoms. It has been derived from the curve 1 by multiplying it with $E_2^{0.5}$. In the range $E_2 < 135$ eV the yield of Cu^+ ions is smaller and in $E_2 > 135$ eV higher than expected. Maximal discrepancy between the experimental and expected values in the first range is smaller than factor 2, and in the second range smaller than 2.5. Curve 4 represents the energy distribution of Al^+ ions obtained in these measurements. A good agreement of this curve with the theoretical law $E_2^{-1.5}$, as well as with curve 3 was noticed. A slight discrepancy of results may be caused by the difference of experimental conditions in those two experiments.

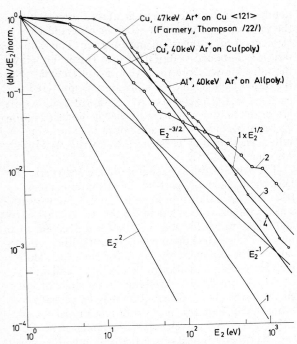

FIGURE 7   Comparison of energy distribution curves for
sputtered copper neutral atoms and ions from targets
bombarded with Ar$^+$ ions.

5 CONCLUSION

The majority of experimental energy distribution curves
of monatomic secondary ions is close to the law $E_2^{-1.5}$.
To check that law experimental data for energy distributions of sputtered neutrals are necessary. Up to now
such data are scarce. The most probable energy of all
the distribution curves of secondary ions is about 1 eV.
Energy distribution curves for positive atomic secondary
ions of target material show a slight periodic dependence
on the atomic number Z. Apart from that, the above
mentioned curves, for the targets from the same group
of the periodic table, have narrower distribution for
the elements with higher atomic number. Sharper drop-
off of distribution curves and smaller halfwidth of the
distribution are characteristic for all kinds of molecular
ions.

REFERENCES

1. D. P. Smith, *J. Appl. Phys.* **38**, 340 (1967).
2. B. V. Panin, *Zh. Eksp. Teor. Fiz.* **42**, 313 (1962).
3. S. Datz and C. Snoek, *Phys. Rev.* **134**, A347 (1964).
4. P. Dahl and J. Magyar, *Phys. Rev.* **140**, A1420 (1965).
5. M. W. Thompson and R. S. Nelson, *Phil. Mag.* **7**, 2015
   (1962).
6. F. Kirchner und H. J. Klein, *Z. Naturforschg.* **22a**, 577
   (1967).
7. Z. Jurela, *Rad. Effects* **13**, 167 (1972).
8. R. S. Nelson, *Phil. Mag.* **11**, 291 (1965).
9. C. A. Andersen and J. R. Hinthorne, *Science* **175**, 853
   (1972).
10. A. Benninghoven, *Z. Physik* **199**, 141 (1967).
11. H. E. Stanton, *J. Appl. Phys.* **31**, 678 (1960).
12. A. A. Adilov, V. I. Veksler i A. M. Reznik, *Fiz. Tverd.
    Tela* **11**, 1779 (1969).
13. J. F. Hennequin, *J. Physique* **29**, 655 (1968).
14. R. F. K. Herzog, W. P. Poschenrieder, F. G. Ruedenauer and
    F. G. Satkiewicz, 15th Ann. Conf. Mass Spectrom. Denver,
    1967, p. 301.
15. Z. Jurela and B. Perović, Proc. Int. Conf. on Phenomena in
    Ionized Gases, Vienna 1967, p. 30; *Canad. J. Phys.* **46**, 773
    (1968).
16. Z. Jurela, Fourth Yugoslav Symposium on Physics of
    Ionized Gases, Herceg Novi 1968, p. 14; Proc. Ninth Int.
    Conf. on Phenomena in Ionized Gases 1969, Bucharest, p. 89.
17. V. I. Veksler, *Fiz. Tverd. Tela* **11**, 3132 (1969).
18. M. W. Thompson, *Phil. Mag.* **18**, 377 (1968).
19. B. W. Farmery and M. W. Thompson, *Phil. Mag.* **18**, 415
    (1968).
20. G. E. Chapman, B. W. Farmery, M. W. Thompson and
    I. H. Wilson, *Rad. Effects* **13**, 121 (1972).
21. L. N. Dobrecov, M. V. Gomojunova, Emisionaja Elektronika,
    Nauka, Moskva 1966, pp. 452–465, (in Russian).
22. E. Ja. Zandberg, N. I. Ionov, Poverhnostnaja ionizacija,
    Nauka, Moskva 1969, (in Russian).
23. Z. Jurela, *Atomic Collision Phenomena in Solids*. Ed.
    D. W. Palmer, M. W. Thompson and P. D. Townsend
    (North-Holland, Amsterdam, 1970) p. 339.
24. A. P. M. Baede, W. F. Jungmann and J. Los, *Physica* **54**,
    459 (1971).
25. G. Staudenmaier, *Rad. Effects* **13**, 87 (1972).
26. E. Dennis and R. J. MacDonald, *Rad. Effects* **13**, 243
    (1972).
27. V. E. Krohn, Jr., *J. Appl. Phys.* **33**, 3523 (1962).
28. G. Hortig and M. Müller, *Z. Phys.* **221**, 119 (1969).
29. G. Blaise and G. Slodzian, *C.R. Acad. Sci.* **266**, B1525
    (1968).
30. G. D. Tanciirev and E. N. Nikolaev, *Zh. Eksp. Teor. Fiz.*
    13, 473 (1971).
31. V. S. Ban and B. E. Knoh, *Int. J. Mass Spectrom. Ion Phys.*
    3, 131 (1969).

DETERMINATION OF THE NEGATIVE ION YIELD
OF COPPER SPUTTERED BY CESIUM IONS

M. K. ABDULLAYEVA, A. K. AYUKHANOV and U. B. SHAMSIYEV

Institute for Electronics, Uzbek Academy of Sciences, Tashkent, USSR.

This paper describes the determination of secondary ion yields for negative ions obtained by bombardment of copper by cesium ions. Stable and reproducible surface conditions are reached by high rate sodium deposition simultaneously with sputtering. An optimum thickness of sodium corresponding to about one monolayer is found. Total negative ion yields K_{Σ}^- are measured by a double modulation technique. Individual negative ion yields K_i^- are then found by mass spectrometrically determining the various negative ion intensities, the sum of which relates linearly to K_{Σ}^-. This method is based on the assumption of an equal angular and energy distribution of all sputtered negative ions. Data are given for K_{Σ}^- and K_{Cu}^- and K_0^-. The dependence of K_i^- on primary ion energy (500 to 2500 eV) is similar to ordinary sputtering which points to the same basic mechanism in both cases.

INTRODUCTION

The mass spectrometric study of negative ion sputtering of some solids during bombardment by positive alkali metal ions has revealed that the presence of an alkali film which is deposited on the surface by the incident ion beam itself or from a special evaporator creates favourable conditions for target sputtering and surface contamination in the form of negative ions.[1-6] It was also shown in those works that during cesium ion bombardment of targets like Al, Zr, Si, Cu, CuBe, Ag, Au, Ni and graphite atomic and molecular negative ions sputtered from the target material have been observed. Special experiments have shown this to be in some cases ordinary cathode sputtering from the crystal lattice and not sputtering of chemical compounds on the surface which is accompanied by dissociation.[2-4]

All these experimental results show that reducing the surface work function partially transforms cathode sputtering of usually neutral atoms and multiple atom complexes into a negative ion state.

The results observed during the study of negative ion sputtering of graphite in the regime of Na film deposition on the sputtered surface show that even though the mass-spectrometric method fails to estimate the complete number of sputtered particles, it nevertheless provides interesting data on the atomic and molecular products of sputtering of the target itself.[6]

This brings us to the necessity to come to a quantitative description of the negative ion sputtering process. In this regard negative ion sputtering is an almost unstudied phenomenon.

Most important is the negative ion yield: $K_i^- = I_i^-/I_{\Sigma}^+$ where I_i^- = intensity of sputtered negative ions of definite mass; I_{Σ}^+ = intensity of primary incident ions of alkali metal. These coefficients would in a certain way characterize the composition of the surface layer in the target and the possibility of obtaining negative ions of one or another type under given conditions on the studied surface and the given parameters of the primary ion beam.

Measurements of the absolute value of K_i^- entail considerable difficulties and there exist only estimates of the yield for some types of secondary ions.[7,8]

This paper describes a method for determining the negative secondary ion yield of solids during their bombardment by positive ions in regime of alkali metal film deposition on the sputtered surface. We also give a result for the negative ion yield for individual ions and the values of the total negative ionization yield of copper during cathode sputtering in the case of its bombardment by positive ions of cesium within the energy range of 500–3000 eV.

EXPERIMENTAL FACTS

The process of determining negative ion yields comprises the following. The target for studies is bombarded by positive ions in a regime of alkali metal film deposition on its surface. The dependence of the total negative ion yield K_{Σ}^- on the incident ion energy is measured by the method of double modulation[9] at various alkali film depositions. The total negative ion yield K_{Σ}^- is determined by the ratio $K_\Sigma^- = \sum_i I_i^-/I_\Sigma^+$, where $\sum_i I_i^- =$ total current of all secondary negative ions which had been sputtered from the surface of the bombarded target; $I_\Sigma^+ =$ total current of primary incident ions.

Measurements of the total yield K_Σ^- are accompanied by observations of the composition of negative ions and determination of maximum heights of each peak in the spectrum for a certain incident ion energy. The height of all peaks is then summarized, and equalled to the total negative ion yield K_Σ^- at same energy. K_Σ^- is then split into the various K_i^- by approportioning according to the intensity for each type of ion within the mass spectrum.

Determination of the K_i^- value by the above-mentioned method is correct only when the angular distribution of all secondary ions is identical.

It is evident therefore that it makes sense to determine secondary ion yields only under given conditions on the bombarded surface. In vacuum of 10^{-6}–10^{-7} Torr conditions are indefinite. Hence the nature of the dependence of ion sputtering on the state of the surface was studied to explore the conditions on the surface of the target which give reproducible results.

After maximum de-gassing of the target the authors of this paper measured the total negative ion yield and composition of negative ions depending on energy of the cesium ions which bombard the copper target at various alkali film deposition rates. An alteration in the film deposition rate under permanent vacuum conditions results in a changing ratio between the number of alkali atoms and residual gas atoms which are deposited with the film. Hence the state of the bombarded target surface is changed deliberately.

The composition of the sputtered ions was determined with a mass-spectrometer, operating with electric scanning of the acceleration voltage, and an ion-electronic converter used to register the analysed particles.[1,6] Measurements of the total negative ion yield were based on the method of double modulation[9] within the sputtering chamber of the same mass spectrometric installation either simultaneously or alternately with observation of the ion composition.

The value of K_Σ^- was determined through independent measurements of the negative ion and the electron emission. The electrons were deflected to a separate collector by a magnetic field.[9] The secondary currents of electrons and negative ions were picked up by separate collectors, screened on two oscillographs and photographed.

A cine-camera was used to register the changing volt ampere characteristics of the summary negative ion emission and section of the mass spectrogram at a high film deposition yield (~15–20 monoatomic layers per second).

RESULTS

Figure 1a illustrates the dependence of values of the total ion yield (K_Σ^-) on the energy of primary cesium ions at various Na film deposition rates. The film deposition rates was estimated by measuring the filament current of the evaporator which corresponds to alkali metal deposition rate. Of course, if the film deposition rate changed it was necessary to conduct measurements at one and the same thickness of the film. However, at present it is not possible to register one and the same thickness of the film in the process of continuous film deposition. It is only possible to register the moment when the optimum thickness of the film is reached at a high-speed deposition rate at which the surface has a minimum work function value.[5] That is why at low deposition rates (~1 monolayer per second) we registered the value of secondary currents at regular time intervals. The value K_Σ^- at low Na film deposition rates was registered 30 seconds after the high-temperature flash of the target. At high-rate Na film deposition it was registered at the maximum value.

Figure 1b shows that with increased film deposition rate K_Σ^- first goes through a maximum value. At high deposition rate it reaches a permanent value which is reproduceable in repeated experiments. This change in the total ion yield reflects the changes which occur in the composition of the sputtered ions.

Figure 2 illustrates curves which characterize the change in the composition of sputtered ions depending on the Na film deposition rate during the bombardment of a copper target by cesium ions at energy of 2.500 eV. It is seen from the figure that alterations in K_Σ^- are accompanied by substantial changes in the composition of the sputtered ions. At low film deposition rates the composition of sputtered ions is very mixed. Figure 2 illustrates the behaviour of only some of them. It is very indicative that as the film deposition rate grows one group of ions disappears monotonously or by a transition through a maximum while another group grows monotonously to a certain stationary value. The first group characterizes the content of residual gases on the surface and in the deposited film while the second group characterizes the sputtering main target. For this reason the composition of the ions is less diverse at high rates of alkali film deposition. It does not change when the rate grows further, is reproduceable and apparently characterizes the maximum achievable stage of film purity in vacuum of 10^{-7} torr. Hence the composition of ions provides the best characteristic of the state of a target having an alkali coating of optimum thickness.[5]

The mass spectrum of negative ions sputtered from copper during bombardment by cesium ions at an energy

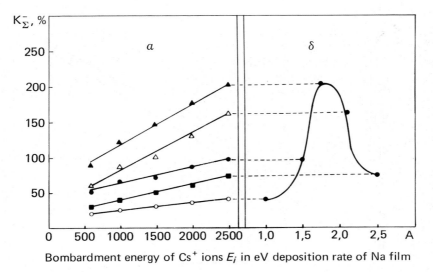

FIGURE 1   Dependence of the total negative ion yield K_{Σ}^{-} on the energy of the incident ions of cesium at various Na film deposition rates on the sputtered surface of copper.

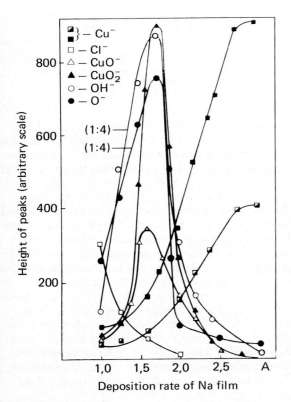

FIGURE 2   Changing composition of sputtered ions depending on the Na film deposition rate on the surface of copper which is bombarded by cesium ions at an energy of 2500 eV.

2000 eV comprises at the moment of optimum Na film thickness basically two very intense peaks belonging to the ions of the two isotopes of copper. Of the impurity ions in the surface the spectrum displays only O^- which constitute a negligible share of the total negative ion sputtering of copper (about 5–6%). Thus as a result of surface activation by high rate alkali film deposition the vacuum background in the spectrum goes down to a very low level and the data then obtained characterize the sputtering of the bulk itself. For these states of the bombarded surface we determined the differential coefficients of negative ions sputtered from copper in the form of Cu^- and O^- ions.

The dependence of the total negative ion yields of copper K_{Σ}^{-} on the energy of the incident cesium ions is registered at the moment when the Na film on the bombarded surface reaches an optimum thickness. The corresponding negative ion yield of K_{Cu}^- and K_O^- are illustrated in Figure 3. It is seen from the figure that the formation of negative ions Cu^- during the deposition of an alkali film on the surface of copper proceeds with great efficiency and grows on a linear basis along with increased energy of the incident ions of cesium. In this case the observed dependence of K_{Cu}^- on the energy of the incident ions of cesium once again shows that the negative ion sputtering of the target material from a surface deliberately polluted by an alkali film shows the same dependence on the energy of the incident ions as ordinary cathode sputtering. The value of the sputtering yield in this state of the bombarded surface

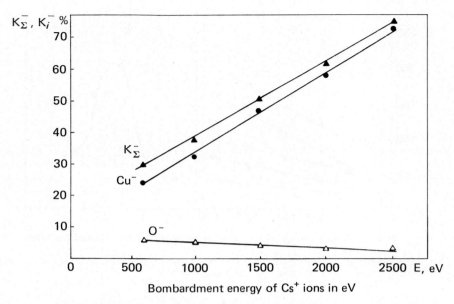

FIGURE 3   Dependence of total negative ion yield K_{Σ}^{-} and individual negative ion yield K_i^{-} on energy of the ions of cesium incident on copper at the moment when the Na film reaches its optimum thickness at high rate film deposition (about 15–20 monoatomic layers per second).

is apparently slightly smaller than that of the sputtering of an atomic pure surface which is due to the partial screening of the surface by the alkali film.

Attention is drawn to the fact that even at such conditions on the surface of the target when all the other negative ions of impurities disappear from the mass spectrum the emission of O^{-} ions continues at all achievable rates of Na film deposition.

Experimental data obtained in studying the influence of parameters of the bombarded target and incident ions on the mass spectrum of the negative ion sputtering of solids show that O^{-} ions occur during the interaction of primary ions with various particles on the surface of the sputtered target. Depending on which particle on the surface is responsible for O^{-} ion emission, these ions escape the surface of the sputtered target with different degrees of difficulty. Apparently it is a particle strongly connected with the surface which serves as the source of negative ion O^{-} emission that is observed in the regime of high rates of Na film deposition. According to Fogel this particle remained on the surface at a temperature of $1460°C.$[11]

The assessment of the relative ionization efficiency

of copper in such a process is highly interesting. For this it is necessary to determine the sputter yield K_{Σ}° of copper under bombardment by cesium ions at the same experimental conditions under which the negative ion yield for the various ions sputtered from copper was determined. Such measurements had not been conducted so far. A rough approximation can be achieved by using the results obtained in[10] by Olson and Smith by the radioactive indicator method. However, the values of the ionization yields thus obtained would be strongly diminished due to the partial screening, in our case, of the copper surface by the Na film.

For comparison Figure 4 illustrates the dependence of the sputter yield of pure copper which had been obtained in[10] and of the negative secondary ion yield for Cu^{-}, on the energy of the incident ions of cesium. The same figure shows the dependence the relative ionization efficiency defined as $\beta^{-} = K_i^{-}/K_{\Sigma}^{\circ}$ (with the above-mentioned reservations) on the energy of the primary ions. It is seen from this figure that β^{-} is dependent only to a small extent on the energy of the primary ions and grows slightly when the energy of the incident ions is reduced.

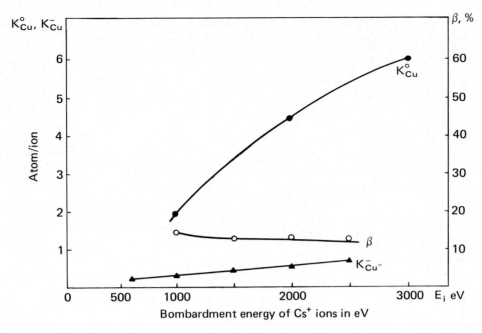

FIGURE 4   Dependence of neutral sputter yield $K^{\circ}_{Cu}{}^{10}$, negative ion yield $K^{-}_{\bar{C}u}$ and relative ionization efficiency $\beta_{\bar{C}u}$ on the energy of the incident ions of cesium ($\beta_{\bar{C}u} = K^{-}_{\bar{C}u}/K^{\circ}_{Cu}$).

REFERENCES

1. A. K. Ayukhanov and M. K. Abdullayeva, *Izv. AN SSSR, Ser. Phys.* **30**, 2000 (1966).
2. M. K. Abdullayeva and A. K. Ayukhanov, *Izv. AN Uz SSR, Ser. Phys. Mat. Nauk.* **4**, 60 (1969).
3. M. K. Abdullayeva and A. K. Ayukhanov, *Radiotechnika i Electronika.* **15**, 1263 (1970).
4. M. K. Abdullayeva and A. K. Ayukhanov, "Vtorichno-emissioniye i strukturniye svoistva tverdikh tel". *Izd. "FAN"*, (Tashkent, 1970).
5. M. K. Abdullayeva and A. K. Ayukhanov, *Izv. AN SSSR, Ser. Phys.* **35**, 404 (1971).

6. M. K. Abdullayeva, A. K. Ayukhanov, U. B. Shamsiyev and M. Z. Gafurova, "Vzaimodeistviye atomnikh chastits v veschestve i na poverkhnosti tverdogo tela". *Izd. "FAN".* (Tashkent, 1972).
7. V. I. Vexler and G. N. Schuppe, *JTF* **23**, 9, 1573 (1953).
8. V. I. Vexler and M. B. Benyaminovich, *JTF* **26**, 1671 (1956).
9. U. A. Arifov, 'Vzaimodeistviye atomnikh chastits s poverkhnostyu tverdoga tela", *Izd. "Nauka"* (Moskva, 1968).
10. N. T. Olson and H. P. Smith, *Raketnaya Tekhnika i Kosmonavtika*, **5**, 186 (1966).
11. Y. M. Fogel, R. P. Slabospitski and I. M. Karnaukhov, *JTF* **30**, 824 (1960).

OBSERVATION OF CLUSTERS IN A SPUTTERING ION SOURCE

R. F. K. HERZOG, W. P. POSCHENRIEDER† and F. G. SATKIEWICZ

GCA Technology Division, Bedford, Mass., USA

This paper reports previously unpublished results which were obtained in 1966. We systematically investigated the dependence of cluster ion intensities on the bombarding gases He^+, Ar^+ and Xe^+ (energies: 4 to 12 keV, current densities: 100 mA/cm^2). Frequently, the observed structures in the relative cluster intensities were quite puzzling, e.g. for Al and Si. Attempts to correlate these structures to crystal configurations failed, nor did any pattern develop from simple valency considerations alone. Initial ion energy distribution measurements from 0 to 1200 eV showed significant differences for atomic ions and cluster ions. This effect is used to reduce interference problems caused by cluster ion peaks in SIMS applied to trace analysis of solids. The results are discussed and compared with those of other investigators, also including cluster formation by vaporization and sparking. Extending known theoretical considerations may possibly afford a general understanding of the intensity structure. The formation and ejection mechanisms of clusters, however, remain unknown. Thermal effects to explain the latter are definitely discounted by the magnitude of the observed initial cluster energy (>10 eV). Also discussed are two phenomena which demonstrate the presence of the bombarding gas in the surface.

INTRODUCTION

The mass spectra obtained with a sputtering ion source frequently show a strong proliferation of polyatomic ions. These "molecular species", which may comprise many atoms, are usually called "clusters". Such clusters have also been found in the vaporization and sublimation of solids,[1-3] and in the rf-spark source.[4-6] The first systematic work on clusters in a sputtering ion source was reported in 1961 by Honig.[7,8] In 1966, during work partly performed under NASA contract No. NAS 5-9254,[9,10] further results were obtained by us. At about the same time the sputtering of cluster ions from Cu was investigated by Woodyard.[11] Unfortunately, neither work has been published in the open literature. The present increasing interest in clusters as part of sputtering phenomena and the still unexplained formation and ejection mechanisms have prompted this belated publication of our results.

EXPERIMENTAL WORK

All our work was done with a laboratory prototype of the GCA IMS 101 sputtering ion mass spectrometer.[12] This instrument, originally designed by Liebl,[13] uses a primary ion beam from a duoplasmatron. The beam can be focused on the target to a spot 0.3 mm in diameter, with a typical current density of 100 mA/cm^2 at a primary energy of 4 to 12 keV. The target surface

† Present address: MPI für Plasmaphysik, D-8046 Garching, Germany.

is bombarded at 45 degrees. The axis of the secondary ion extraction system forms an angle of 90 degrees with the primary ion beam. It should be noted, however, that the application of a secondary acceleration voltage (1000 volts) results in a large cone of acceptance for secondary ions of low initial energy. Only for energetic recoils, above 100 eV, it is reasonable to define a recoil angle of approximately 90 degrees. The secondary ions are mass analyzed with a stigmatically and double focusing mass spectrometer. An electron multiplier and a linear-logarithmic electrometer constitute the ion detection system.

A typical cluster ion spectrum, obtained by bombarding polycrystalline Al with 12 keV Xe^+, is shown in Figure 1. The interesting feature of this spectrum is not only the appearance of large clusters comprising up to 18 Al atoms, but the intensity structure. Al_7^+ is more intense than Al_6^+. There is then a fairly sharp drop, and a group of roughly equal intensities is observed from Al_8^+ to Al_{14}^+, with Al_9^+ and Al_{14}^+ being somewhat more prominent. From Al_{15}^+ to Al_{18}^+ there follows another group of about equal but again weaker intensities. At first sight there seems to be some magic number 7 involved here. The sputtering of Si yielded the spectrum in Figure 2. In spite of its well-known chemical affinity, Si shows clusters only up to Si_{11}^+. Here an intensity inversion of $Si_6:Si_5$ is observed and there is a suggestion of inversion of $Si_4:Si_3$. Sputtering of Mg showed Mg_2^+ but hardly any Mg_3^+. It seems reasonable to associate these intensity patterns with different crystal structures, e.g. aluminium is f.c.c., silicon is diamond-like, and

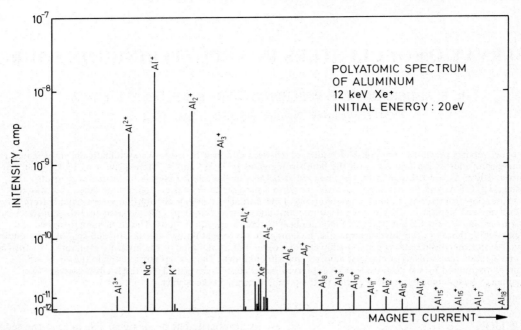

FIGURE 1   Typical cluster ion spectrum of polycrystalline aluminum sputtered by a high density beam of Xe⁺ at 12 keV. This and the following spectra were well reproducible in all details.

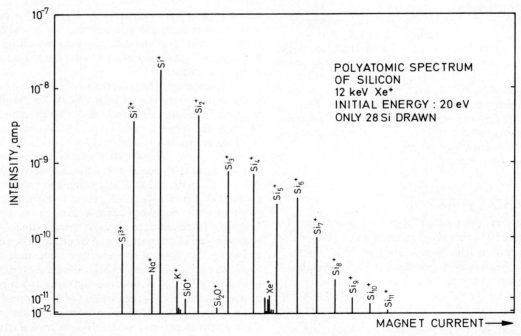

FIGURE 2   Typical cluster ion spectrum of silicon (single crystal) sputtered by a high density beam of Xe⁺ at 12 keV.

magnesium is h.c.p. However, sputtering of allotropic materials such as graphite, diamond, or amorphous carbon only revealed a comparatively minor effect. A comparison of materials from the same groups of the periodic system did not reveal any simple pattern either.

The mass of the bombarding ion turned out to have a very strong effect on the population of the cluster ion spectra. This is demonstrated in Figure 3, which

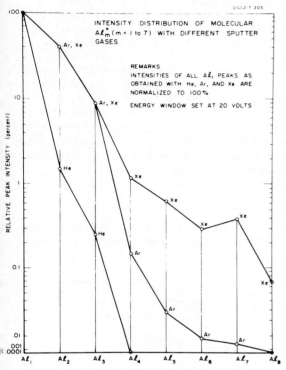

FIGURE 3   Relative cluster ion intensities versus number of Al-atoms per cluster, for He$^+$, Ar$^+$ and Xe$^+$ bombardment. Primary energy was 12 keV.

shows the intensity distribution of Al clusters for He$^+$, Ar$^+$ and Xe$^+$ as sputtering gas. Sputtering with Ar$^+$ gave considerably less heavy clusters, and the typical structure observed with Xe$^+$ is obscured. With He$^+$ the intensity drop is even sharper and nothing was detected beyond Al$_4^+$. In contrast, reducing the energy of Xe$^+$ from 12 keV to 4 keV had no effect on the relative cluster intensities.

An inherent feature of a double focusing mass spectrometer is that it also permits determination of ion energies. To this avail, the width of the energy slit was reduced to transmit an energy spread of $\Delta E = 4$ eV, which is about 10% of the usual energy acceptance.

With the analyzer set to a fixed mean energy, the energy spectrum can be scanned by simply changing the target potential, i.e. the secondary acceleration voltage. The initial energy distribution of the secondary ions sputtered by 12 keV Ar$^+$ from an Al–Mg alloy is shown in Figure 4. Noteworthy is the very different

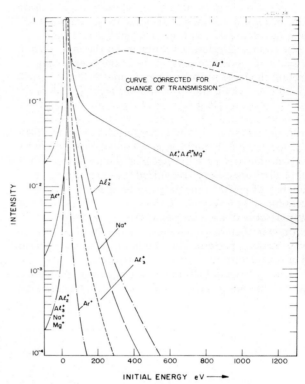

FIGURE 4   Normalized energy distributions of the various atomic and polyatomic ions sputtered from an Al–Mg alloy by 12 keV Ar$^+$.

energy distribution of the atomic ions Al$^+$ and Mg$^+$ compared with Al$_2^+$ and Al$_3^+$. All species have a most probable energy around 20 eV in common, but the atomic ions have a very extended tail on the high energy side, whilst the intensities of cluster ions show a rapid fall-off which increases with cluster size. (Energy distributions obtained in the way described are qualitative at best, because of the energy dependent transmission characteristic of the whole mass spectrometer. The dotted curve in Figure 4 was deduced by calculating the theoretical transmission change based on the—incorrect—assumption of isotropic ion emission at all energies. Thus, the second maximum might well be an artifact, but the true curve should still be closer to the "corrected curve", especially at higher energies.)

Important is the notion that the general picture of Figure 4 was hardly affected by using He^+ or Xe^+ in place of Ar^+. Reduction of the primary ion energy from 12 to 4 keV only caused a change in the high energy slope of atomic Al^+ and Mg^+.

The different initial energy distribution of atomic and polyatomic ions suggests an interesting application. Cluster ions, interesting as they are from one point of view, are a nuisance in trace analysis because they cause serious interference problems in the mass spectrum. If the analyzer accepts the initial energy range around the most probable energy, sputtering of Al with 12 keV Ar^+ yields the spectrum in Figure 5. This is a typical polyatomic spectrum of a contaminated surface, showing a peak at about every mass number. After shifting the range of initial energy acceptance to 250 eV, the spectrum in Figure 6 was obtained. Now almost all polyatomic ions have disappeared and a number of trace impurities are clearly recognized. Frequently, reliable labeling of peaks in the polyatomic spectrum is only possible after investigating the atomic spectrum in the way described. For compounds the polyatomic spectrum may then help to determine stoichiometry in the way cracking patterns are used in organic mass spectrometry to determine molecule structures.

Many of the polyatomic ions frequently observed are not known to exist as stable neutral molecules. An example is shown in Figure 7 representing the spectrum of a sample of aluminum bombarded with 12 keV Ar^+. Here an enormous proliferation of $Al_m Ar_n$ ions is seen. Rare-gas-metal ions sometimes also showed up with other samples, but such an exceedingly high intensity was only found for this specific sample. The special conditions leading to the occurrence of these ions are not known to us. The spectrum of Figure 7 was well reproducible, however.

Another marginal result further underlined the presence of the sputtering rare gas in the surface. It was found that the intensity of the halogens F^+ and and Cl^+—a usual contaminant in Al—in the mass spectrum strongly depended on the nature of the primary ion. This is demonstrated in Figure 8. A comparison of the ionization potentials of the primary and secondary ions in question (He^+: 24.6 V, F^+: 17.4 V, Ar^+: 15.7 V, Cl^+: 13.0 V, Xe^+: 12.3 V) suggests a correlation. A possible explanation is that Cl and F are ionized by charge exchange or Penning ionization, while passing through a layer of primary ions or excited primary metastables respectively.

DISCUSSION

Our investigations clearly failed to show any simple correlation between the crystal structure and the

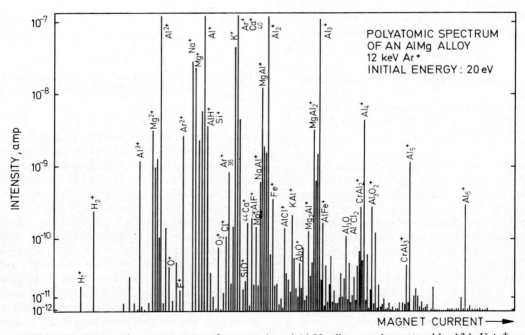

FIGURE 5   Polyatomic mass spectrum of a contaminated Al–Mg alloy sample sputtered by 12 keV Ar^+.

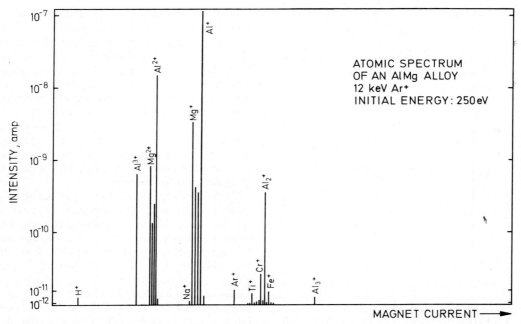

FIGURE 6   Atomic mass spectrum of the same sample (Figure 5). Strong suppression of all poly-atomic species is obtained by using only secondary ions emitted with high initial energy (250 eV).

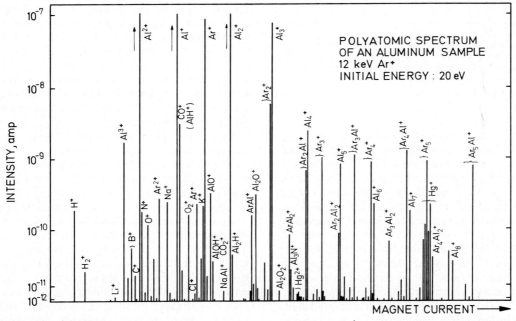

FIGURE 7   Mass spectrum of an aluminum sample bombarded by 12 keV Ar^+. Here, an unusual strong proliferation of $(Al_mAr_n)^+$ is seen.

observed ion cluster spectrum. For the cluster ions of graphite Dörnenburg and Hintenberger[5] had already failed to link the intensity pattern observed in a rf-spark source to the crystallographic structure. In a rf-spark it is difficult to say whether clusters are produced by sputtering, evaporation or association in the plasma.

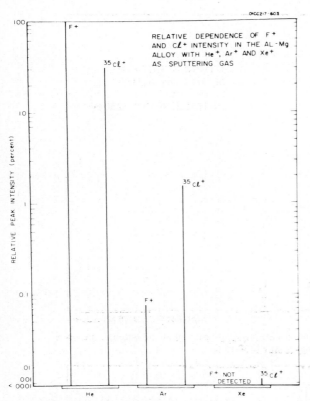

FIGURE 8   The secondary ion yield of elements with high electron affinity (F, Cl) show a very strong dependence on the ionization potential of the bombarding rare gas (He[+], Ar[+], Xe[+]). Intensities were normalized to Al[+]. Al was main constituent of the Al–Mg matrix.

Association is less likely, however. Even in the evaporation and sublimation of solids, corroborated by some evidence, it is assumed that cluster formation takes place during the evaporation process itself[2] and not in the equilibrium vapor phase. The same is certainly true of sputtering, though we are here confronted with an ejection mechanism essentially different from evaporation. However, it had been pointed out by Honig[7] that the carbon cluster spectra from his sputtering ion source showed a striking resemblance to spectra obtained in earlier work[2] by sublimation from a hot graphite filament. These results were subsequently explained by Pitzer and Clementi[14] in a theory based upon a quantum-mechanical concept of the electronic states of the carbon atom. This model gives the cluster intensity distribution in the vapor at equilibrium as a function of the filament temperature. At higher temperature the relative intensity of larger clusters is shown to increase. In the sputtering process we are far

from an equilibrium condition. Still, the use of a heavier bombarding ion qualitatively has an effect similar to increasing the temperature. This trend was also shown by Honig[7,8] for the C[−] clusters. Dörnenburg, Hintenberger and Franzen[5] found very reasonable agreement between the intensity structure of carbon clusters in the rf-spark source and conclusions drawn from the theory of Pitzer and Clementi. Their findings confirmed the assumption of linear chain molecules for clusters of up to 8 C atoms. Beyond this number, ring-structured monocyclic molecules become more likely. It all suggested that this theory could be successfully extended to other elements.

Unfortunately, an explanation of the cluster intensity structure provides little help towards understanding the formation and propagation of such clusters. There is some speculation that previous surface erosion is a prerequisite for cluster ejection. But then, it is difficult to see a linear molecule such as C_8 result in this way. The clusters observed in evaporation raise further doubts. Obviously, clusters do not directly reflect the crystallographic structure and may not be considered to be miniature chips from the bulk material. Are then typical cluster configurations already formed on the surface, e.g. by sputtering induced surface mobility, or are they the result of dissociation of larger aggregates after ejection? We can only say that any dissociation would have to happen very close to the surface, otherwise it would be revealed by the initial energy distribution of cluster ions. The question of the ejection mechanism itself also remains unanswered. The notion of overlapping collision cascades or of "chiselling" by a deflected primary ion which entails sufficient momentum to some members of a larger aggregate in a fast sequence of collisions can neither be supported nor discounted by our observations.

The differential initial energy distributions of atomic and cluster ions, however, is easily explained. A larger aggregate is not likely to survive a more violent collision process which deposits an energy significantly above its dissociation energy. Thus, the half-width of the energy distributions of a cluster ion is correlated with its stability. In the mass spectrum, however, the most stable aggregate does not necessarily correspond to the strongest peak. Sublimation of graphite, for instance, shows C_3^+ to be the most abundant species in accordance with theory.[14] Sputtering, in contrast, shows C[+] to be the dominant peak. Obviously, what we actually observe is the superposition of two phenomena. That is (a) an effect associated with the electronic shell structure, governing the cluster stability and defining a heat of formation, and (b) a mechanism leading to the ejection of clusters with a probability which depends on the cluster size, nature and state of the matrix. In sputtering, direct interaction with primary ions is also possible. Our results indicate that effect (b) is described by a function which decreases steadily with cluster size. The lighter the bombarding ion, the sharper is the decrease (Figure 3). Thus, not in all cases can effect (a) override effect (b), so that intensity inversion is observed, e.g. for $C_2^- : C^-$,[7] or $Al_7^+ : Al_6^+$.

The presence of a significant amount of primary gas is illustrated by the results shown in Figures 7 and 8. We do not know the general role of these atoms in the ejection of clusters, but the presence of many $(Al_n Ar_m)^+$ clusters (Figure 8) indicates a significant interaction. It has already been pointed out that such a proliferation of intense $(Al_n Ar_m)^+$ peaks is unusual and was only observed on one specific sample. Since Ar does not dissolve in Al, it can only have accumulated in voids. It is known that radiation induced voids in Al can occupy at least 10% of the crystal volume. The strong peaks of Ar_2^+ through Ar_5^+ seen in this spectrum indicate the build-up of a very high gas density. A more extensive investigation of this phenomenon is planned.[15]

CONCLUSION

Our discussion has been limited to experimental and theoretical data available in 1967. We could state that the method used by Pitzer and Clementi might well lead to a general understanding of structures in the cluster ion intensity. This has recently been affirmed by the work of Joyes.[16] Further systematic studies of clusters sputtered from tungsten were reported by Staudenmaier.[17] The general trend of his measurements agrees well with our results. But all the experimental and theoretical data available up to now do not resolve the details of the formation and ejection of clusters.

It is interesting to compare cluster intensities obtained by sputtering, vaporization and sparking. We have to keep in mind, however, that the high initial energy of sputtered cluster ions in the 10 eV range, as compared with 0.3 eV in vaporization,[3] precludes any thermal ejection mechanism. Still, dense thermal spikes might well be an important factor in the formation of stable cluster aggregates. It is difficult to imagine a simple binary collision event which transfers an energy of several 10 eV to a larger aggregate without inducing

dissociation. A more collective phenomenon, involving many members, is certainly a more reasonable assumption.

Further investigations are clearly needed to answer those questions. These include the dependence of relative cluster intensities (a) on primary ion mass, including molecular ions and self-sputtering; (b) on primary ion energy, from the threshold upwards to several 10 keV; (c) on primary ion dose and current density, and (d) on the impact angle. In addition, more accurate data on the initial energy and angular distributions of secondary ions should be available.

With a more complete and accurate knowledge of these parameters, the smaller effects observed in the sputtering of allotropic systems, of alloys and compounds might reveal more details of the spatial propagation of collision cascades in actual materials.

REFERENCES

1. R. E. Honig, *J. Chem. Phys.* **21**, 573 (1953).
2. R. E. Honig, *J. Chem. Phys.* **22**, 126 (1954).
3. W. A. Chupka and M. G. Inghram, *J. Chem. Phys.* **21**, 1313 (1953).
4. E. Dörnenburg and H. Hintenberger, *Z. Naturforschung* **14a**, 765 (1959).
5. E. Dörnenburg, H. Hintenberger and J. Franzen, *Z. Naturforschung* **16a**, 532 (1961).
6. J. Franzen and H. Hintenberger, *Z. Naturforschung* **16a**, 535 (1961).
7. R. E. Honig, *Advances in Mass Spectrometry*, Vol. 2, edited by R. M. Elliott (Pergamon Press, London, 1963), p. 25.
8. R. E. Honig, *Proc. 5th Intern. Conf. on Ionization Phenomena in Gases*, Munich 1961 (North-Holland, Amsterdam), p. 106.
9. R. F. K. Herzog, W. P. Poschenrieder and F. G. Satkiewicz, NASA Contract No. NAS5-9254, Final Report, GCA-TR-67-3N (1967).
10. R. F. K. Herzog, W. P. Poschenrieder, F. G. Rüdenauer and F. G. Satkiewicz, presented at the 15th Ann. Conf. on Mass Spectrometry and Allied Topics, Denver, Colorado, May 1967.
11. J. R. Woodyard, presented at the 15th Ann. Conf. on Mass Spectromedry and Allied Topics, Denver, Colorado, May 1967.
12. R. F. K. Herzog, H. J. Liebl, W. P. Poschenrieder and A. E. Barrington, NASA Contract No. NASw-839 GCA Technical Report No. 65-7-N (1965).
13. H. J. Liebl and R. F. K. Herzog, *J. Appl. Phys.* **34**, 2893 (1963).
14. K. S. Pitzer and E. Clementi, *J. Amer. Chem. Soc.* **81**, 4477 (1958).
15. F. G. Satkiewicz, private communication.
16. P. Joyes, *J. Phys. Chem. Solids* **32**, 1269 (1971).
17. G. Staudenmaier, *Rad. Effects* **13**, 87 (1972).

ANGULAR DEPENDENCE OF CLUSTERS SPUTTERED FROM A TUNGSTEN SINGLE CRYSTAL SURFACE

G. STAUDENMAIER

Sektion Physik, University of Munich, Munich, Germany

The anisotropy of positively charged clusters sputtered with 150 keV Ar^+ ions from a tungsten single crystal has been investigated with a mass and energy spectrometer. Not only the monomer W^+ but also the clusters W_2^+, W_3^+ and W_4^+ (so far investigated) show maxima of emission along close packed directions. The maximum cluster yield along a $\langle 111 \rangle$ direction normalized to the yield along a random direction increases with increasing cluster size and cluster energy, although the absolute cluster yield decreases. An interpretation of the experimental results on the basis of a simple statistical model of cluster formation kinetics is presented which explains the yield profiles of the anisotropic contribution fairly well.

1 INTRODUCTION

In a previous investigation[1] the influence of projectiles on the energy distribution of charged clusters sputtered from polycrystalline tungsten was studied. The experiments have led to the conclusion that the energy distribution is independent of the mass and the species of the projectiles (either rare gas or metal ions) is used. However, there were various parameters open, which might have influenced the emission of clusters, e.g., the surface structure and the surface orientation.

Therefore the investigations have now been extended to a study of cluster emission from a tungsten single crystal, in particular from a (110) surface. In addition the angular dependence of the anisotropy of cluster ejection has been studied, since this quantity seems to be less influenced by any unknown parameter involved in secondary ion emission.

2 EXPERIMENTAL

The experimental set-up is shown in Figure 1. Positively charged particles emitted from the target are mass and energy analyzed in a double focusing spectrometer described in Ref. 1. A beam of 150 keV Ar^+ ions, collimated to an angle of divergence of 0.5 degrees is incident at angle ϕ with respect to the surface normal. The angle between the incident beam at ϕ and the direction of observation at ψ was always fixed to 60 degrees for experimental reasons.

The tungsten single crystal with a (110) surface can be rotated around a $\langle 110 \rangle$ axis. To avoid the incident beam and the direction of observation to pass exactly through low index planes the crystal was tilted so that the surface normal and the plane containing the incident beam and the direction of observation were kept inclined by an angle of 10 degrees, see Figure 1.

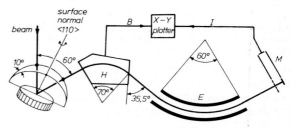

FIGURE 1   Experimental arrangement. 150 keV Ar^+ ions strike a (110) tungsten surface. Sputtered charged particles are analyzed with the mass and energy spectrometer consisting of magnetic field H and cylindrical condenser E and are detected by the windowless multiplier M. The angle between incident beam and direction of observation is fixed to 60 degrees; the angle between the plane set by the incident beam and the direction of observation and the surface normal $\langle 110 \rangle$ is fixed to 10 degrees. The axis of rotation is parallel to the surface normal $\langle 110 \rangle$.

3 RESULTS

In Figure 2 and Figure 4 the anisotropic contribution to the emission of clusters is plotted against the angle of emission. In the plots the cluster yields have been normalized to the yield obtained when both the direction of incidence and the direction of observation are high index directions, which, by the way, was found to be identical with the yield obtained from polycrystalline tungsten.

For a secondary particle energy of 30 eV maxima of emission in the $\langle 111 \rangle$ direction are obtained not only for the monomer W^+ but also for the clusters W_2^+, W_3^+, and W_4^+, Figure 2. Note that the relative peak

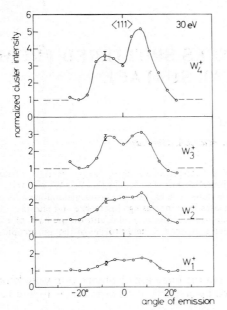

FIGURE 2   Cluster yield normalized to the yield in a random direction of a ⟨111⟩ spot vs. angle of emission for a cluster energy of (30 ± 0.8) eV. The shape of the yield profiles is partially determined by the experimental procedure.

height increases with increasing cluster size, though the absolute yield decreases drastically with increasing cluster size. The complicated shape of the emission curves is due to the already mentioned fact that by varying the angle of observation ψ we inevitably pass through planar directions of beam incidence, see Figure 3.

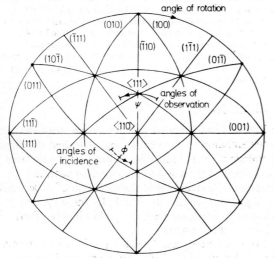

FIGURE 3   Stereogram of a (110) tungsten surface. The range of angle of incidence ϕ and of observation ψ are indicated. $\phi + \psi$ was always kept fixed to 60 degrees.

The peaking effect of cluster yields becomes even more pronounced for a cluster energy of 60 eV, Figure 4. Though the absolute yields are decreasing

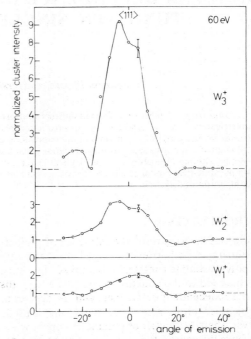

FIGURE 4   Cluster yield normalized to the yield in a random direction of a ⟨111⟩ spot vs. angle of emission for a cluster energy of (60 ± 1.6) eV. The shape of the yield profiles is partially determined by the experimental procedure.

with increasing energy, the normalized cluster yield increases with increasing energy. Again the relative anisotropic contribution is the higher the larger the clusters are. As far as experimental data are available at present, the energy dependence of the normalized yield of W^+ seems to have a maximum at an energy of about 45 eV. At energies below 60 eV there are no maxima observed for W_2^+ and W_3^+. A summary of experimental values is given in Table I.

By bombarding tungsten with 1200 eV Hg^+ ions Ashmyanskii et al.[2] found in the anisotropic contribution of the W^+ emission a small maximum at about 30 eV, assuming that the emission in ⟨110⟩ direction is not essentially different from the emission in a random direction.

MacDonald[3] observed in the sputtering of a copper single crystal a maximum of the anisotropic contribution of Cu^+ for an energy of about 17 eV for emission in ⟨110⟩ direction.

The half widths at half maximum of the normalized yields are about 10 degrees for the monomer W^+ and also for the clusters. This value is comparable with

TABLE I
Normalized yield maximum $R_i^{\langle 111 \rangle}(E_i)$ of clusters of energy E_i and of i atoms emitted along a $\langle 111 \rangle$ direction measured for three energies, 30, 45 and 60 eV. The cluster yield relative to the W^+ yield for the emission maximum is given for comparison.

E (eV)	30	45	60
$R_1^{\langle 111 \rangle}(E)$	1.8 ± 0.1	2.6 ± 0.2	2 ± 0.1
$R_2^{\langle 111 \rangle}(E)$	2.5 ± 0.2	2.7 ± 0.2	3.2 ± 0.2
$R_3^{\langle 111 \rangle}(E)$	3.1 ± 0.2	3.9 ± 0.2	9.2 ± 0.6
$\dfrac{I(W_2^+)}{I(W^+)}$	35%	30%	19%
$\dfrac{I(W_3^+)}{I(W^+)}$	6%	3.2%	1.8%

typical half widths of spot patterns obtained in single crystal sputtering. For comparison with theoretical estimates we mention that the Lehmann and Sigmund model predicts a $\langle 111 \rangle$ spot width of 8.2 degrees for tungsten.[4]

4 DISCUSSION

From the results shown we conclude that an enhanced emission not only for the monomer W^+ but also for the clusters W_2^+, W_3^+ and W_4^+ occurs along the close packed direction $\langle 111 \rangle$ of bcc tungsten. Preliminary data indicate similar results for the emission along the $\langle 100 \rangle$ direction. With polycrystalline tungsten targets we have also performed measurements with various angles of incidence and directions of observation. No enhanced emission of clusters relative to the W^+ yield was observed in any case, although the absolute yields varied by an order of magnitude.

It is intriguing to interpret the present experimental results on the basis of a simple kinetic model of cluster formation. We assume that clusters are produced by collision cascades which penetrate through the surface layer. There may be then an appreciable probability that two or more adjacent atoms receive about the same momentum at about the same instant. Then these atoms are likely to leave the surface as a cluster, a process which needs less expenditure of energy than the emission of separate atoms. Therefore, as a simple estimate, the probability for the formation of a cluster of kinetic energy E_n containing n atoms is

$$P_n(E_n) \propto [I_1(E_1)]^n \qquad (1)$$

where $E_n = nE_1$ and I_1 is the yield of W^+ at energy E_1. Applying this picture of cluster formation to the experimental results obtained in single crystal sputtering, in particular to the anisotropic contribution, for reasons we shall discuss later, we obtain for the relative yield of clusters the relationship

$$[R_n^{\langle 111 \rangle}(E_n)]^m = [R_m^{\langle 111 \rangle}(E_m)]^n \qquad (2)$$

where $R_i^{\langle 111 \rangle}(E_i)$ is the normalized maximum yield along $\langle 111 \rangle$ of clusters of energy E_i and of i atoms. In Table II the available two combinations are listed. The

TABLE II
Comparison of the normalized cluster yield in $\langle 111 \rangle$ direction according to Eq. (2). The original numbers of the $R_i^{\langle 111 \rangle}(E_i)$ values are listed in Table I.

$[R_1^{\langle 111 \rangle}(30 \text{ eV})]^2$	$\longleftrightarrow$	$R_2^{\langle 111 \rangle}(60 \text{ eV})$
3.2 ± 0.4		3.2 ± 0.2
$[R_2^{\langle 111 \rangle}(30 \text{ eV})]^3$	$\longleftrightarrow$	$[R_3^{\langle 111 \rangle}(45 \text{ eV})]^2$
15.6 ± 3.8		15.2 ± 1.6

agreement of the numerical results with Eq. (2) is obvious.

The simple model exploited can only be considered to be a first approximation, since there are a number of so far disregarded parameters which, however, must have an influence on the cluster emission yield.

i) Because of partial neutralization the observed W_n^+ intensity can be only an energy dependent fraction of the total W_n^+ intensity at the surface.

ii) On the same basis, not only the measured W^+ ion intensity but the total intensity of neutral atoms must be considered to contribute to cluster formation.

iii) For the formation of clusters the strongly fluctuating number of knocked-on surface atoms per unit area[5] should be more important than the measured average number emitted from the total surface.

iv) The conditions concerning the assumption for the W^+ emission to occur at the "same time" and with the "same momentum" are stringent:

a) Time. A pair of tungsten ions with each the same energy E leave the surface in the same direction but with a time difference Δt. The maximum difference in time which still may lead to a bound tungsten cluster then is

$$\Delta t < a_0 \sqrt{\frac{M}{2E}} \simeq \frac{0.5 \cdot 10^{-13}}{\sqrt{E \text{ (eV)}}} \ (s)$$

(M is the mass of the tungsten ions, $a_0 = 0.53 \cdot 10^{-8}$ cm) This leads to about 10^{-14} s, a time difference which should be compared with the duration of a collision cascade of presumably smaller than 10^{-13} s.[6]

b) Momentum. Assuming that two adjacent particles W^+ are emitted at the same time with the same magnitude of momentum but in different directions, the two particles can form a cluster only if the angle 2α between the two directions of emission is smaller than a critical value $2\alpha_c$ with $\sin \alpha_c \propto \sqrt{E_b/E}$ where E_b is the binding energy. Because of this the probability for the formation of a W_2^+ cluster, which is proportional to the solid angle $\Omega/2\pi = (1 - \cos \alpha_c) \approx \alpha_c^2/2$ (for small α_c), modifies Eq. (1) to $P_2(E_2) \propto [I_1(E_1)]^2/E_1$. In case of larger clusters not too serious deviations from this estimate are expected.

In conclusion the factors mentioned in (i) to (iv), the values of which are still uncertain at present, prevent a quantitative evaluation of the formation kinetics of clusters produced on polycrystalline surfaces.

However, in the case of the anisotropic contribution to the cluster yields the agreement is much better, probably because many uncertainties are eliminated with the normalization.

The enhancement of cluster formation along low index directions is likely to be a direct consequence of directional sputtering due to focusons or according to the surface model proposed by Lehmann and Sigmund.[7] In these cases not only dense knock-on regions are generated, but also the neighbouring knock-on atoms receive closely the same momentum.

The energy distribution of anisotropic cluster emission is under investigation. This can give additional information about the emission mechanism. A close examination of the experimental data will be performed to prove or disprove the present simple assumptions of cluster formation kinetics.

ACKNOWLEDGEMENTS

The author would like to thank Prof. R. Sizmann for encouragement and valuable discussions. The technical assistance of Mr. H. Bradatsch is appreciated. Part of this research was supported by the Bundesministerium für Bildung und Wissenschaft, Germany.

REFERENCES

1. G. Staudenmaier, *Rad. Effects* **13**, 87 (1972).
2. R. A. Ashmyanskii, M. B. Benyaminovich and V. I. Veksler, *Sov. Phys.—Sol. State* **7**, 1314 (2965).
3. E. Dennis, and R. J. MacDonald, *Rad. Effects* **13**, 243 (1972).
4. P. Sigmund, *Collision Theory of Displacement Damage, Ion Ranges, and Sputtering*, Int. Summer School on "Interaction of Radiation with Matter" (Predeal, Roumania, 1971).
5. J. E. Westmoreland and P. Sigmund, *Rad. Effects* **6**, 187 (1970).
6. G. Leibfried, *Bestrahlungseffekte in Festkörpern* (Teubner, Stuttgart, 1965), p. 262.
7. Chr. Lehmann and P. Sigmund, *Phys. stat. sol.* **16**, 507 (1966).

SECONDARY EMISSION OF MOLECULAR IONS FROM LIGHT-ELEMENT TARGETS

MIREILLE LELEYTER and PIERRE JOYES

Laboratoire de Physique des Solides,†
Bâtiment 510, Université Paris-Sud, 91405 Orsay, France

We present some experimental results and calculations on ion clusters produced by primary ion bombardment of light element samples (lithium, beryllium, graphite).

Alternations according to the parity of the atom number in the clusters are found in their stabilities, ionization potentials and electron affinities: $Li_{2n\pm1}^+$ and C_{2n}^- ions are found more stable than Li_{2n}^+ and $C_{2n\pm1}^-$ ions respectively. On the other hand, stability of Be_n^+ decreases when n increases.

These alternating properties are studied with CNDO computations.

We also compare our results to those of other authors.

1 INTRODUCTION

The purpose of this paper is to study some ion clusters produced by primary ionic bombardment of light solid elements.

Some experimental results on cluster formation by ion bombardment are already known for light elements which are not in solid state at room temperature: for instance, frozen hydrogen bombarded with alkali-metal A^+ ions[1] gives rise to $[A(H_2)_n]^+$ ions. Films of frozen water or ammonia, or methyl (or ethyl) alcohol give by ion bombardment clusters of the types: $[H(H_2O)_n]^+$, $[H(NH_3)_n]^+$ or $[H(R{-}CH_2OH)_n]^+$ (with R = H or CH_3) with large values of n (up to $n \simeq 30$).[2] In these cases, the targets are molecular crystals.

Here we are studying the secondary emission of molecular ions M_n^+ or M_n^- ($1 \leqslant n < 10$) from pure targets made of an element M (Li, Be or graphite) and bombarded by a few keV Ar^+ ions.

As it is likely that the emission intensity of polyatomic ions is related to the properties of these aggregates, we theoretically study their stabilities, their ionization potentials, and electron affinities. A similar study was previously done for noble metal polyatomic ions (Cu_n^+, Ag_n^+, Ag_n^-) by Joyes.[3]

2 CALCULATION METHODS AND EXPERIMENTS

We use a semi-empirical quantum chemistry calculation, the Complete Neglect of Differential Overlap method developed by Pople et al.[4] and known as CNDO. It is a more elaborate procedure than the Hückel method because it is self-consistent, though Hartree–Fock equations are solved with several approximations (this method is less accurate than ab-initio

† Associated to the C.N.R.S.

calculations, but needs less computation time): only valence electrons are considered, one-electron integrals are evaluated from experimental data, such as ionization potentials and electron affinities of the component atoms, and two-electron integrals are calculated from the Mulliken approximation formulae.[5]

The CNDO program has been written for the first three rows of the periodic table and allows the study of light elements to which we limit ourselves here. Computations were done on an IBM 370-165 computer.

The intensities $I(M_n^+)$ or $I(M_n^-)$ of secondary polyatomic ions were recorded at room temperature with the Castaing–Slodzian ion microanalyzer[6] which includes a double homogeneous magnetic prism associated with an electrostatic mirror. In the figures, we shall plot normalized intensities $I(n)/I(1)$.

3 RESULTS

3.1 Lithium

First let us examine the lithium case. The solid line (Figure 1) represents the emission intensities[7] of clusters as a function of n. It shows an oscillating behaviour quite similar to spark source results.[8] The dashed line is the variation of the negative value of ionization energies Δ_n^+ for linear lithium polyatomic ions, with $\Delta_n^+ = E(Li_n^+) - E(Li_n)$. E is the total computed energy of the neutral or ionized cluster.

As we can see, the two curves have a similar behaviour; for $n > 2$ we have:

$$I(2n) < I(2n \pm 1)$$
$$-\Delta_{2n}^+ < -\Delta_{2n\pm1}^+ \qquad \text{(linear clusters)}$$

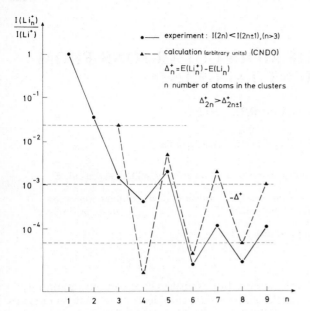

FIGURE 1   Normalized intensities $I(n)/I(1)$ of clusters sputtered by 10 keV Ar^+ ions from lithium.vs. number of atoms n in a cluster. The negative values of the ionization energies $-\Delta_n^+$ calculated for linear clusters with the CNDO approximation are given for comparison (energy scale: 0.04 au between two successive horizontal dashed lines).

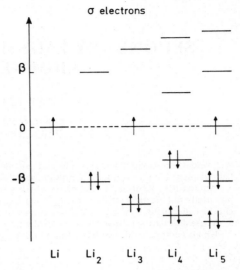

FIGURE 2   Energy levels and occupation of these levels for Li_n clusters in the Hückel approximation. The origin of the energies is shifted by an amount of α.

For electron affinities $\Delta_n^- = E(Li_n^-) - E(Li_n)$, the calculation gives the same inequalities:

$$-\Delta_{2n}^- < -\Delta_{2n\pm1}^- \qquad \text{(linear clusters)}$$

Thus the calculation shows that it is easier to get n-odd ions than n-even ones, a result in agreement with experimental data.

We also studied some kinds of non-linear lithium clusters[9] and came to similar conclusions.

In fact these results are already contained in the following simple one-electron Hückel scheme: a Li_n neutral cluster has n valence electrons and n external energy levels which can accommodate two electrons each. In Figure 2, we have plotted these external levels obtained with Hückel's approximation for linear shape: energy levels ϵ_q of n-atom molecules are given by

$$\epsilon_q = \alpha + 2\beta \cos \frac{q\pi}{n+1} \qquad \begin{array}{l} q = 1, 2, \ldots, n \\ n > 2 \end{array}$$

α and β being the two usual integrals of the Hückel theory[10]. A n-even neutral cluster has its $n/2$ first levels filled. In the $n + 1$ odd cluster, the extra electron must go to the $[(n/2) + 1]$th level which thus is half filled.

As it can be seen in a one-electron scheme, the energy needed to remove the external electron $(Li_n \rightarrow Li_n^+ + e^-)$ is smaller and the energy gained by filling the first unoccupied level with one extra electron $(Li_n + e^- \rightarrow Li_n^-)$ is larger for n odd than for n even. It is therefore easier to get n-odd ions than n-even ones.

We see that one of the main features of the calculation is the fact that each atom brings one valence electron to the cluster. Therefore this result can be extended to other cases:

i) H_n clusters: in hydrogen plasma, n odd H_n^+ ions are more encountered than n even ones.[11]

ii) LiH_n clusters: secondary ion emission experiments quoted above[1] show only the presence of $[Li(H_2)_m]^+$; positive clusters with an odd number of hydrogen atoms do not seem to exist. Indeed, in our simple scheme, LiH_n has $(n + 1)$ valence electrons and is equivalent to Li_{n+1}, easier to ionize for $(n + 1)$ odd (or n even).

iii) Cu_n, Ag_n, Cr_n clusters: experiment shows a more abundant emission of Cu_n^+, Cu_n^-, Ag_n^+, Ag_n^-, Cr_n^+ when n is odd, and this has been already explained by Hückel's method.[3]

Let us also note that, always in Hückel's framework, by adding all the individual electron energies, we get the electronic part of the binding energy.[3] This calculation shows that $(LiH_{2m})^+$ ions are more stable than $(LiH_{2m\pm1})^+$, a result confirmed by recent ab initio calculations.[12,13,14] Furthermore these authors also deal with BeH_n^+ clusters and show that $(BeH_{2m+1})^+$ ions have a larger stability than $(BeH_{2m})^+$ and $(BeH_{2m+2})^+$, a conclusion which can be found again easily with the Hückel scheme: the Be atom brings to the system two valence electrons instead of one, which reverses the alternations found in Li_n^+.

3.2 Beryllium

On Figure 3, we are dealing with positive beryllium ions (we did not find any Be_n^- ions because of too small intensities). Intensity of emission $I(n)$ regularly decreases when n increases: $I(2n-1) > I(2n) > I(2n+1)$ We have CNDO calculation results for Δ_n^+ with $n < 4$, which actually show a decreasing behaviour. This

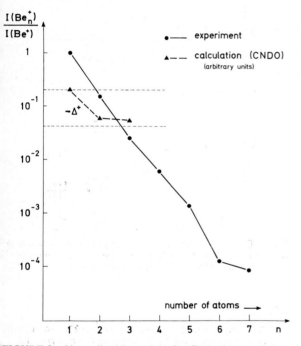

FIGURE 3   Normalized intensities $I(n)/I(1)$ of clusters sputtered from beryllium vs. number of atoms n in a cluster. The negative values of the ionization energies $-\Delta_n^+$ calculated with CNDO are given for comparison (energy scale: 0.01 au between the two horizontal dashed lines).

monotonous variation can be understood from Hückel's method; in Be_n neutral clusters, we have $2n$ valence electrons which completely fill the n energy levels; so we have always closed shells whatever n is. Hence we see that no alternation in the properties of Be_n^+ can be expected. Furthermore, it seems very difficult to get negative ions because the extra electron would have to go into upper energy levels which are no longer σ-levels.

Let us recall that all the transition elements (except Cr), which have the same valence shells as Be, show also a monotonous decreasing behaviour.

3.3 Graphite

In Figure 4 is drawn the spectrum of negative polyatomic C_n^- ions of graphite with alternations according to the parity of n: $I(2n) > I(2n \pm 1)$. We notice the existence of carbon ions with one extra hydrogen atom $(C_nH)^-$, showing also alternations with the parity of n.

Our experimental results (full line in Figure 5) are similar to those of Honig for the secondary ion emission of graphite[15] or silicon carbide[16] or for the sublimation of graphite[17] (in particular, a very strong peak for C_2^- seems to show that this diatomic ion is extremely stable). The dashed line (Figure 5) represents the negative value of the electron affinities Δ^- computed by CNDO for linear carbon clusters. It shows the expected alternations: $-\Delta_{2n}^- > -\Delta_{2n\pm1}^-$; thus, for C_n^-, it is easier to get n-even ions than n-odd ones, which is the inverse behaviour of lithium.

Here again, we can explain the phenomena from Hückel's scheme. For carbon, there are four valence electrons: $2s^2\,2p^2$. In a n-atom neutral cluster, there are then $4n$ valence electrons. Pitzer and Clementi[18] proposed the following electronic structure for $n > 2$: the lowest energy levels are $(n-1)$ σ-bonding states which can accommodate $2(n-1)$ electrons. Above these states exist two "surface states" (or Shockley states) filled with 2×2 electrons, and above them, there are n π-states in which $2(n-1)$ electrons must be accommodated.

In Figure 6, we represent these π-levels calculated by using as for lithium σ-levels, the Hückel method: a n-atom cluster presents an external π-shell, half filled for n even (except for C_2 which is a special case) and filled for n odd. We see that we can apply a reasoning similar to the lithium case, which leads to the conclusion that it is easier to get n-even negative ions than n-odd ones.

Alternations in emission intensities are also found in spark source experiments[19,20] for C_n^+ which are the most abundant when n is odd; neutral C_n clusters are probably formed at first and then ionized. This mechanism is supported by our CNDO results which show that neutral C_n are more stable when n is odd.

4 CONCLUSION

It may be concluded that the intensity distribution of secondary emission of polyatomic ions seems to be closely related to the electronic properties of the clusters.

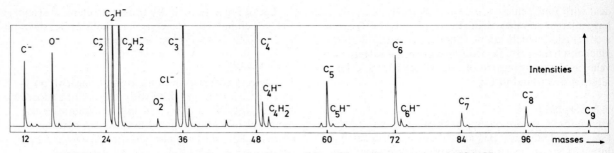

FIGURE 4   Spectrum obtained on bombarding graphite with positive argon ions (~10 keV).

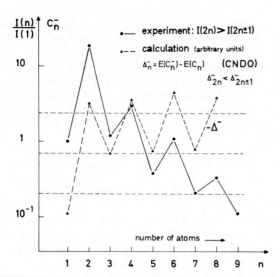

FIGURE 5   Negative clusters of graphite; results of CNDO calculations are given for linear clusters (energy scale for the electron affinities Δ_n^-: 0.08 au between two successive horizontal dashed lines).

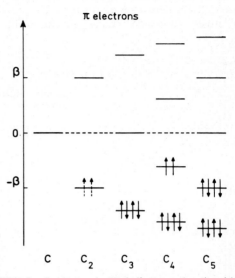

FIGURE 6 C_n clusters in Hückel's approximation (the origin has been shifted by α). In fact the higher occupied level of C_2 is a σ-state, completely filled with two electrons.

It is interesting to notice that when n increases, some solid properties become apparent from the calculation. Ionization potential and electron affinity calculated for lithium tend towards the same limit which is the work function of the bulk for infinite n, whereas for carbon, they seem to have different limits which are connected to a forbidden band. Then, already in those small aggregates, lithium shows a metallic character and carbon tends to behave as an insulator.

Secondary ion emission appears to be a new means to check theoretical calculations of small metallic clusters about which very little information was available until now. Thus it is promising for attaining a better understanding of the metallic bond, through investigations of the stability criteria of small clusters.

REFERENCES

1. R. Clampitt and D. K. Jefferies, *Nature* **226**, 141 (1970).
2. G. D. Tantsyrev and E. N. Nikolaev, *JETP Letters* **13**, no. 9, 337 (1971).
3. P. Joyes, *J. Phys. Chem. Solids* **32**, 1269 (1971).
4. J. A. Pople and D. L. Beveridge, *Approximate Molecular Orbital Theory* (McGraw Hill, New York, 1970).
5. R. S. Mulliken, *J. Phys. Chem.* **56**, 295 (1952).
6. R. Castaing, J.-F. Hennequin, L. Henry and G. Slodzian, in *Focusing of charged particles*, **2**, edited by A. Septier (Academic Press, New York, 1967).
7. P. Joyes and M. Leleyter, *C.R. Acad. Sci, Paris* **274 B**, 75 (1972).
8. R. Bourguillot, A. Cornu, R. Massot and J. Pellet, rapport GAMS 1964, Centre d'Etudes Nucléaires, Grenoble, France.
9. M. Leleyter and P. Joyes, to be published.
10. L. Salem, *The Molecular Orbital Theory of Conjugated Systems* (W. A. Benjamin, New York, 1966).
11. References (3), (4), (5) in reference (7) of the present paper.

12. J. Easterfield and J. W. Linnett, *Nature* **226,** 142 (1970); *Chem. commun.,* 64 (1970).
13. R. D. Poshusta, J. A. Haugen and D. F. Zetik, *J. Chem. Phys.* **51,** 3343 (1969).
14. S. W. Harrison, L. J. Massa and P. Solomon, *Chem. Phys. Lett.* **16,** 57 (1972).
15. R. E. Honig, *Adv. Mass. Spectr.* **2,** 25 (1962).
16. R. E. Honig, Proceedings Fifth International Conference on Ionization Phenomena in Gases, Munich 1961, 106.
17. R. E. Honig, *J. Chem. Phys.* **22,** 126 (1954).
18. K. S. Pitzer and E. Clementi, *J. Am. Chem. Soc.* **81,** 4477 (1959).
19. E. Dörnenburg and H. Hintenberger, *Z. Naturforsch.* **14a,** 765 (1959).
20. E. Dörnenburg, H. Hintenberger and J. Franzen, *Z. Naturforsch.* **16a,** 532 (1961).

NEUTRAL BEAM SPUTTERING OF POSITIVE ION CLUSTERS FROM ALKALI HALIDES

J. RICHARDS and J. C. KELLY

Materials Irradiation Laboratory, Physics,
University of New South Wales, Sydney 2033, Australia

Neutral argon atom beams of 15 keV energy have been used to sputter alkali halides and the ejected positive ions have been analysed in energy, mass and angular distribution.

The use of a neutral beam, rather than an ion beam, minimizes surface charge and the deflection of ejected ions by electrostatic interaction with a charged incident beam.

A cluster component of the form K_2Cl^+, $K_3Cl_2^+$ and higher members of the series is found for all alkali halides studied.

1 INTRODUCTION

The ion bombardment of alkali halides is complicated, compared with the bombardment of metals, by the build up of surface charge, by the ejection of both positive and negative ions and by a high yield of secondary electrons. The simple collection of current is hence inadequate as a measure of the ejection ion current in a given direction. We have used a system which analyses ejected positive ions for energy and mass at almost any angle of ejection from the surface. The random sputtered component, of lower energy, is more effected than is the directed component, by electrostatic interactions with both the incident ion beam[1,2] and with target surface charges. The directed component peaks at 55° to the (100) plane and being of greater energy is less deflected after leaving the crystal surface. The peak is close to the ⟨112⟩ direction which is at 54.8°. The large angular width of the peaks, the predominance of low energy ions and the lack of a strong temperature effect support a surface correlated mechanism[3] rather than a focuson model[4] but the evidence is not yet conclusive.

2 APPARATUS

An ion accelerator[5] was modified by the addition of a sodium vapour charge exchange beam neutralizer. The cross sections for charge exchange between argon (and indeed many other gases) and alkali vapour are large and have been tabulated.[6] The high cross section is probably associated with an almost resonance exchange, producing argon atoms which are neutral but in excited states. This is unlikely to effect the sputtering as the excitation energy (~3 eV) is small compared with the beam energy of 15 keV. The charge exchange

oven which surrounded the beam was maintained at a temperature of 250°C giving a sodium vapour pressure of 2×10^{-3} torr which is a number density, n, of 3.6×10^{13} atom/cm³. The theoretical efficiency is between 60 and 85% of the ion beam neutralized, depending on the cross section, σ, used in

$$N = N_0 \exp(-n\sigma L)$$

Efficiencies of 50–65% were achieved in practice for our oven with $L = 5$ cm. Remaining ions were removed by electrostatic deflector plates mounted just down the beam line from the oven.

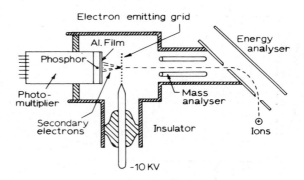

FIGURE 1   Analysis system of electrostatic parallel plate energy analyser and short quadrupole mass analyser with detector used for sputtered positive ions. The entire assembly is supported on a yoke and can be moved to scan in elevation.

For the results reported here the argon atom beam was incident normally on the target crystal and rotation of the target mount in azimuth and the detector mount in elevation enabled ion ejection over most of the hemisphere above the target to be measured. The

191

detector and analyser assembly is shown in Figure 1. It has
been made as small as possible for easy manoeuvreability.
The sputtered particles enter a parallel plate energy
analyser[7,8] with a resolution of 2.5%. They pass through
a 5 cm long quadrupole mass analyser[9] designed for a
resolution of 1 amu up to energies of 50 eV thus
enabling argon, 40 amu, to be separated from sputtered
potassium, 39 amu. The detector, based on a design by
Mason[10], accelerates the positive ions onto a 90% transmission stainless steel grid held at −10 kV which
produces secondary electrons which are accelerated
onto an earthed aluminium coated phosphor and the
resultant light pulse is detected by a photomultiplier.
The detector is not easily contaminated, will detect
currents of less than 10^{-13} amp and has a noise level of
less than 10^{-16} amps. The need to make the assembly
compact and manoeuvrable limited the noise figures
that could be achieved but 10 pulses per second could
be resolved and this was adequate for our purpose. The
900 Angstrom thick aluminium coating prevented
surface charging of the phosphor but allowed transmission of the 10 keV electrons. By reflecting light
back into the photomultiplier the aluminium doubled
the efficiency of the detector.

The relative response of the detector to Na^+ ions is
shown in Figure 2. The curve is flat up to an incident
sodium ion energy of 35 eV and falls smoothly at
higher energies. Similar curves were obtained for the
other alkali metal ions used.

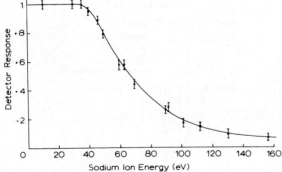

FIGURE 2   Response of the detector to incident Na^+ ions
with an energy range of 0–160 eV.

3 CATION BINDING ENERGIES

The surface charge produced by bombardment can be
measured from the movement of the peak yield which,
with increased incident ion current, moves to high
values as shown in Figure 3 which is for a thin NaCl
target at 270°C bombarded with 15 kV argon ions.
The linear region gives an effective conductivity of

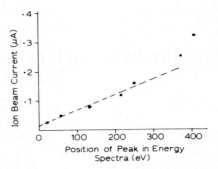

FIGURE 3 Na^+ ion spectra energy peaks vs. incident 15 keV
argon ion beam current for 150 μm thick NaCl at 270°C.

$(4 \pm 0.5) \times 10^{-8}$ ohm^{-1} m^{-1} which is close to that of
Lehfeldt,[11] 3.6×10^{-8} ohm^{-1} m^{-1}. The large magnitude
of this surface charge under ion bombardment is the
reason for the use of 15 kV neutral argon atoms for
most of our observations.

The general features of the angular distribution of
sputtered positive ions is shown in Figure 4 for KBr at
20°C. The azimuth scan was taken by fixing the
detector at 45° elevation to the (100) crystal surface
and rotating the crystal about a ⟨100⟩ direction. The
elevation scans were taken in (100) and (110) planes
with the crystal fixed. The yield increases with increasing
elevation and shows a fourfold symmetry in the (110)
planes. A series of azimuth scans at a number of
elevations enables the directions of maximum yield to
be located. Such a series is shown in Figure 5 for
positive ions sputtered from NaCl at 340°C by 15 keV
Argon atoms, i.e. atoms with the same energy as 15 keV
singly charged argon ions. Preferred ejection occurred
in the ⟨112⟩ direction for all crystals studied. Energy
curves for sputtered cations were typically of the form
shown in Figure 6 for ejection in a random direction.
The linear rise is to be expected from random cascade
theory,[12] and extrapolated back to zero yield, gives the
surface potential which acts to displace the peak to
higher energy values than are found for metals. When
correction is made for surface potential, the energy on
the high side of the peak falls as E^{-2} for energies above
11–15 eV and out to 100 eV depending on target
material for the alkali halides LiF, NaCl, KCl, KBr and
KI which have been measured.

If we assume from random cascade theory that the
surface binding energy, E_b, produces a peak at $\frac{1}{2}E_b$,
and that the surface potential merely shifts the energy
spectrum to higher energies without serious distortion
near the peak region the binding energies in Table I
emerge.

There appears to be no other direct experimental
method of obtaining cation binding energies and we

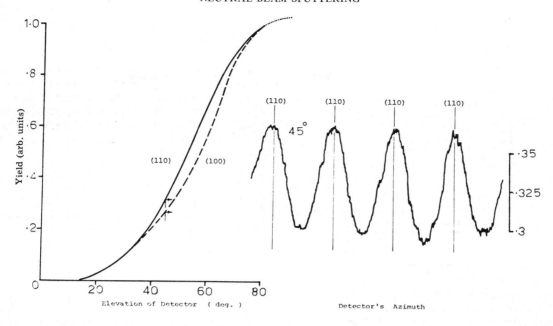

FIGURE 4   Angular distributions of positive ions sputtered from KBr at 20°C by 15 keV argon atoms. The elevation scans are measured with respect to the (100) surface of the crystal. The azimuth scan is for the detector at a fixed elevation of 45°.

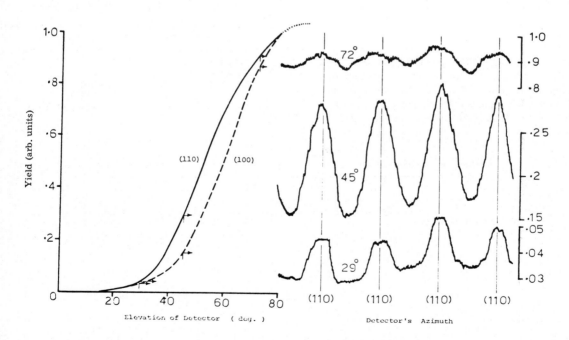

FIGURE 5   Angular distributions of positive ions sputtered from NaCl at 340°C by 15 keV argon atoms.

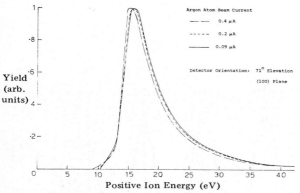

FIGURE 6   Yield of sputtered cations in a random direction at an elevation of 71° to (100) surface, target material KCl. The high elevation was chosen to minimize errors due to surface potential effects.

know of no other experimental results with which these results can be compared. For metals the sublimation energy is used for this comparison. However for alkali halides sublimation produces not cations but neutral molecular species.[13] The results of Table I follow the same trend as the bond energies in these alkali halides.

The peak position in Figure 6 is relatively independent of bombarding atom intensity, unlike the ion bombarded case shown in Figure 3, since the surface charge is smaller and much less variable for atom than for ion irradiation.

4 CLUSTERS

In addition to the alkali metal ions the mass filter revealed the ejection of a series of heavier positive ions of the general form K_2Cl^+, $K_3Cl_2^+$ for the alkali halides LiF, NaCl, KCl, KBr, and KI. Presumably a series of analogous negative ions occur but our detector was

not equipped to find these. The ejection directions of these heavier ions appear to be random but the sensitivity of our present equipment would not have allowed us to find distribution differences of less than a few percent. The mass distribution did not change over the range of target temperatures from 20 to 400°C. The cluster ions, expressed as a percentage of the total ion current collected by the detector in the same position, varied from 4% K_3Cl_2 and 7% K_2Cl^+ to 23% Li_2F^+.

ACKNOWLEDGEMENT

We are indebted to the Australian Research Grants Committee and the Australian Institute for Nuclear Science and Engineering for support.

REFERENCES

1. P. Bryce, J. Richards and J. C. Kelly, *J. Phys. C: Solid St. Phys.* **4**, L263 (1971).
2. P. Bryce and J. C. Kelly, *J. Phys. C: Solid St. Phys.* **5**, 1064 (1972).
3. C. Lehmann and P. Sigmund, *Phys. Stat. Solidi* **16**, 507 (1966).
4. M. W. Thompson, *Defects and Radiation Damage in Metals* (Cambridge University Press, 1969).
5. G. E. Chapman, M.Sc. Thesis, Univ. of New South Wales (1967).
6. J. R. Peterson and D. C. Lorents, *Phys. Rev.* **182**, 152, (1969).
7. G. C. Yarnold and H. C. Bolton, *J. Sci. Instr.* **26**, 38 (1949).
8. G. S. Harrower, *Rev. Sci. Instr.* **26**, 850 (1955).
9. W. Brubaker, *Proc. Intl. Instr. and Methods Conf. Stockholm* (Academic Press, New York, 1961).
10. D. W. Mason, *J. Nucl. Eng.* **6**, 553 (1964).
11. W. Lehfeldt, Z. Y. *Physik* **85**, 717 (1933).
12. M. W. Thompson, *Phil. Mag.* **18**, 377 (1968).
13. T. A. Milne and H. M. Klein, *J. Chem. Phys.* **33**, 1628 (1960).

TABLE I
Binding energies of alkali halides for cations sputtered by 15 keV argon atoms

	LiF	NaCl	KCl	KBr	KI
E_b	5.9 ± 1.8	5.8 ± 0.4	6 ± 0.3	5.2 ± 0.3	4 ± 0.5
No. of measurements	3	12	17	13	3

DEFLECTION OF SPUTTERED IONS BY SURFACE CHARGES

J. RICHARDS, R. L. DALGLISH and J. C. KELLY

Materials Irradiation Laboratory, Physics, University of New South Wales, Sydney, 2033, Australia

The trajectories of ions sputtered from an insulating surface carrying a surface charge have been computed. The results show that yields measured in particular directions can be in serious error as the trajectories of low energy sputtered ions depend critically on surface charge and point of origin of the ion. The distorted yield curves for 1.2 eV ions and 10 eV ions leaving a surface with a ten volt surface potential are calculated.

1 INTRODUCTION

Measurements[1] of the energy spectra of ions sputtered from alkali halides by heavy ions indicate that considerable surface charges are always present and that the charge distribution is not uniform. For neutral atom irradiated surfaces the surface potentials are only of the order of several volts and almost uniform over the area. The computation described here calculates the deflection of sputtered ions due to a uniform circular distribution of positive charge over the irradiated area, falling abruptly to zero at the beam edge. The trajectories prove to be fairly insensitive to potential variation across the charged surface which suggests that the errors involved in our assumptions do not seriously affect the results.

2 THE ELECTRIC FIELD ABOVE THE SURFACE

The field above the surface was computed by adding the contributions from segments of the charged disc as in Figure 1. An image charge 0.4 mm below the surface (corresponding to a metal backed crystal 0.2 mm thick) was included and the image charge calculated from[2]

$$\sigma_{image} = \sigma 2\epsilon_1/(\epsilon_1 + \epsilon_2)$$

taking the dielectric constant, ϵ_2, of the target crystal as six and that above the crystal as one. The inclusion of the image charge proved to have little effect on the trajectories.

A computer plot of equifield lines above the circular uniform charge distribution is shown in Figure 2 in contours of 0.05 times the maximum field. A surface potential of five volts, corresponding to a charge density of 1.25×10^{-7} coulombs/m^2 gives a maximum field in the centre of the distribution of 3000 V/m which is clearly able to distort the trajectories of sputtered ions.

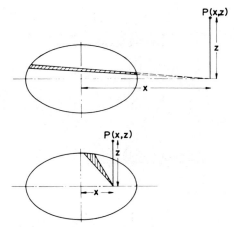

FIGURE 1   Procedure for deflecting field calculation.

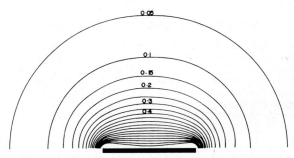

FIGURE 2   Equified lines above a circular uniform charge distribution in steps of 0.05 times the maximum field.

3 COMPUTED TRAJECTORIES

An iterative technique was used with the field calculated at the centre of each line segment and the length of the line segments increasing as the ion moved away from the crystal. The paths were followed up to 5 cm from the crystal. As all trajectories were smooth, interpolation methods were used between selected trajectories calculated in detail.

Some results for ions ejected in an axial plane with energies from 0.4 to 40 eV are shown in Figure 3 for a 1.5 mm diameter disc with a uniform positive surface charge of ten volts. The ions remain in the same plane because of symmetry and these are thus the simplest

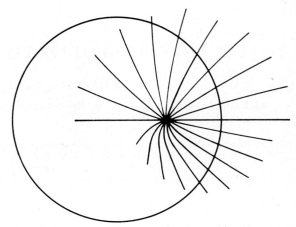

FIGURE 4   Plan view of the trajectories of ions launched with an initial elevation of 50° and energy 5.7 eV in planes 18° apart from a point 0.5 R from the disc centre. All trajectories are 3 R long. Those which are rising most steeply from the surface appear the shortest due to foreshortening.

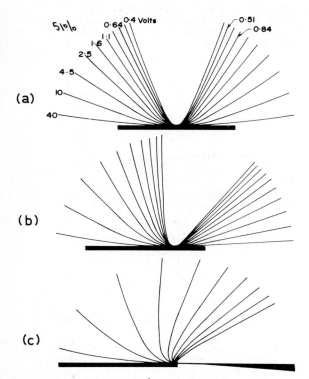

(a)

(b)

(c)

FIGURE 3   Computed trajectories of ions ejected near grazing incidence in an axial plane from a circular disc held at a potential of 10 volts as a function of ion energy from 0.4 to 40 eV. (a) launched from the disc centre, (b) from half way along a disc radius and (c) from the edge of the disc. The trajectories are viewed in elevation and the disc diameter is 1.5 mm. The same set of energies are used in each case.

elevation of 50° in planes 18° apart. All trajectories are three disc radii (R) in length. The apparently different length indicating the trajectory elevation angle. The point of launching is half way along a radius. In Figure 5 a similar series of ions are launched at 70° elevation, energy 1.2 eV from a point 0.75 R from the centre. For all but those launched in an axial plane,

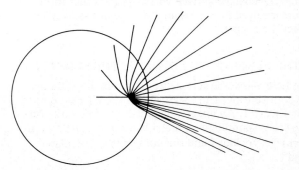

FIGURE 5   Plan view of ions with an initial elevation of 70° and energy 1.2 eV launched in planes 18° apart from a point 0.75 R from the disc centre. All trajectories are 5 R long.

trajectories. However even in this simple case it is clear that low energy ions launched initially at grazing incidence from a disc edge, Figure 3(c), away from a detector can be deflected sufficiently to enter the detector, and very large differences in final direction result from small differences in initial energy.

Ions launched in a non axial plane will follow spiral paths under the influence of the surface charge. Figure 4 shows a plan view of 5.7 eV ions launched at an initial

complex paths result. A fixed detector at any angle will collect many ions not initially launched in that direction and in the extreme case of low energy, may collect none of those it is assumed to be collecting.

4 ISOTROPIC YIELD DISTORTIONS

An initially isotropic yield becomes seriously distorted. We have calculated the shape of this distorted yield curve for the case of 1.2 eV ions leaving a surface with a 10 volt surface potential. The individual curves are shown in Figure 6 for particular initial angles of elevation together with the shape of the total yield curve. The total yield curve is the sum of the individual curves weighted by the cosine of the ejection angle to compensate for the projected area. The distorted yield is seen to be less than a cosine distribution from about 0–30° and to have a pronounced peak near 65° elevation. Similarly computed curves for 10 eV ions leaving a 10 volt surface are shown in Figure 7 where the peak has moved to an elevation of 80°.

In an actual ion beam sputtering experiment the distributions will be more complicated as the ejection is not all isotropic and a wide mixture of energies are found. In addition, electrostatic interactions with the incident beam[3,4] may be significant.

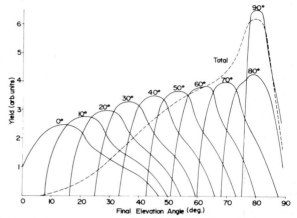

FIGURE 7   Yield vs. angle of elevation for 10 volt ions emitted under the same conditions as in Figure 6.

ACKNOWLEDGEMENTS

This work has been supported by the Australian Institute for Nuclear Science and Engineering and the Australian Research Grants Committee.

REFERENCES

1. J. Richards and J. C. Kelly. To be published.
2. V. V. Batygin and I. N. Toptygin, *Problems in Electrodynamics* (Academic Press, New York, 1961).
3. P. Bryce, J. Richards and J. C. Kelly, *J. Phys. C: Solid St. Phys.* **4**, L263 (1971).
4. P. Bryce and J. C. Kelly, *J. Phys. C: Solid St. Phys.* **5**, 1604 (1972).

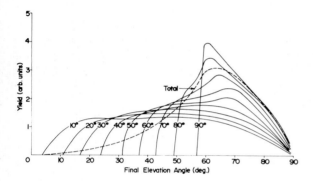

FIGURE 6   Yield vs. angle of elevation for 1.2 eV ions emitted isotropically and deflected by a uniform 10 volt surface potential. The individual curves are for ions initially isotropically emitted at the indicated angle of elevation and the total yield curve is obtained by weighting the individual curves by the cosine of the angle of elevation.

SURFACE DAMAGE AND TOPOGRAPHY CHANGES PRODUCED DURING SPUTTERING

R. S. NELSON and D. J. MAZEY

Metallurgy Division, A.E.R.E., Harwell, Berks

Ion bombardment of solids creates radiation damage in the form of interstitials and vacancies which agglomerate into dislocation structures, so becoming visible in the electron microscope. Furthermore, ion bombardment causes sputtering of atoms from the surface of solids which can give rise to changes in surface topography. The paper will briefly review the salient features of these two phenomena and in particular will point out their interrelation.

1 INTRODUCTION

It is well known that ion bombardment of solids results in the creation of radiation damage in the form of interstitials and vacancies. In many circumstances, these defects are mobile at the temperature of bombardment, and the resulting damage is a manifestation of their agglomeration. Furthermore, ion bombardment causes sputtering of atoms from the surface of solids which often gives rise to changes in surface topography. In this paper we will briefly review the salient features of these two phenomena, in particular pointing out their interrelation. It should be emphasized that a plethora of information exists on these subjects, and it is impossible to list all the numerous contributions in the field which have been published over the years. We shall therefore simply point out the observed features and present a discussion of the physical processes involved in a rather general way.

2 SURFACE DAMAGE

During heavy ion bombardment of metals at temperatures where both vacancies and interstitials are mobile radiation damage is manifest in the formation of defect clusters to a depth equal to the range of the incident ions (Figure 1). Such clusters grow rapidly into a complex dislocation network, which intersects the bombarded surface. It should be remembered however, that a substantial number of both vacancies and interstitials are lost by recombination either by mutual annihilation or at the free surface. This latter point infers that the surface receives a substantial flux of both defects and essentially acts as an infinite sink. Furthermore, as a consequence of processes such as dislocation slip and climb the dislocation network is in continuous motion during irradiation. This behaviour is readily seen during

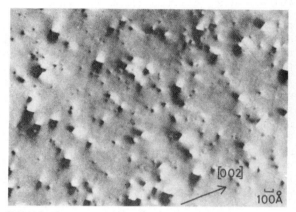

FIGURE 1   Isolated dislocation structures in low dose ion bombarded Cu.

observations of damage production in sophisticated experiments performed in combined ion accelerator-electron microscopy facilities.

Dislocations are associated with comparatively long range elastic strain fields which interact in a complex way with each other and with the free surface. The dislocation networks formed during ion bombardment will therefore be associated with excess elastic strain energy, and in a dynamic system, such as during irradiation, the network will move in such a way as to minimize the total free energy. This is manifest by the loss of dislocations to the surface due to strong image forces which are set up, and by the tendency for dislocations to align themselves in specific crystallographic directions (Figure 2). However, the fact that dislocations can only move in certain crystallographic directions as dictated by their specific Burgers vectors also plays a role in the detailed nature of any alignment. In this context, the orientation of the surface plays a critical role in determining the number of dislocations which

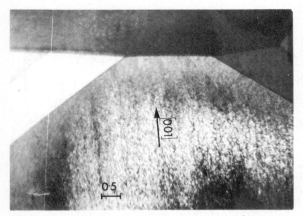

FIGURE 2   Dislocation network in high dose Xe⁺ bombarded
Cu {110} showing the tendency for dislocation line up.

can slip-out under the surface image force, and in
determining the exact nature of any tendency towards
dislocation line up. For instance, a simple edge dislocation in f.c.c. copper has an $a/2\langle 110\rangle$ Burgers vector
perpendicular to the dislocation line and can therefore
only slip in $\langle 110\rangle$ directions. The array which develops
from edge dislocations will therefore depend on the
number and relative disposition of the various $\langle 110\rangle$
directions lying parallel to the surface, see Table I. It
should be pointed out that each time a dislocation
moves to the surface, material is effectively transferred
from the bulk to form localized regions of extra atomic
planes on the surface.

TABLE I
f.c.c. lattice

Surface	No. of $\langle 100\rangle$ directions parallel to surface	Remarks
(100)	2 at 90°	Rectangular array
(111)	3 at 60°	Triangular array
(110)	1	Linear array
(310)	1	Linear array

During irradiation, atoms are continually removed
from the surface by sputtering, so leaving surface
vacancies. However, for every surface vacancy so
created, at least an order of magnitude more vacancies
and interstitials migrate to the surface from the bulk.
Surface vacancies and interstitials are thought to be
highly mobile and a substantial fraction recombine. A
consequence of this is that as well as the loss of atoms
by sputtering, the bombarded surface is in a highly
mobile state. Under such circumstances any gradients
in chemical potential will influence the flow of matter
both towards and over the surface. For instance, a

surface pit will tend to fill in so as to lower the total
surface free-energy.

The above discussion is clearly very generalized,
and the detailed behaviour of a particular metal will
depend on a variety of parameters, such as: crystal
structure, surface cleanliness, activation energies for
defect migration, temperature, sputtering ratio etc.
For instance, in the case of copper at room temperature, the vacancy is just mobile. This means that
recombination within the metal will significantly
restrict the flux of defects to the surface. On the other
hand, the dislocation density will build up to a relatively
high value so causing significant elastic strains in the
surface region.

3 THE INFLUENCE OF THE BOMBARDING SPECIES

The above discussion relates to the ideal situation which
might exist during so called self-ion bombardment,
where no complicating effects of foreign atoms can
perturb the situation. In practice, the majority of
experimental studies have used ions which are readily
available, such as the inert gases. In this section, we
will therefore review briefly the physical state of solids
which have been bombarded with impurity ions, and
consider how the implanted atoms can influence the
formation of surface damage topography changes.

During the course of a typical sputtering experiment, an appreciable quantity of the bombarding ions
become trapped within the target. For instance, the
saturation concentration is given approximately as the
inverse of the sputtering ratio, and this is usually of the
order of between 1 and 10 atoms/ion for heavy
particles; the average saturation concentration can
therefore be as high as between 10 and 100%. In general,
the final configuration of the injected atoms depends
on chemical and physical parameters, such as solubility and temperature. It is well known from the phase
diagrams of simple binary alloy systems, that once the
solid solubility limit has been exceeded, precipitation
occurs. Depending on the particular constituents, the
second phase so formed can consist of either just
implanted atoms, or a compound containing both
implanted and target atoms. We therefore have three
possibilities of the state of the alloy produced during
bombardment:

a) atomic solid solution, e.g. Au–Ag

b) precipitation of a compound second phase, e.g.
Cu–Al, oxides, nitrides or hydrides.

c) precipitation of an elemental second phase, e.g.
Pb–Al, or Au–Cu.

In the first case, if Ag^+ is the bombarding ion, no matter what concentration is attained, the injected atoms will always sit in sites indistinguishable from target atoms. In the second case, the alloy will consist of atoms in solution, together with small compound second phase precipitates, e.g. $CuAl_2$, ZrH_2 or SiO_2 (see Figure 3). It is worth pointing out that under

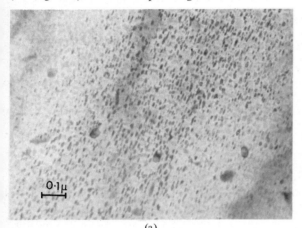

(a)

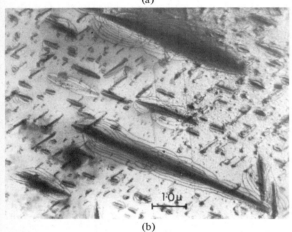

(b)

FIGURE 3   (a) $CuAl_2$ precipitates in Cu^+ bombarded Al. (Courtesy P. A. Thackery.) (b) Zr H_2 precipitates in proton bombarded Zr. (Courtesy G. P. Walters.)

these conditions of ion bombardment, precipitation can occur at very much lower temperatures than usually experienced during isothermal heating. This is because the defects produced during bombardment can create an enhanced diffusion of both solute and solvent atoms. In the third case, where either element is essentially insoluble in the other, precipitation leads to isolated second phases of the injected atoms. A classic example of this is the inert gases, which even in liquid metals have solubilities below

10^{-10}. At temperatures where thermal or irradiation enhanced diffusion occurs such gas atoms will rapidly agglomerate. Such agglomeration forms small pockets of gas within the solid which behave very similarly to gas bubbles in a liquid, e.g. they grow to a size defined by the surface energy of the solid, they can migrate through the solid by diffusion processes, they may grow by coalescence, and will burst when they intersect a free surface. Such agglomerates in solids are therefore called gas bubbles, and their behaviour is readily studied by techniques using equipment such as a combined ion accelerator and electron microscope (see Figure 4).

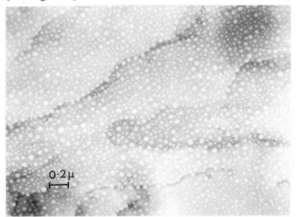

FIGURE 4   Argon gas bubbles in A^+ bombarded Cu at 400°C.

We must now discuss the implication of precipitate formation on the evolution of radiation damage structures and on the state of the sputtered surface. First of all, it is well known that precipitates, solid or gaseous, act as pinning points for moving dislocation lines and can thus modify the nature of the dislocation networks which form during bombardment. However, perhaps the major implication stems from the intersection of the surface with such precipitates as it steadily recedes under sputtering. In the case of bubbles, these will burst as the surface approaches so leaving small craters; whether such craters persist or heal will then depend on the particular environment e.g. bombardment temperature etc. However, in the case where solid precipitates intersect the surface, the subsequent behaviour will depend on the relative sputtering ratios of the precipitate and the surrounding matrix, and in some instances on the vapour pressure of the precipitate at the bombardment temperature. For instance, if the precipitate has a low sputtering ratio, then the matrix will recede at a faster rate, so leaving the precipitates projecting from the surface; however, this will be considered later.

4 SURFACE TOPOGRAPHY

The changes in surface topography which occur as a consequence of sputtering have been studied for many years. However, only recently has any significant understanding emerged. The detailed structures depend on variables such as sample purity, surface cleanliness, grain size, crystal orientations and the bombarding ion species.

The majority of theoretical work[1,2] has been concerned with what may be termed amorphous solids and takes no account of phenomenon such as channelling, dislocations, diffusion processes or of the variation of binding energy with orientation. Some of these studies have been aimed at explaining the formation of cones which appear on a variety of sputtered surfaces, see for example Figure 5. Such cones generally

FIGURE 5   Cones developed on the surface of a tin crystal after A[+] ion bombardment (courtesy M. W. Thompson).

appear after the removal of several microns of material and have their axis parallel to the direction of incidence of the bombarding ions. It is generally assumed that dust particles, small inclusions or precipitates are responsible for the formation of cones. For instance, if such a precipitate is situated in a surface which is being eroded by ion bombardment, the material beneath the precipitates will be shielded from the bombarding ions. Furthermore if the sputtering ratio

of the precipitate or particle is significantly less than the surrounding material a protrusion will develop as a consequence of the erosion process. Eventually the precipitate will be removed, leaving behind the protrusion. It is then the subsequent changes in shape of such protrusions which are thought to give rise to cone formation. For instance it is well known that the sputtering ratio of a polycrystalline or amorphous solid varies with angle of incidence approximately as the secant of the angle, passes through a maximum, $S(\hat{\theta})$ between about 60–80° and falls to zero at 90°. It is then easily shown that as sputtering continues, a protrusion will develop facets corresponding to the development of those faces with the highest sputtering ratio. A hemispherical protrusion will for instance degenerate into a cone of semi-angle $[(\pi/2) - \hat{\theta}]$ with its axis parallel to the ion beam direction.

The simple explanation of cone formation outlined above appears quite adequate to account for the more general features. However, closer examination of the background material in Figure 5 shows a rather interesting fine scale furrowed structure. Such structures are in fact quite commonly observed after ion bombardment of clean high purity specimens irradiated with ions of the inert gases. The detailed nature of the structure depends intimately on the orientation of the particular specimen or grain within a specimen, see for instance the selection of photographs in Figure 6 and 7. In this context, we should recall that differential erosion from grain to grain in a polycrystalline sample, can occur as a consequence of two phenomena. Firstly

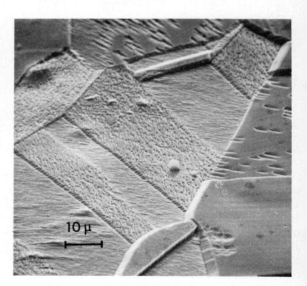

FIGURE 6   Surface of Xe[+] bombarded polycrystalline Cu after removing several microns.

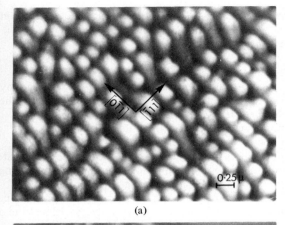

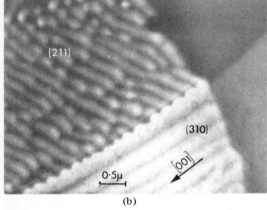

FIGURE 7 Transmission micrographs of Xe⁺ bombarded Cu, after removing about 500 Å, (a) (211) orientation, (b) mixture of orientations showing delineation of structure at grain boundary together with an area shielded during bombardment.

if the ions are incident along a low index direction ion-channelling will determine the energy density deposited in the surface layers, and hence the energy available for sputtering. Secondly, the sputtering yield is strongly dependent on the surface binding energy; different crystal faces have different values of binding energy, which will therefore influence the sputtering from differently orientated grains. It is thus clear that each grain in a polycrystalline sample will sputter at a different rate (depending on orientation) with the result that an irregular structure varying in depth from grain to grain will develop on the surface. However, let us return to a discussion of the evolution of the periodic structures as mentioned above. It is perhaps useful to list some of the major features of the observations.

1) Sputtering structures form on flat, unetched surfaces, at a periodicity which does not significantly alter as it develops during subsequent prolonged bombardment.

2) The nature and periodicity of the sputtering structures bear a close relation to the irradiation induced dislocation networks which form under the surface.

3) There is some evidence which suggests that regular structures develop more readily when the ion beam is incident along a channelling direction. However, structures do develop under random incidence ion bombardment.

4) Identical structures are formed for instance during A⁺, Xe⁺, Pb⁺ and Cu⁺ ion bombardment of copper.

It has been suggested that it is the tendency to form regular dislocation arrays which is the underlying reason for the development of regular surface structures.[3] The question then arises: how can the dislocation arrays influence the sputtering structure? Two possible mechanisms have been suggested.

1) *Dechannelling at Dislocation Lines* It is well known that if the bombarding ions are incident along a channelling direction, defects below the surface can result in dechannelling. Thus in the vicinity of dislocation lines the local sputtering rate can be increased relative to the surrounding material. Hermanne and Art[4] have therefore suggested that for the case of channelled bombardments, the existence of regular dislocation arrays below the surface can give rise to periodic sputtering structures as a consequence of localized bands of increased sputtering.

2) *Modification of Surface Binding* We have already stated that the surface binding energy depends on the orientation of the specimen surface. However, local variations in binding energy over a surface can arise from variations in elastic stress. The sputtering ratio is to a first approximation, inversely proportional to the binding energy, so local variations will readily produce differential sputtering. It is well known that significant variations in binding energy occur in the vicinity of grain boundaries or in the immediate vicinity of a dislocation which intersects a surface. Such variations can give rise to grain boundary grooving or localized etch-pits. However, we must consider the implications of regular dislocations array below the surface, and estimate the elastic forces which these produce in the surface itself. For simplicity if we assume a parallel array of dislocations, of separation h, located a distance, a, from the surface elastic theory allows us to calculate the stress distribution over the surface. Using standard techniques Bullough[5] has

calculated the following expression for the surface stress.

$$p_{yy}(o, y) = - \frac{2b\mu}{h(1 - \gamma)}$$

$$\times \left[\frac{\sinh{(2\pi a/h)}}{\cosh{(2\pi a/h)} - \cos{(2\pi y/h)}} \right.$$

$$\left. + \frac{2\pi a \{1 - \cosh{(2\pi a/h)} \cos{(2\pi y/h)}\}}{h\{\cosh{(2\pi a/h)} - \cos{(2\pi y/h)}\}^2} \right]$$

where y defines the position in the surface, b is the Burgers vector of the dislocation, μ is the shear modulus and γ is Poisson's ratio. The stress in the surface is therefore highly oscillatory with a periodicity defined by the dislocation spacing. If we take typical values for the various constants pertinent to copper, and choose h and a equal to 2000 Å and 20 Å respectively, we calculate a maximum stress of $\sim 5 \times 10^9$ dynes/cm^2. This corresponds to a reduction in surface binding energy of ~ 0.5 eV. The equilibrium binding energy for copper is of the order of 3.5 eV and it is therefore reasonable to expect a periodic sputtering structure to evolve during erosion as a consequence of the periodic variation in binding energy.

However, the periodic elastic forces which appear below and in the surface, can give rise to other effects which may play a part in the formation of regular sputtering structures. Firstly, the variation in chemical potential in the surface will influence the diffusional migration of both vacancies and interstitials over the surface. We recall that for every atom sputtered from the surface, many more defects flow to the surface from the interior of the metal. We expect a substantial number of these to annihilate, but even so, material will flow across the surface in an attempt to lower the total free energy of the system. In other words, we might well expect a periodic structure to develop if this has the overall effect of reducing the total strain-energy of the system. Secondly, dynamic observations of the changes in dislocation networks which occur during irradiation suggest that substantial amounts of material are transferred within the vicinity of the surface by extensive dislocation slip and climb. We have already pointed out that a large number of dislocations can slip out from the surface under the influence of their image force; however, those edge dislocations which have their Burgers vector parallel to the surface can in fact only slip in a direction parallel to the surface. Experiment has shown that during irradiation such dislocations are influenced by the

periodic stresses and move in such a way as to account for significant material transfer into the regions of the sputtering structures.

We therefore have several mechanisms by which regular structures can be initiated on crystalline surfaces during ion bombardment. However, whether any of these are in themselves sufficient to account for the observed structures which develop after continued sputtering, is not clear. If we invoke the mechanism proposed for cone formation, then we would expect sputtering structures to expose those surfaces associated with the maximum sputtering ratio. In the case of single crystal surfaces, or individual grains, the angular dependence of sputtering ratio is quite different from that expected for amorphous or polycrystalline solids.

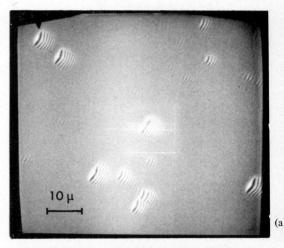

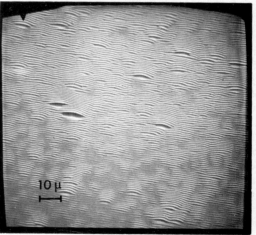

FIGURE 8   Xe$^+$ bombarded Si (courtesy J. A. Cairns), (a) low dose showing early stage of structure, (b) higher dose showing complete structure.

Local maxima and minima appear depending on channelling effects etc. and these would give rise to features somewhat different from those to be expected based on amorphous target results. However, the observations are in general consistent with such a mechanism.

The periodic structures so far discussed appear in materials such as copper which remain essentially crystalline even after extensive bombardment, further-more, the periodicity of such features is about 2000 Å and is independent of the angle of incidence. However, materials such as glass or silicon (which is rendered amorphous during ion bombardment at room tempera-ture), remain relatively smooth during bombardment at normal incidence. On the other hand, if silicon is bombarded at an angle of say 45°, then periodic structures can in fact develop—see for instance Figure 8. It is interesting to note that in contrast to the features described previously, in this case, the periodi-city is of the order of 1 μm. Furthermore it has no specific orientation relative to the original crystal structure, but depends only on the ion beam direction, the striations lying perpendicular to the ion beam. The gradual development of the structure from discrete regions, as illustrated in Figure 8 is also quite different from that found in metals. As yet, no unambiguous explanation of this phenomenon has been provided; however, it is possibly associated with regions of

segregated impurity within the silicon, which steadily becomes exposed as the surface is eroded. Effects of strain caused by the reduction in density of the silicon during amorphisation may also play a role; but apart from these rather vague suggestions, we can say nothing definite at the present time to explain such structure as those observed on silicon.

5 SUMMARY

In this review we have briefly described the experi-mental evidence for surface damage and the develop-ment of surface topography during ion bombardment. During recent years the realisation that damage structures and topography changes are closely related, has provided a better understanding of the underlying mechanisms responsible for the development of surface structure. However, although some advances have been made, more critical research work is needed before we have an unequivocal understanding.

REFERENCES

1. A. D. G. Stewart and M. W. Thompson, *J. Mat. Sci.* **4**, 56 (1969).
2. C. Catana, J. S. Colligon and G. Carter, *J. Mat. Sci.* **7**, 467 (1972).
3. D. J. Mazey, R. S. Nelson and P. A. Thackery, *J. Mat. Sci.* **3**, 26 (1968).
4. N. Hermanne and A. Art, *Rad. Effects* **5**, 203 (1970).
5. R. Bullough, Private Communication (1972).

Appendix

CINE FILM OBSERVATIONS

The presentation of this paper at the Conference was accompanied by a short 16 mm ciné film which demon-strated some of the irradiation damage phenomena discussed in the paper. The irradiation experiments were carried out in a combined ion accelerator and electron microscope facility which permits the direct observation of irradiation effects during ion bom-bardment of specimens in the electron microscope. The facility consists of a 120 kV heavy-ion accelerator connected by a flight tube to electro-static deflection plates which directs the ion beam at an angle of 45° into the specimen chamber of a 200 kV JEOL electron microscope. The film was produced from video tape recordings of dynamic changes occurring in thin metal foils whilst under ion bombardment.

Initial sequences showed formation of dislocation

loops, dislocation networks, and voids in aluminium during bombardment with 100 keV Al^+ ions at a dose rate of 1×10^{13} ions cm^{-2} sec^{-1}. The next sequence showed the progressive stages of surface sputtering of copper during 60 keV Xe^+ ion bombardment (2.5×10^{13} ions cm^{-2} sec^{-1}) at 25°C. The damage consisted initially of point-defect clusters followed by the rapid formation of dislocation loops which grew and interacted to produce dislocation alignment. Sputtering structure consisting of an aligned surface undulation was seen to evolve parallel to the dislocation alignments. It was then shown that prolonged bom-bardment of aluminium at room temperature with 100 keV Ar^+ ions (1×10^{14} ions cm^{-2} sec^{-1}) leads to the formation of inert gas bubbles in the metal, and blisters between the metal surface and its oxide.

The final sequence showed examples of formation,

growth, coalescence, and bursting of argon bubbles in copper foils during bombardment with 100 kV argon ions (5×10^{13} ions cm^{-2} sec^{-1}) at 450°C.

In most of the processes demonstrated in the film the irradiation damage structure resulting from atom displacements was characterized by the continuous formation and movement of point-defect clusters and dislocation loops created by the flux of ions. The film served to provide an indication of the physical changes produced in metal surfaces by high-dose ion bombardment, and the value of a facility which permits observation and recording of such dynamic phenomena.

A NECESSARY CONDITION FOR THE APPEARANCE OF DAMAGE-INDUCED SURFACE-TOPOGRAPHY DURING PARTICLE BOMBARDMENT

N. HERMANNE

Physique des Surfaces, Fac. Sci. Université Libre de Bruxelles

A first approach to the establishment of an "order-of-magnitude" criterion for the initiation of damage-induced surface topography on clean and well-polished metal single crystals is proposed. It is based on a model proposing a mechanism of preferential erosion of parts from the sample surface due to a local increase of sputtering yield above "extensive defects". These "extensive defects", resulting of the migration of the point-defects, created during the collision-cascade started by the incoming ions, can indeed influence the sputtering yield if they are located close under the surface, i.e. when the surface has regressed, due to sputtering, to the level where these "extensive defects" have been formed.

The condition implies that the formation of "extensive defects" must have taken place before the surface reaches them. The condition is written in terms of the irradiation parameters and quantities characteristic for the migration of point-defects due to the presence of a potential gradient related to the "extensive defects" under formation.

INTRODUCTION

It is well known that ion, or more generally particle irradiation introduces damage in the bulk of the irradiated crystal and also, under some experimental conditions, changes in surface structure. In the context of this paper two types of surface structure should be distinguished according to their origin.

One type consists of features of different aspects (e.g. cones, spikes, craters, protusions) appearing on amorphous, poly- or single-crystalline surfaces, whose origin (Figure 1) is related[1,2] to impurities or irregularities present in or under the surface before the beginning of the irradiation. It can for example happen that a regular type of structure exists on amorphous or polycrystalline surfaces crossing the different grains. It appears that they are growing above a system of traces remaining after the pre-bombardment treatment, particularly polishing.[3] Important is that the origin of this first type is independent of implantation and of the radiation damage created by the incoming ions and that, if desired, its presence can be avoided by adequate cleaning and polishing and use of pure samples.

The second type is a more or less regular pattern of arrays or any of the previous features appearing (Figure 2) on single-crystal surfaces or on the surface of only some of the grains of polycrystalline samples. As these also[9] appear on surfaces, very clean and smooth before bombardment, one has to conclude that their origin is related to radiation damage, i.e. the defects created and impurities implanted during the bombardment (Figure 3).

Rather confusing is that similar-looking features can have different origins, e.g. regularly distributed protusions (blisters) can be of the second type,

FIGURE 1 Scanning micrography showing typical aspects of surface topography of the first type of origin: cones have been left over under parts of the surface shadowed by impurities present before the initiation of the bombardment (Cu sphere irradiated with 6 keV A$^+$ ions).

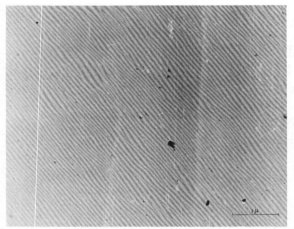

FIGURE 2   Replica electron micrography of a typical aspect of a regular structure of the second type of origin: ribbons grown by a mechanism of preferential sputtering above dislocation lines due to the irradiation. (Cu irradiated with 1.6 mA/cm^2, 5 keV A$^+$ ions); by courtesy of A. Art.

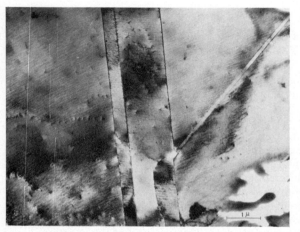

FIGURE 3   Transmission micrography displaying the previous type of surface structure on different grains of a polycrystalline Cu thin-foil, showing that its direction is related to the crystal orientation. This type of structure is originated due to the bombardment (CuGe 8% atomic, irradiated with 5 keV A$^+$ ions at d.r. = 10^{18} ions/sec · cm^2, during 70 sec).

whereas irregular arrays are not necessarily of the first type.[12]

Some surfaces display together features which could be of both types of origins,[4,5] when both mechanisms act without competing.

As there is an influence of surface-roughness on measurements done on irradiated samples, e.g. yield,[6] resistivity,[7] back-scattering, scattering,[8] etc.; it is necessary to know more about the origin of the second type of surface topography. It was observed

previously by several authors using different methods, that its origin is dependent on the irradiation parameters (doses, energy of ions, target type, incidence direction, crystal orientation of the sample, target temperature, ion type, etc.) as reviewed[13] by R. Behrisch and later on by G. Carter and J. C. Colligon. The aim of this paper is to establish an "order-of-magnitude" criterion for the possible appearance of these damage-induced surface structures, taking into account all experimental parameters.

A model for the mechanism of formation of this type of surface topography starting from the creation of point defects in the collision cascade of the incoming ions was proposed previously.[9] It is based on the migration of point-defects in clusters (this supposes a relatively high concentration of radiation induced damage). When these clusters become bigger by absorption of more point-defects they grow out to one of the configurations of extensive defects; dislocations for example.[10,11] If the surface which is regressing due to sputtering approaches one of these extensive defects, these cause a local decrease of penetration for the ions entering the crystal in their neighbourhood. The consequent local increase of sputtering yield is the origin of pits in the surface above the extensive defects, which, from then on, grow by a Stewart–Thompson mechanism[1] adapted for single-crystals if necessary,[9] until reaching their equilibrium configuration if the irradiation goes on.

The regularity or randomness of this type of surface structure depends on the configuration of the extensive defects, when the surface reaches their level.[12] For example, if the clusters have become loops which open when the surface reaches them to form dislocation lines, the dislocation lines interact to spread out over the whole width of the grain and thus form a regular network. The surface structure growing above such a network is very regular and its symmetry depends on the crystal orientation. On the other hand, if the extensive defects do not reach a form where they can interact before the surface reaches their level and thus keep randomly spread, the surface structure developing above these will be randomly spread too.

A necessary condition for the development of surface structure is thus the formation of extensive defects at a certain level under the initial surface, before the surface regresses, due to the sputtering to this level. It is necessary that the defects are extensive, for isolated point-defects do not affect the penetration of a sufficiently large amount of incoming ions to cause a sputtering yield increase. (Later on, we will take as "extensive enough" the size of agglomerates of defects overgrowing the maximum size of stability of three-dimensional clusters, i.e. the size of collapsing into (two-dimensional) loops. This size (30 A for copper), appears to be extensive enough to be at the origin of surface formations[14]).

At this point we will restrict ourselves to the topography induced by extensive defects, grown by the migration of point-defects of target atom type. Of course the formula established below can be enlarged to surface structures which find their origin in the agglomeration of implanted projectiles (e.g. precipitates or gas bubbles), after minor modifications concerning the migration speeds of these particles through the substrate (in Section 2). In the first case, the condition can, according to the model,[9] be expressed as: time (*formation of extensive defects*) < time (*regression of the surface to their level*). This means that a sufficiently large amount of defects must have migrated towards the position where an extensive defect will grow in the time that the surface regresses through the distance separating its initial level from the level of high damage concentration, i.e. approximately the depth of the peak of damage distribution. In terms of migration speed averaged over all the defects participating in the formation of an extensive defect and of the average distance they have to cover, this condition (as it can be seen from Figure 4, the velocities are parallel respectively to d_d and d_l and we treat from now on

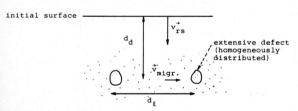

FIGURE 4   Schematic representation of the damage produced by incident ions.

with the corresponding scalars) can be written in the following form:

$$\frac{d_d}{v_{rs}} > \frac{\frac{1}{4}d_l}{v_{migr.}} \tag{1.1}$$

where:

v_{rs} is the regression speed of the surface, perpendicular to itself.

d_d is the depth of damage peak under the initial surface of the sample.

$v_{migr.}$ stands for an average migration speed towards

the extensive defect of all the defects which will contribute to the formation and growth of it.

d_l is the distance between two neighbouring extensive defects. The defects contributing to the growth of each extensive defect are initially spread randomly in a volume limited by $(\frac{1}{2}d_l)^3$. The distance to be covered by the defects to reach the extensive defect is on an average (approximately) $\frac{1}{4}d_l$. It is here supposed that all the defects present in the zone participate to the growing of the extensive defects, which is a very crude approximation and will be corrected for later on.

All these quantities depend on the values of several experimental parameters, except $v_{migr.}$ which only depends on the target temperature as experimental parameter and on quantities typical for the target. As it is also the only quantity in (1.1) depending on temperature, it is going to play a special role in the following calculations. Condition (1.1) can also be written as:

$$v_{migr.} > \frac{d_l \times v_{rs}}{4d_d} = v_{min} \tag{1.2}$$

For convenience, let us first find (Section 1) an expression in terms of the experimental (irradiation) variables of the right-hand side of (1.2), which will be represented by v_{min}, i.e. a minimum value for the migration speed of the defects if we want to satisfy the necessary condition. Later on we will express (Section 2) the temperature dependence of the migration speed of the defects through the lattice towards the sinks in terms of parameters typical for the target.

1 Expression for v_{min} Related to Irradiation Parameters

a) The speed at which the sample surface regresses, due to sputtering, is:

$$v_{rs} = \frac{Y \times d.r.}{c} \tag{1.3}$$

where dose rate (d.r.) is given in ions/sec. surf. unit.

yield is expressed in atoms/ion and is noted Y.

c = atomic density of target (atoms/vol. unit).

b) The distance d_l, appearing in (1.2), depends on the number of extensive defects formed per unit volume and thus on the amount of point-defects present at the concerned level. Therefore d_l will be calculated in terms of the resulting damage concentration at the concerned level. Indeed, as shown below (Figure 5), from a certain moment on, the total amount of point-defects at a given level becomes almost independent of time, i.e. of total dose.

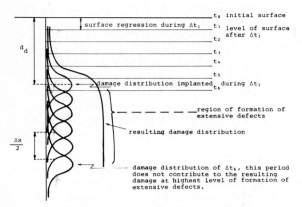

FIGURE 5   Resulting damage distribution.

A statement that will be used here, is that, for the reasons explained previously we will take as condition for having extensive defects that the maximum size for stability of the 3-dimensional configuration of aggregates of defects is overgrown and loops are formed. To overgrow the stable size of 3-dimensional clusters, a minimum amount of defects is necessary: n_{min}.

The number of loops formed per volume unit ρ_l is related to the concentration of defects (ρ_l) by:

$$\rho_l = \frac{\rho_d}{n_{min} \times \text{unit vol.}}$$

The average volume concentration of defects ρ_d is found by calculating the total quantity of defects created per irradiated surface unit divided by the depth (same units) over which these defects are spread (i.e. the distribution width, noted Δx):

$$\rho_d = \frac{\text{dose} \times v_d}{\Delta x}$$

where the mean defect production rate v_d is the average number of defects created by each incoming ion in the crystal during its collision cascade, and the dose is calculated over the time which is needed to tend to a stable resulting damage distribution at the level we are concerned with. During irradiation the surface regresses and, at constant depth under it, the damage distribution too (Figure 5). (This is not exactly realistic, for during the bombardment, the penetration decreases due to the presence of damage in the crystal). The number of defects created at a certain level under the surface increases continually beginning from the moment where its distance to the surface is equal to the maximum penetration, until the surface and the corresponding damage distribution have shifted downwards (at speed v_{rs}) so far that its contribution is negligible (e.g. from t_1 to t_5 on Figure 5). The defects created in this period are spread over a width approximately equal to the width of the damage distribution, thus:

$$\text{dose} = \text{d.r.} \times \frac{\Delta x}{v_{rs}}$$

The number of loops per vol. unit created per bombarded surface unit and spread over the distribution width of damage is given by:

$$\rho_l = \frac{\text{d.r.} \times v_d}{v_{rs} \times n_{min}}$$

The distance d_l between closest neighbouring loops can be approached by the expression for a homogeneous distribution:

$$d_l \simeq \frac{1}{\sqrt[3]{\rho_l}} = \sqrt[3]{\frac{v_{rs} \times n_{min}}{\text{d.r.} \times v_d}} \qquad (1.4)$$

a) The mean defect production rate v_d can be expressed by the formula[15] for the number of defects created by an ion:

$$v_d = E_i/2E_d$$

where E_d is the displacement energy of an atom in the lattice and E_i is the energy of the incoming ion.

b) The radius associated with the volume occupied by n_{min} defects is approximated by the following relation[16]:

$$n_{min} = \frac{4\pi r_0^3}{3} c, \quad \text{or:} \quad n_{min} = 4.2 r_0^3 c$$

where:
   c is the atom density of the target.
   r_0 is given by $\mu b^2/\gamma$.
   μb^2 is the formation energy of dislocation line per unit length.
   γ is the energy/surface unit.

as far as the assumption that the density inside this volume is equal to the density in the unperturbed lattice is valid.

c) The depth d_d is approximated by the mean projected range, expressed (for amorphous targets) by:[17]

$$d_d \sim R(A) \simeq \frac{130 \, E_i}{g} \times \frac{1 + M_2/M_1}{Z_1^{2/3}} \times \frac{1}{1 + M_2/3M_1}$$

where:

g is the density of the target expressed in g/cm^3
M_2 is the target atom mass
M_1 is the ion mass
Z_1 is the atom number of the ion
E_i is the energy of the incoming ion, expressed in keV.

From (a) and (b) expression (1.2) becomes now:

$$v_{min} = \frac{1}{4} \times \frac{(v_{rs})^{4/3} \sqrt[3]{n_{min}}}{\sqrt[3]{v_d} \times d_d \times \sqrt[3]{\text{d.r.}}}$$

and combined with (1.3):

$$v_{min} = \frac{1}{4} \times \frac{(n_{min})^{1/3} \times Y^{4/3} \times \text{d.r.}}{(v_d)^{1/3} \times d_d \times c^{4/3}} \quad (1.5)$$

In Eq. (1.5), we see the role of the different experimental parameters. Indeed, ion energy appears in the sputtering yield and depth of damage distribution d_d; type of target influences n_{min}, sputtering yield, atom density c, mean defect production rate and d_d; whereas the type of ion acts on yield, mean defect production rate and d_d.

To express v_{min} in explicit terms of more practical quantities, let us introduce the following approximations, although some of them are very crude, but satisfactory for this order-of-magnitude estimation.

It is evident that if reliable experimental values are available for mean defect production rate or depth of damage distribution for the particular case one is interested in, it is far better than using the previous approximations, to substitute these values in the expression (1.5).

After substitution of the previous approximations and taking account of $g = cM_2/N_{Av}$, expression (1.5) becomes now:

$$v_{min} = 22 \cdot 10^{-20} \times r_0 \times Y^{4/3} \times \text{d.r.} \times \frac{E_d^{1/3}}{E_i^{4/3}}$$
$$\times Z_1{}^{2/3} \times \frac{M_2(3M_1 + M_2)}{M_1 + M_2} \quad (1.6)$$

2 Migration Speed of Point-Defects in the Stress-Field of a Sink

The migration of point-defects in a stress-field being a very complex problem, we will restrict ourselves to a simple approach for this work. It is assumed here that dose has no influence on diffusion and that normal diffusion theory is applicable. This supposes that the implanted layer is either of low concentration, so that no bubble formation or precipitation takes place (i.e. high damage production rate); either not lying at the same level as the damage layer (i.e. high energy projectiles), or, that the total dose is small.

The potential gradient due to the presence of a sink in a crystal induces a migration of each of the defects present in the crystal towards or away from the sink (depending on type of point-defect and sink), to be distinguished from the normal diffusion (random motion).

The mean migration velocity $\vec{v}$ in a potential V is given by:[18]

$$\vec{v} = -\frac{D}{kT}\overrightarrow{\nabla V} \quad (2.1)$$

where D is the diffusion coefficient, T the absolute temperature and $\overrightarrow{\nabla V}$ stands for the present potential gradient.

The diffusion coefficient D, appearing in (2.1) is temperature dependent as given below:[19]

$$D = D_0 \times e^{-E/RT} \quad (2.2)$$

with D_0 as temperature independent diffusion coefficient, E as activation energy (per mole) and R as gas constant.

This migration speed will vary with the distance separating the sink and the defect, corresponding to the distance-dependence of the potential gradient. We will therefore first find an expression for the migration speed of a defect located at a given distance from the sink (Section 2.1) and will later on see how the average migration speed of defects towards sinks can be expressed in terms of the latter (Section 2.2).

2.1 As the formation and growing of the sinks, which will cause the migration of the point-defects, only starts with the beginning of the bombardment, the stress-field due to the former and even their type will vary, starting from very small agglomerates of point-defects up to extensive defects. We will thus treat the problem for a sink without specification.

The important factors determining the potential of interaction between two stress-fields (one surrounding the point defect and the other due to the presence of a sink) are:[20]

a) the elastic interaction of the two stress-fields

b) the electrical interaction arising from the redistribution of the electrons

c) the interaction with the non-elastic distortions very close to the sink.

The last one can be neglected here in comparison to the first one because of the fact that even if the defect is submitted to a very strong interaction during the short period when being very close to the sink, this will not influence the average migration speed with which it covers the "long" distances from its point of creation to the close neighbourhood of the sink. The interaction potential $V(r)$ of a defect with a stress-field can in some cases be written as:[13]

$$V(r) = -|A|/r \qquad (2.3)$$

where $|A|$ is a constant depending on the type of sink, and r the distance between point-defect and sink.

Let us consider positive r away from the sink and from now on, we consider only the absolute value of the mean migration velocity.

Due to the form of the potential, the potential gradient $\overrightarrow{\nabla V}$ can here be written as $\partial V/\partial r$, where V is the potential due to the sink, acting on a point-defect. The potential gradient thus becomes:

$$\frac{\partial V}{\partial r} = +|A|/r^2 \qquad (2.4)$$

Let us now assume that the distance from each position of relative dip of potential energy to the next one is covered at constant speed, i.e. a transition of the continuous varying migration speed to a discrete variation; and on the other hand, that the migration of the defects happens along preferential crystallographic directions along which the distance between two next equilibrium positions for the defect is equal to the crystal parameter a. We can express the distance r as entire multiples of the crystal parameter $a (r = na)$.

After combination of (2.1), (2.2) and (2.4), the migration speed of a defect (located at distance $r = na$ from the sink without initial migration speed) moving from this position $r = na$ to position $(n-1)a$, is given by:

$$|v(na)| = \frac{D_0}{kT} \times \frac{|A|}{n^2 a^2} \times \exp\left(-E/RT\right) \qquad (2.5)$$

A defect, during its migration approaching the sink, crosses a series of positions of relative dips in potential energy (i.e. the equilibrium positions in an unstressed crystal), where it undergoes (in our discrete approach) subsequent changes in migration speed. The migration speed with which a defect moves from position ia to position $(i-1)a$, is given by the sum of the migration speeds with which a defect created in each position

(ia) is leaving this position, from ia to na, where na is the position where it was actually originated:

$$|v_{ia \to (i-1)a}| = |v(ia)| + |v((i+1)a)| + \ldots$$
$$+ |v(na)| \qquad (2.6)$$

The average migration speed $\langle v_{na} \rangle$ of a defect from its position of creation na towards the sink has to be expressed as the sum of the speeds with which the n successive steps between position of creation of defect and sink are covered divided by n:

$$|\langle v_{na} \rangle| = \frac{1}{n} \{ |v_{na \to (n-1)a}| + \ldots + |v_{ia \to (i-1)a}|$$
$$+ \ldots + |v_{1a \to 0a}| \}$$

or after replacing each of the $v_{ia \to (i-1)a}$ by the expression (2.6) and after regrouping the terms:

$$|\langle v_{na} \rangle| = \frac{1}{n} \left\{ \sum_{n=1}^{n} i |v(ia)| \right\} \qquad (2.6')$$

2.2 To obtain the migration speed $v_{\text{migr.}}$ as appearing in (1.2), we have to average over all the defects participating in the elaboration of the extensive defect, i.e. over all possible distances between positions of creation of the point-defects and the sink. This means we have to average $|\langle v_{na} \rangle|$ over all values of the distance na ranging from 0, $1a$, $2a$, ... to $na_{\max} = d_l/2$; taking into account that the point-defects can be created at distances varying from 0 to $\frac{1}{2} d_l$ from the closest sink. We suppose here that point-defects created further than $\frac{1}{2} d_l$ away from any sink, do not contribute to the growth of the extensive defects.

As in each spherical shell of radius na different quantities of defects are created corresponding to the differences in volume of these shells, the previous average has to be weighted by the quantity α_n of contributing defects created in each of the shells:

$$|v_{\text{migr.}}| = \frac{\displaystyle\sum_{n=1}^{n_{\max}} \alpha_n |\langle v_{na} \rangle|}{\displaystyle\sum_{n=1}^{n_{\max}} \alpha_n}$$

As the initial defect distribution is homogeneous (i.e. constant volumic concentration of defects) the amount α_n of defects created in a spherical shell at distance na of the sink and thickness a, can be written as:

α_n = amount of volume units in the spherical shell x volumic concentration of defects

= vol. conc. defects x $4\pi a r^2{}_{\text{shell}} = \beta n^2 a^3$

with: β = vol. conc. defects x 4π

We thus obtain:

$$|v_{\text{migr.}}| = \frac{\beta \times a^3 \times \displaystyle\sum_{n=1}^{n_{\max}} n^2 |\langle v_{na}\rangle|}{\beta \times a^3 \times \displaystyle\sum_{n=1}^{n_{\max}} n^2}$$

$$= \frac{\displaystyle\sum_{n=1}^{n_{\max}} n^2 |\langle v_{na}\rangle|}{\displaystyle\sum_{n=1}^{n_{\max}} n^2}$$

or, after replacing $|\langle v_{na}\rangle|$ by the expression obtained in (2.6'):

$$|v_{\text{migr.}}| = \frac{1}{\displaystyle\sum_{n=1}^{n_{\max}} n^2} \times \sum_{n=1}^{n_{\max}}$$

$$\times \left\{ n^2 \times \frac{1}{n} \times \left\{ \sum_{i=1}^{n} i\,|v(ia)| \right\} \right\}$$

and after replacing $|v(ia)|$ by the expression (2.5) for $r = ia$:

$$|v_{\text{migr.}}| = \frac{D_0}{kT} \times \frac{|A|}{a^2} \times \exp(-E/RT)$$

$$\times \frac{\displaystyle\sum_{n=1}^{n_{\max}} n\left\{ \sum_{i=1}^{n} \frac{1}{i} \right\}}{\displaystyle\sum_{n=1}^{n_{\max}} n^2}$$

The final equation for $|v_{\text{migr.}}|$ is thus:

$$\boxed{\begin{aligned} |v_{\text{migr.}}| &\simeq 6 \times \frac{D_0}{kT} \times \frac{|A|}{a^2} \times \exp(-E/RT) \\ &\times \frac{\displaystyle\sum_{n=1}^{n_{\max}} n\left\{ \sum_{i=1}^{n} \frac{1}{i} \right\}}{n_{\max} \times (n_{\max}+1) \times (2n_{\max}+1)} \end{aligned}}$$

(2.7)

Equation (2.7) can be used for all types of sinks, as far as the potential is of the type $V(r) = -|A|/r$.

The necessary condition, for the possible initiation of surface structure of the type of which the origin is related to radiation damage, is satisfied if the averaged migration speed of all the defects participating in the elaboration of the extensive defects, calculated by expression (2.7), is larger than the minimum required by expression (1.6).

3. Comparison with the Experiments and Conclusions

Three remarks have to be made before comparison with results:

a) After initiation of topography, it is necessary to hold on the bombardment to reach a sufficient total dose, to erode the surface preferentially according to an adapted Stewart–Thompson mechanism, to obtain finally a stable configuration of the surface, regressing parallel to itself if the bombardment is continued.

During this evolution the characteristic height of the surface structure will, at each moment, be proportional to the total dose since its initiation and to the local differences in sputtering yield. The observability (and thus the required total dose for observability) depends on the observation technique used, but the surface structure can already influence measurements for total doses far below the one giving observability with an observation technique of low resolution (the characteristic length is of the order of 10^3 Å in some experiments, as low as a few hundreds Å in other cases). It is for this reason that a criterion concerning the initiation of surface topography was established and not one for the obtention of the final equilibrium configuration.

b) For each material there exists[18] a temperature above which the surface-diffusion tends to flatten out the surface structure due to irradiation. In this temperature range thermal facetting[21] can start to play a role, which can be distinguished from the surface structure due to bombardment, by the fact that the facets of the latter are not planes of low crystal indices, whereas the previous are.

c) The temperature one can measure experimentally is an overall target temperature, evidently different from the local temperature of the region of the sample where the migration takes place, i.e. the region of highest damage production and thus of highest energy deposit. As the migration speed is very temperature dependent, this difference influences very much the results.

Among all the experimental results concerning surface effects available from the literature[13] and from private communications, as well as from own experiments, of course only those data which are concerned with topography of the second type of origin are considered. Indeed, checking the criterion in its present form (established for migration of point-defects) makes no sense if the samples display cone or crater formations due to the presence of impurities or inclusions before the beginning of the irradiation, or even craters or blisters due to inclusions (precipitates or gas bubbles) formed during the bombardment.

TABLE I
Some comparisons of the predictions by the criterion and the observations

Target	Ion	Energy keV	Dose rate ions/sec cm^2	Experimental temperature	Surface structure + references
Cu	A$^+$	6	$2.5 \cdot 10^{17}$	917	predicted and observed (own experiments)
Cu	H$^+$	120	$2 \cdot 10^{13}$	1074	predicted and observed (B. M. U. Scherzer: private communication)
Ag	A$^+$	40	$6.25 \cdot 10^{14}$	400	excluded and not observed (Z. Jurela: private communication)
Au	A$^+$	6	$6.25 \cdot 10^{14}$	300	excluded and not observed (P. H. Schmidt: private communication)

Therefore we started to do very systematic irradiations combined with microscope observations of single crystal spheres and thinned foils, under variation of only one of the parameters appearing in the criterion, during the same series of experiments. Until now our experiments were done on Cu, Ag, Pt and Au, with ions such as He$^+$, A$^+$ and Sb$^+$ of energies ranging from 5 keV up to 90 keV, under target temperatures varying from room temperature up to 1000°C and under dose rates varying from $2 \cdot 10^{10}$ ions/sec $\cdot$ cm^2 up to $0.8 \cdot 10^{18}$ ions/sec $\cdot$ cm^2; total dose covering the corresponding range.

A first comparison (see Table I) with our results and a few collected from private communications[4] show that, if no surface structure was observed, the condition was indeed not satisfied, i.e. the migration speed was smaller than the minimum value required by (1.6) in terms of the experimental parameters, and on the other hand, that if structure of the second type was observed, the criterion was indeed satisfied.

This necessary condition concerns all types of surface structure growing on surfaces, above extensive defects, large enough to perturb the penetration of the incoming ions: a regular network will grow above a regular alignment of dislocation lines, but randomly spread and isolated craters will grow above loops or tetrahedra, which have not been interacting yet.

After thorough comparison with more complete experimental data, it will probably appear that the theory (until now only an "order-of-magnitude" estimation) needs refining. Indeed, established as above it contains the following approximations:

a) The depth of the damage peak is approached by an expression for the mean projected range established for amorphous targets;[17]

b) The mechanism of the migration of the point-defects into clusters is very simplified; we suppose for example interindependence of the point-defects during their migration towards the sink and on the other hand the possibility of establishing, due to their migration, a concentration gradient of point-defects is neglected;

c) The nucleation of clusters was supposed to be due to a high probability of encounter of point-defects in a level of high damage density. This was supposed to happen in times sufficiently short to be neglected in comparison to the long times needed to grow from small clusters towards "extensive defects." On the other hand the dependence of the potential of interaction in terms of time, i.e. in function of the state of advancement of their growth, was neglected.

Nevertheless, the criterion for the possible initiation of damage-induced surface topography on clean and well-polished crystal surfaces during particle bombardment, as established above, can be usefully applied as a necessary condition. It should indeed enable experimenters, if desired, to choose the values of their irradiation parameters in such a way that the origination of surface topography is excluded.

NUMERICAL TREATMENT OF ONE COMPARISON WITH EXPERIMENTAL OBSERVATIONS

Regular structure was observed (Figure 2) on the surface of Cu samples irradiated with 5 keV A$^+$ ions at a dose rate

$$\text{d.r.} = 10^{16} \text{ ions/sec} \cdot \text{cm}^2.$$

a) For calculating v_{min} by (1.6) we need:

$$M_1 + M_2 = 103.4; \frac{3M_1 + M_2}{M_1 + M_2} = 1.79:$$

$$r_0 \simeq 30 \text{ A}^{16}; Y = 7 \text{ at/ion}$$

$$E_i/2E_d = 100$$

We obtain then $v_{min} = 0.23$ A/sec.

b) To calculate the migration speed in terms of the temperature, we need:

$$D_0 = 11 \text{ cm}^2/\text{sec}^{[19]}$$

$$a = 3.61 \cdot 10^{-8} \text{ cm}$$

$$E = 57.2 \text{ kcal/mole}^{[19]}$$

in the case of a dislocation acting[9] as a sink we have:

$$\mu = 4 \cdot 10^{11} \text{ dyn/cm}^2 \text{ and } b = 2.5 \cdot 10^{-8} \text{ cm},$$

thus giving[20]

$$|A| = \frac{2\mu a^3 |b|}{3\pi} = 10^{-19} \text{ erg. cm}$$

the value of d_l measured on the micrographs: $d_l = 70$ A.

The migration speed $|v_{\text{migr.}}|$ is now given by:

$$|v_{\text{migr.}} (T)| \simeq \frac{74 \cdot 10^{10}}{T} \text{ cm}^\circ \text{ K/sec}$$

$$\times \exp \left[- \frac{2.9 \cdot 10^4 \, ^\circ\text{K}}{T} \right]$$

On a plot of $|v_{\text{migr.}}|$ in terms of temperature in this case, can be seen that the temperature under which surface topography is excluded according to our criterion, is 714°K or 441°C, which is indeed lower than the target temperature during the experiment.

ACKNOWLEDGEMENTS

The author wishes to express gratitude to Professor Albert Art for many helpful discussions and criticism of the manuscript, and to all those who have been willing to send data about their observations of surface topography. Thanks are also expressed to Professor J. Jedwab for kindly helping with scanning electron microscopy and to Dr. A. Johansen for irradiating some of our samples, as well as to the reviewers of the paper for many interesting comments.

REFERENCES

1. A. D. G. Stewart and M. W. Thompson, *J. of Mater. Sci.* **4**, 56–60 (1969).
2. M. J. Nobes, J. S. Colligon and G. Carter, *J. of Mater. Sci.* **4**, 730–733 (1969).
3. I. Teodorescu, Fourth European Regional Confer. on E.M. pp. 371–372 (1968).
4. G. Staudenmaier: Private communication.
5. Ref. 1. p. 57, Figure 1.
6. J. J. Ph. Elich, H. E. Roosendaal, H. H. Kersten, D. Onderdelinden, J. Kistemaker and J. D. Elen, *Rad. Effects,* **8**, 1–11 (1971).
7. H. Schwarz and R. Lück, *Mater. Sci. Eng.* **5**, 149–152 (1969/70).
8. S. H. A. Begemann, Ph.D. Thesis, Univ. of Groningen (1972).
9. N. Hermanne and A. Art, *Fizika* **2**, Suppl. 1, 72–80 (1970).
10. D. J. Mazey, R. S. Nelson and P. A. Thackery, *J. Mater. Sci.* **3**, 26 (1968).
11. N. Hermanne and A. Art, *Rad. Effects* **5**, 203 (1970).
12. N. Hermanne, *Rev. Roumaine de Physique* **17**, 747–749 (1972).
13. For review: G. Carter and J. C. Colligon, *Ion Bombardment of Solids* (Heinemann Educat. Books, 1968); and R. Behrisch, *Ergibn. der Exakten Naturwissensch.* **35**, 295–443 (1964).
14. P. Haymann, *Thèse* (Univ. of Paris, Editions Métaux, 1962).
15. D. S. Billington and J. H. Crawford, *Radiation Damage in Solids*, p. 36 (Princeton University Press, 1961).
16. M. W. Thompson, *Defects and Radiation Damage in Metals.* p. 77, (Cambridge University Press, 1969).
17. J. W. Mayer, J. A. Davies and L. Eriksson, *Ion Implantation in Semiconductors* (Academic Press, p. 13, 1970).
18. P. G. Shewmon, *Diffusion in Solids*, Eq. 1–44 (McGraw-Hill, 1963).
19. *Metals Reference Book*, C. J. Smitthells editor, p. 556 (Intersci. Publ. N.Y. 1955), 2nd edition
20. A. C. Damask and G. J. Dienes, *Point Defects in Metals,* pp. 69–70 (Gordon and Breach, 1963–1971). See also: J. W. Christian, *The Theory of Transformations in Metals and Alloys,* p. 257 (Pergamon Press, 1965) and Ref. 18, pp. 25 and 27.
21. A. J. W. Moore, Metal Surfaces, Structures, Energetics and Kinetics, ASM. AIME Seminar, New York (1962).

THE TOPOGRAPHY OF SPUTTERED SEMICONDUCTORS

I. H. WILSON

Department of Electronic and Electrical Engineering,
University of Surrey, Guildford, Surrey, England.

The surfaces of semiconductor single crystals were observed using a scanning electron microscope after bombardment at normal incidence at room temperature with doses between 0.5×10^{18} and 3×10^{18} of 40 keV argon ions/cm^2.

The usual topography of the sputtered surface and the perturbing effects of dirt and contamination are reported, and observations of the variation of topography with dose are reported for GaAs. An explanation for the features observed is discussed with particular reference to current models of cone formation and secondary processes such as flux enhancement at the foot of steep slopes.

The behaviour of Si, Ge, GaAs and InP was consistent with an amorphous surface. In the case of CdTe, CdS and GaP it appeared that the crystal structure was retained resulting in a topography of the sputtered surface that is sensitive to orientation. InSb appears to undergo a chemical change during ion bombardment.

I THOSE SEMICONDUCTORS SHOWING BEHAVIOUR CONSISTENT WITH AN AMORPHOUS SURFACE; Si, Ge, GaAs AND InP.

1.1 INTRODUCTION

The topography of sputtered surfaces has become of great interest because of the large influence this can have on sputtering yields, ejection patterns, cluster formation and analysis techniques such as secondary ion mass spectroscopy and Rutherford backscattering.

Two classes of topography can be defined:

a) That which results from the effects of ion channelling and radiation damage and is sensitive to the crystal orientation. An example of this is the grain relief that develops on metals[1,2] due to sputtering rate varying with orientation. Another example are the facets that are observed on single crystal metals that possibly result from the formation of periodic dislocation networks.[3,4,5]

b) That which develops from the perturbation of the surface caused by dirt, contamination, inclusions or the initial irregular topography. In this category are the observations of ion etching of flaws[6] and mounds[7] in glass and the formation of cones on metal surfaces.[2,8,9]

It is the latter class of topography that concerns us here.

A lot can be learned of the mechanisms of sputtering and the influence of the angle of incidence of the beam by observing the manner in which the perturbed surface settles down to equilibrium.

The cones formed on metals have exhibited features that cannot be explained by the models proposed for cone formation[8,10] on amorphous surfaces. For example it has been observed[2] that cones of two angles can exist on the same grain, and that in fact they are not true cones but are of hexagonal cross-section. It is thought that these phenomena are the result of orientation effects.

In this work are reported the results of observations on semiconductors bombarded with argon to high doses where, hopefully, the surface is amorphous with an equilibrium topography in the form of a flat plane normal to the ion beam. Thus the effects of perturbation of the surface can be studied in isolation from orientation effects.

Of the semiconductors studied those that came nearest to this ideal were Si, Ge, GaAs and InP. The results for these materials are reported in this section. Observations on the other materials studied, namely: GaAs, CdS, CdTe and InSb are reported in section II.

1.2 EXPERIMENTAL

All specimens were prepared from slices cut, with a diamond saw, from single crystals; they were nominally 1 mm thick, greater than 5 mm diameter. The gallium arsenide specimens were then given a heavy free etch, a polish-etch and finally a light chemical etch. The other materials were lapped, mechanically polished and then cleaned ultrasonically in methanol followed by hot distilled water.

No gross changes were observed in sputtered topography if a chemical etch was introduced after the final polish. This contrasts with the effects seen on metal polycrystals.[2,3]

The nominal orientation of the surface normal was as follows; Si and Ge ⟨111⟩, GaAs ⟨110⟩ and InP ⟨100⟩.

The specimens were bombarded at normal incidence in a conventional linear accelerator[11] with 40 keV mass analyzed argon singly charged ions at a current density of about 20 μA/cm^2, with a beam divergence of about 1°.

A d.c. bias on an electrode between the beam defining aperture and target was used to suppress secondary electrons. The specimens were attached by means of a film of colloidal silver to a substantial mount; the power input to them was less than 0.2 watts and it is estimated that their temperature was not greater than 50°C during a run. Residual pressure was ~5 x 10^{-7} torr using an oil diffusion pump with a liquid nitrogen cold trap.

The observations were made using a Cambridge Instruments scanning electron microscope with a resolution of 25 nm. The viewing angle used with respect to the surface normal was either 45° or 75°.

1.3 OBSERVATIONS

1.3.1 The Unperturbed Sputtered Surface
At a dose of ~2 x 10^{18} ions/cm^2 the surface of Si and Ge consisted of a series of bumps approximately 0.15 μm in diameter (Figure 1). This was also the case for

GaAs at a dose of 0.5 x 10^{18} ions/cm^2 but at doses of greater than 1.5 x 10^{18} ions/cm^2 the surface appeared smooth.

Very small truncated cones of similar diameter and periodicity to these bumps were observed on InP after a dose of 2 x 10^{18} ions/cm^3.

1.3.2 Perturbation of the Surface by Foreign Bodies
A particle on the surface which has a low sputtering rate acts as a mask such that a column of material is left below it of a height equivalent to the thickness removed by sputtering (Figures 2a and b). No significance is attributed to the angle of slope of the side of the column as this depends on beam parameters and details of the masking particle. It is frequently seen that when a large foreign particle (say greater than 2 μm) is sputtered away complex projections of the target material are left (Figure 2c).

If a small area is masked (say less than 1 μm) single cones are formed of the type shown in Figure 3 (a, b and c). These cones have a half angle of 3° with a rounded tip.

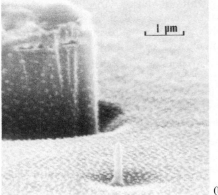

(a)

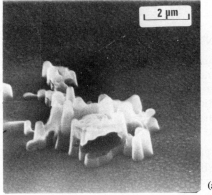

(b)

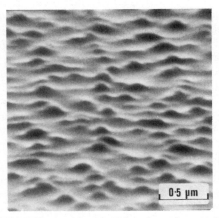

FIGURE 1   The normal sputtered surface of germanium (dose 2.0 x 10^{18} ions/cm^2) viewed at 75°.

FIGURE 2

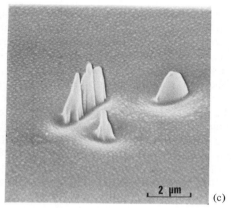

(c)

FIGURE 2   Perturbation of the surface by large dirt particles.
(a) Silicon (45° viewing angle, dose 1.8 x 10^{18} ions/cm^2). (b)
InP (75° viewing angle, dose 2.0 x 10^{18} ions/cm^2). (c) GaAs
(45° viewing angle, dose 1.4 x 10^{18} ions/cm^2).

It is interesting to note there there appears to be a
build-up of material at the foot of the cone in the case
of silicon, whereas a pit is observed in the other materials
which is deeper for GaAs and InP than for Ge.

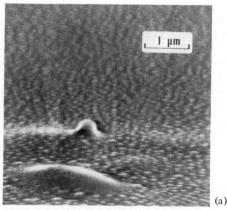

(a)

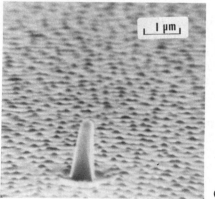

(b)

FIGURE 3

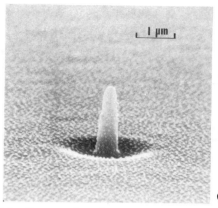

(c)

FIGURE 3   Cones resulting from perturbation of the surface
by small dirt particles (viewing angle 75°). (a) Silicon (dose
2.1 x 10^{18} ions/cm^2). (b) Germanium (dose 2 x 10^{18} ions/
cm^2). (c) InP (dose 2.0 x 10^{18} ions/cm^2).

1.3.3 Variation of Profile with Increasing Dose

A specimen of GaAs was masked with a wire grid
during the first bombardment. This provided a
reference so that one could map the surface and
return to the same features after successive doses.
More than 30 features were followed in this way and
all showed a similar variation to the pair illustrated in
Figures 4 (a, b, c, d, e, and f). A diagram summarizing
all the variations observed is shown in Figure 5.

For low ion doses the surface sputters away leaving
a column beneath any contaminant particle with a
deep pit around the base. The column apparently
increases in height until the foreign particle is finally

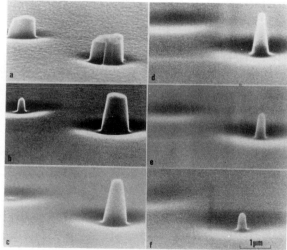

FIGURE 4   GaAs; variation of a surface feature with increasing dose (viewed at 75°). Dose x 10^{18} ions/cm^2, (a) 1.0,
(b) 1.5, (c) 1.8, (d) 2.0, (e) 2.3, (f) 2.5.

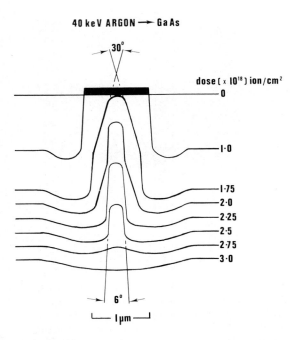

40 keV ARGON → GaAs

FIGURE 5   GaAs diagram of the stages observed in the variation of the profile of the surface initially perturbed by a foreign body (note: this is drawn to scale).

removed by sputtering from the edges inwards. This leads to the formation of a conical top to the column of half-angle 15°. This cone then sweeps down the column, the rate of removal being much greater than the normal surface. A cone of much narrower half-angle (3°) is left which in turn sputters from the tip, disappearing at a rate in between that of the normal surface and that of the wider angled segment.

Eventually a shallow pit is left which slowly flattens out during still further bombardment.

1.4 DISCUSSION

1.4.1 The Unperturbed Sputtered Surface
The undulating surface may arise from

a) Formation of argon bubbles.

b) Microcrystals formed under the influence of radiation enhanced diffusion.

c) Perturbation of the surface topography in the early stages of bombardment.

Any explanation in terms of (a) and (b) will have to account for the disappearance of the undulations on GaAs after more than 2 μm of material has been

removed. This is difficult as the projected range[12] of A^+ in GaAs is 0.03 μm, so that changes in topography due to changes in concentration of gas and damage in the surface will have ceased at doses well below those necessary to remove the undulations.

The most reasonable explanation would seem to be that the bumps develop from a varying sputtering rate in the early stages of bombardment. The undulations formed would then remain until a considerable amount of material had been removed as there are only small departures from normal incidence of the ion beam and therefore only minor variations in sputtering rate. Possible reasons for perturbation of the sputtering rate of the microscopically smooth surface are;

a) A discontinuous oxide film,

b) Mechanical damage, or

c) Low sputtering yield material (such as aluminium sputtered from the beam-defining aperture onto the specimen surface.

It is not possible to reject any of these possibilities from the experimental data presented here.

Chemical analysis by electron spectroscopy (ESCA) of Ge and GaAs[13] reveals that a considerable proportion of the surfaces of the semi-conductors are covered by oxides (or both gallium and arsenic in the case of GaAs after the normal preparation.

Great care was taken with the GaAs specimens to see that the absolute minimum of mechanical damage remains, as confirmed by etch pit counts, electron channelling patterns, Rutherford backscattering and transmission electron microscopy.[13]

From a comparison of Ge and GaAs it would appear that effects due to mechanical damage cannot be discounted. Whereas for etched (damage free) GaAs a smooth surface is observed after more than 2 μm of material is removed, the undulations are still present on a mechanically polished Ge surface.

The effects of atoms sputtered from the beam defining aperture are seen on InP but only near the edges of the bombarded area, where one would expect the flux of these atoms to be greatest. In this region a very high concentration of cones is seen (Figure 6), of the type first observed by Wehner and Hajicek.[9] Also with InP, bumps are observed on the sides of columns and cones, which must have been generated after the slopes were established and therefore possibly come from nuclei formed from sputtered atoms or inclusions of impurities. The undulations observed on the other materials are all similar in nature and extend uniformly over the entire surface so an explanation in terms of sputtered atoms, inclusions or contamination seems improbable.

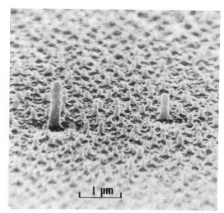

FIGURE 6 InP, a region with a high density of small cones (near the beam edge). Viewing angle 75°, dose 2.0 × 10^18 ions/cm².

If sputtered aluminium, from the defining aperture is the cause of the dense region of cones found on InP then this implies that the surface diffusion of aluminium over InP must be rapid enough to favour formation of nuclei which resist re-sputtering and therefore generate small cones.

1.4.2 Flux Enhancement

The build up or erosion of the surface at the foot of steep slopes is probably the result of secondary processes not considered in the theoretical models.[8,10] It is proposed that such effects are due to the bombardment of the surface at the foot of a slope by ions reflected and atoms sputtered from the sloping side of a column or cone.

If we consider a surface element at the foot of the slope then in addition to the direct ion beam it receives a flux of reflected ions proportional to $R(\theta, E_0)$ the fraction of incident ions of energy E_0 reflected from the slope whose normal is at angle θ to the direction of ion incidence. As these ions will lose little energy and even after reflection are at nearly normal incidence to the foot of the slope they will have a similar sputtering yield to the direct beam $S(O, E_0)$.

Atoms sputtered from the slope will also be incident on the surface at the foot of the slope and flux will be proportional to the sputtering yield of the slope $S(\theta, E_0)$. Three terms determine whether this will result in removal or deposition of material, the sputtering yield for atoms incident at energy E_s, $S(E_s)$, the fraction of sputtered atoms with this energy, $N(E_s)$, and the fraction of these that remain trapped on the surface $F(E_s)$.

The approximate forms of $N(E_s)$ and $S(E_s)$ can be deduced from results for other materials and are illustrated in Figure 7a. The only available data on the form at $N(E_s)$ is for single crystal gold at $\theta = 45°$.[14] This peaks at about 2 eV and drops as E^{-2} at higher energies, $S(E_s)$ is > 1 atoms/ion for $E_s > 500$ eV in the case of Ge and GaAs[15] (> 0.5 atoms/ion for Si) and approached a threshold at ~ 20 eV. A guess at the form of $F(E_s)$ has been made assuming that at low energies atoms are re-emitted from the surface unless they find nuclei and therefore $F(E_s) < 1$ and decreasing with increasing energy till the displacement energy is reached. At higher energies incident atoms can be buried in the surface and retained and $F(E_s)$ begins to rise to unity in the keV range of energies. The term

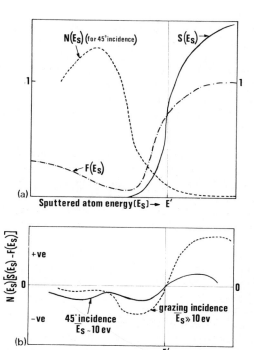

FIGURE 7 Variation of number $N(E_s)$, sputtering rate $S(E_s)$ and fraction retained on surface $F(E_s)$ as a function of energy E_s of atoms sputtered from a slope, incident on the surface at the foot of the slope.

$N(E_s) [S(E_s) - F(E_s)]$ describes the overall effect of sputtered atoms and may take the form illustrated in Figure 7b.

The mean energy of sputtered particles $\bar{E}_s$ will move from the 45° value of ~10 eV to higher energies when the ions are incident at more acute angles and a dis-proportionate number of the more energetic sputtered atoms are directed at the foot of the slope. This could have the effect of increasing the chances of removing material from the foot of the slope.

All the factors mentioned can be combined to show that the overall removal of material at the foot of a slope will be proportional to

$$S(\theta, E_0)\left\{\int_0^{E_0} N(E_s) \ [S(E_s) - F(E_s)] \ dE_s\right\} +$$

$$+ R(\theta, E_0)S(0, E_0) \qquad (1)$$

$S(\theta, E_0)$ which is a maximum at $\theta = \hat{\theta}$ will fall as θ approaches 90°. However, $\bar{E}_s$ shifts to higher energies which may balance this out to a certain extent. $R(\theta, E_0)$ increases with increasing θ from 0 at ~45° to unity at $\theta = 90°$. Therefore the reflected ions are probably dominant in causing erosion for $90° > \theta > \hat{\theta}$ but the effects of sputtered atoms may be significant and become more important for $\theta \leqslant \hat{\theta}$. Both terms predict that the effects of flux enhancement will be sensitive to sputtering rate. This is confirmed by the results where deposition occurs for Si, slight erosion for Ge and marked erosion for GaAs and InP, the sputtering rates for 40 keV argon being Si, 1.9; Ge, 3.0; GaAs, 6.4 and InP, 5.5 atoms/ion.[16]

It would seem that for Si the sputtering rate is low enough to result in the first term in Eq. (1) being negative and larger than the second term. For the other materials, there is a removal of material at the foot of the slope, so it may be assumed that the first term is either positive or if negative is now smaller than the second term.

1.4.3 The Formation of Cones
In the early stages of erosion of GaAs the dirt particle sputters inwards from the edge which is explained by the fact that the maximum sputtering rate will be achieved where the beam is near grazing incidence. This therefore results in the formation of a conical or rounded top to the column immediately after all the dirt is removed. The shape at this point would depend on the sputtering rate of the dirt particle. The top of the column then forms a cone of 15° half angle, possibly by the Stewart-Thompson[8] mechanism where the angle is such that the maximum sputtering rate is achieved (i.e. beam incidence angle $\theta = \hat{\theta}$). This expla-nation is supported by the fact that the cone sweeps rapidly down the column.

The narrow angled cone (3° half angle) that always remains, after the changes outlined above, is observed on all four semiconductors. It would seem to be close to the topography predicted as stable in the model of Catana, Colligon and Carter,[10] where the sputtering rate for the beam incident at θ' on the side of the cone equals that for normal incidence i.e. $S(\theta', E_0) = S(0, E_0)$. In fact the cones observed have rounded tips, probably as a result of transmission sputtering adding to the effects of reflection sputtering at the point increasing the erosion rate. (The wider angled cones seen on metals, where transmission sputtering would not ex-tend so far, have a much smaller rounded region at the tip.[2]) The cones therefore are not stable and sputter most rapidly at the tip presumably because the ion beam is incident at angles of $\theta < \theta'$ so that

$$S(\theta, E_0) > S(\theta', E_0).$$

The narrow angled cone could be generated at a slight concavity formed at the tip of the wide angled cone. If the concavity was such that the ion beam would strike at angles $\theta > \hat{\theta}$ then theory[6] predicts that the surface would rotate towards 90°, but if stable at $\theta = \theta'$ a cone of half angle 90–θ' would be formed. Dirt remaining after the wide angled facet was formed could create the concavity. Another possibility would be radiation enhanced surface diffusion. There is some evidence[17] that surface diffusion is enhanced in the direction of the incident ion; if this reaches a maximum value at the rounded tip for $\hat{\theta} > \theta > 0$ a concavity could be formed.

1.5 CONCLUSIONS
Si, Ge, GaAs and InP approach the behaviour consist-ent with a completely amorphous surface, namely formation of a flat surface normal to the ion beam. The surface relief that does develop can be removed by continued sputtering although as much as 5 μm may have to be removed. InP appears to be sensitive to deposition of metallic atoms sputtered from masks and apertures, leading to the development of small cones.

The shape of cones formed from small particles of dirt (< 1 μm) and the changes in shape during bom-bardment do not fit either of the models proposed so far. This probably arises from the fact that the models do not take into account effects such as increased yield at sharp points, the role of the perturbing dirt particle, and the enhancement of flux due to reflected ions and sputtered atoms at the foot of steep slopes. This latter effect is shown to be very sensitive to sputtering rate.

II GaP, InSb, CdS, AND CdTe

2.1 INTRODUCTION

In Section I results are reported of a scanning electron microscope study of the topography of Si, Ge, GaAs and InP after bombardment with 40 keV argon at normal incidence. These materials approach a topography consistent with an amorphous surface after doses in excess of 1×10^{18} ions/cm². Similar observations of GaP, InSb, CdS and CdTe, which do not show this simple behaviour, are reported here. Experimental conditions were similar to those reported in Section I. The ion dose used was $2 \pm 0.2 \times 10^{18}$ ions/cm² in all cases. The CdS was obtained as cleaved wafers, the GaP, CdTe and InSb were mechanically polished slices cut from single crystals. All specimens were cleaned ultrasonically in methanol followed by hot distilled water. The nominal orientation of the surface normal was ⟨111⟩ for GaP and CdTe and ⟨0001⟩ for CdS. The orientation of the InSb specimen was unknown. The microscope viewing angle was 45° for all micrographs. The observations are reported separately for each material studied.

2.2 THE TOPOGRAPHY OF CdTe

After bombardment the unperturbed surface appeared to be flat with none of the undulations seen on the semiconductors studied in Section I. A large number of circular pits were observed over the surface which were almost semi-circular in cross-section. These presumably are generated by flux enhancement from the sides of columns or cones which have been subsequently removed by sputtering. CdTe has a high sputtering rate (9.5 atoms/ion for 40 keV argon[16]), and this means that any cones that are formed would only exist on the surface for a very short time during bombardment. In fact no isolated cones were found. Another consequence of the high sputtering rate seems to be that the flux enhancement effect is very exaggerated. This is illustrated in Figure 8 where a very deep pit can be seen to have formed around cone-like projections. It is probable that these deep pits would take a long time to remove by sputtering and this would explain why so many are seen. Dense areas of cones are observed in regions where presumably the contaminant has been removed by sputtering. As the cones are rounded from base to tip no angle can be accurately defined. However, the approximate half angle was found to be inversely proportional to cone height, the angle ranging from 6° for the tallest cone to 17° for the smallest. This observation can be explained if a wide angled cone was formed by the Stewart Thompson mechanism[18] at the

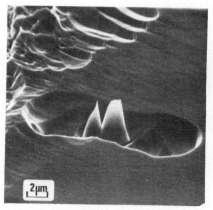

FIGURE 8   CdTe; cone-like projections with a very large pit at the base.

top of the column, thus effectively forming a rounded cone. The cone on top would then sputter at a greater rate than the rest of the surface resulting in the apparent increase in angle with decreasing height. This effect was observed in GaAs with the difference that a narrow angled cone was left behind after removal of the wide angled cone.

In some regions of the surface long lines of overlapping pits have been seen and this effect may be due to a high dislocation density resulting from mechanical damage.

A band of very small cones is formed around the edges of the bombarded area. This phenomenon is similar to that observed on InP and is attributed to atoms, sputtered from apertures, which form nuclei on the surface able to resist re-sputtering.

2.3 THE TOPOGRAPHY OF CdS

The sputtered surface was found to become very rough, being entirely covered by facets with one plane nearly parallel to the ion beam and one apparently perpendicular to this (Figure 9a). As in nearly all the other materials studied the presence of contaminant particles results in the formation of columns and complexes of cones.

The facet-covered surface is reminiscent of the topography that develops on metals for certain orientations. Thus it appears that the crystal structure is retained in the surface and this view is supported by the fact that cleavage steps, seen in masked regions of the specimen, continue into bombarded regions as a change in topography (Figure 9b). The lack of radiation

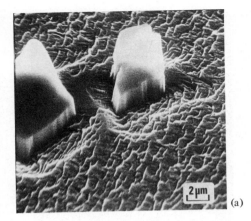

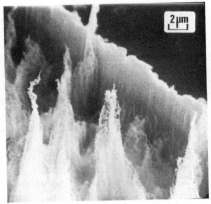

FIGURE 10   InSb; the top right-hand corner has been masked by a wire grid.

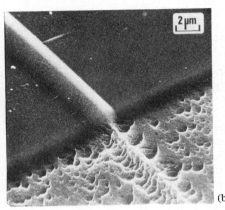

FIGURE 9   CdS. (a) Sputtered surface with contaminant masked projections. (b) A cleavage step. The upper half of the area has been masked from the beam.

damage in the surface is confirmed by other workers concerned with analysis by Rutherford backscattering of ion implanted CdS.[18]

2.4 THE TOPOGRAPHY OF InSb

The bombarded regions have a matt black appearance to the eye. In the microscope one sees a dense mass of whiskers (Figure 10) which are very irregular in shape and grow well above the parts of the surface that have been masked from the beam. The whiskers appear to be readily deformed as many near the edge of the bombarded region have bent over to lie flat on the masked surface.

An indication of the structure of the specimen was seen at the very edge of the bombarded area which is defined by an aperture 10 mm from the surface. Here, due to shadowing, the surface is subjected to a much lower dose than in the bombarded region proper. A

dendritic structure appears which is decorated by whisker growth at the edge of the dendrites. This structure is very similar to that observed in a two phase (InSb + In) thin film,[19] where dendrites of InSb have solidified leaving an indium rich matrix.

The whiskers formed by ion bombardment appear to grow from the indium rich areas. They could result from a growth of oxide or a phase change resulting from preferential sputtering of one constituent. Radiation enhanced diffusion could have the effect of accelerating these changes.

2.5 THE TOPOGRAPHY OF GaP

The sputtered surface was found to be almost completely flat and smooth with no resolvable features other than the projections formed by the masking effect of contaminant particles.

Unlike, say, Ge this semiconductor shows only a very small pit at the foot of any projection. The sputtering rate of GaP is 3.8 atoms/ion[16] under these bombarding conditions. As this is greater than Ge one would expect that if GaP were behaving like the materials studied in Section I then quite sizeable pits would be formed as a result of flux enhancement.

2.6 DISCUSSION AND CONCLUSIONS

The materials studied exhibited none of the features, such as an undulating surface and narrow angled cones, found to be typical of Si, Ge, GaAs and InP.

The most probable explanation of this is that CdTe, CdS and GaP retain their crystal structure at the surface and therefore the topography that develops after sputtering depends sensitively on beam orientation which makes interpretation difficult. The evidence that supports this idea is the existence of wide angled cones and lines of pits on the CdTe surface; faceting, wide angled cones and memory of cleavage steps on the CdS surface, and the lack of flux enhancement effects or any cone formation on the GaP surface.

InSb appears to undergo a chemical change in the surface.

ACKNOWLEDGEMENTS

The InP and InSb was kindly provided by John Shannon of Mullard Research Laboratories (Salfords), the CdS by Bill Grant of the University of Salford, the GaAs, GaP and CdTe by Peter Hemment, David Titley and Michael Gettings of this department. The specimen preparation was done by Mrs. Vera Hinton. I would like to thank them all and especially our group leader Ken Stephens for encouragement and advice.

REFERENCES

1. W. Hauffe, *Phys. Stat. Sol. (a)* **4**, 111 (1971).
2. I. H. Wilson and M. W. Kidd, *J. Mat. Sci.* **6**, 1362 (1971).
3. I. A. Teodorescu and F. Vasiliu, *Rad. Effects* **15**, 101 (1972).
4. R. S. Nelson, Int. Conf. Ion-Surface Interaction, Garching, Paper M1 (1972).
5. N. Hermanne, Int. Conf. Ion-Surface Interaction, Garching, Paper M2 (1972).
6. A. R. Bayly, *J. Mat. Sci.* **7**, 404 (1972).
7. I. S. T. Tsong and D. J. Barber, *J. Mat. Sci.* **7**, 687 (1972).
8. A. D. G. Stewart and M. W. Thompson, *J. Appl. Phys.* **42**, 1145 (1971).
9. G. K. Wehner and D. J. Hajicek, *J. Appl. Phys.* **42**, 1145 (1971).
10. C. Catana, J. S. Colligon and G. Carter, *J. Mat. Sci.* **7**, 467 (1972).
11. P. J. Cracknell, M. Gettings and K. G. Stephens, *Nucl. Inst. and Meth.* **92**, 465 (1971).
12. W. S. Johnson and J. F. Gibbons; Projected range statistics in semiconductors.
13. P. L. F. Hemment, University of Surrey, private communication.
14. G. E. Chapman, B. W. Farmery, M. W. Thompson and I. H. Wilson, *Rad. Effects.* **13**, 121 (1972).
15. G. Carter and J. S. Colligon, *Ion Bombardment of Solids* (Heinemann, 1968), p. 323.
16. I. H. Wilson, Sputtering, focused ion beams and bombardment of thin films, University of Surrey report (1972).
17. G. K. Wehner, private communication.
18. Professor G. Carter, University of Salford, private communication.
19. N. M. Davis and H. H. Wieder, *Proc. 8th Electron and Laser Beam Symp.* (Univ. Michigan, April 1966), p. 385.

SURFACE DAMAGE AND TOPOGRAPHY OF ERBIUM METAL FILMS IMPLANTED WITH HELIUM TO HIGH FLUENCES†

R. S. BLEWER

Sandia Laboratories, Albuquerque, New Mexico 87115

Topographical and expansion effects which occur as a result of implanting erbium thin films with helium up to fluences of 1.5×10^{18} He$^+$/cm^2 are described. There exists an inverse relationship between critical dose and annealing temperature with respect to the formation of surface bubbles. Post implantation annealing at or below 400°C is found to strongly reduce implantation induced expansion for doses less than 3.5×10^{17} He$^+$/cm^2, but is observed to result in increased expansion above this dose. At temperatures above 400°C, expansion is increased for all doses investigated. Details of bubble development in the implanted layer are discussed and the manner in which surface bubbles develop from enlarged subsurface bubbles is illustrated.

1 INTRODUCTION

It has now been well established that implanting metals with high fluences of inert gas atoms results in the formation of surface bubbles and blisters at a critical dose if the rate of surface removal by sputtering is sufficiently low to permit large concentrations of gas atoms to accumulate.[1,2] A detailed experimental study into the manner in which the microscopic gas bubbles in the implanted layer combine and enlarge to a size capable of producing surface bubbles and blisters has not been reported. In this paper the topography and expansion effects observed on erbium thin films as a function of dose and anneal temperature will be reviewed and the role of subsurface bubble development on the formation and morphology of surface bubbles will be explored.

2 EXPERIMENTAL DETAILS

As shown schematically in Figure 1 thin films of erbium 2.5 μm thick were deposited on 1 cm diameter polished sapphire disks 1 mm thick. The periphery of the films was shielded during implantation by a copper annulus placed in contact with the film surface for thermal grounding purposes. Likewise, crossed metal ribbons over the film surface preserved unimplanted regions along perpendicular diameters to provide adjacent areas in which relative expansion and topography effects could be compared. A water-cooled holder was used together with low beam currents (<10 μa/cm^2) to assure that beam heating did not cause premature gas migration effects since surface

† This work was supported by the U.S. Atomic Energy Commission.

SCHEMATIC VIEW of SPECIMEN

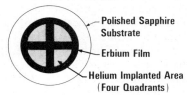

- Polished Sapphire Substrate
- Erbium Film
- Helium Implanted Area (Four Quadrants)

FIGURE 1   Schematic view of film samples used in the implantations.

bubble formation has been observed to be very sensitive to sample temperature (see Section 3). Sample temperatures did not exceed 5°C above room temperature during implantation. Since bubbling effects are also strongly fluence dependent, the accelerator beam was diffuse focused and then x and y scanned to achieve ±4% lateral uniformity.

A series of Er film samples was implanted ($E = 160$ keV) at room temperature with from 4 to 12 well defined fluences per sample. Implantation-induced expansion was measured by interferometry in each fluence area (2×10^{16} to 1.5×10^{18} He ions/cm^2). The calculated projected range for He in Er at 160 keV is 0.77 μm with a $\pm 1\sigma$ spread of ± 0.345 μm.[3] Several samples were then subjected to a linear 5°C/min temperature rise (25°C to 600°C) in a scanning electron microscope (SEM); the speed and character of changes in surface topography were observed in the rapid scan mode and recorded on videotape.[4] Other samples were annealed isothermally in a gas mass spectrometer at 400°C and 500°C and then characterized using the SEM in the normal scan mode. A selection of films from both groups above were later prepared for cross sectional SEM viewing of subsurface bubble development by carefully cutting a deep trough into the rear side of the sapphire substrate using a miniature air cooled diamond wheel and then fracturing the substrate by applying pressure along the notch.[5] Magnification calibration was determined for each sample by comparing the film thickness indicated on the micrographs with that determined interferometrically.

3 RESULTS AND DISCUSSION

3.1 Surface Phenomena

Of 15 erbium film samples implanted at fluences ranging up to 1.5×10^{18} He$^+$/cm^2, none exhibited surface bubbling at room temperature. High magnification SEM micrographs of adjacent implanted and unimplanted areas exhibited a total absence of implantation-induced topographical features. This result for thin films is in contrast to the behavior of a variety of helium implanted bulk metals[6] (including erbium)[7] where surface bubbles have been observed above fluences of about 5×10^{17} He$^+$/cm^2 at room temperature.

Implantation-induced film expansion perpendicular to the film basal plane was previously observed to increase linearly with fluence up to 3×10^{17} He$^+$/cm^2 and to increase at a faster rate at higher fluences.[8] After vacuum annealing several samples at 400°C and above, expansion was observed to decrease toward zero for doses up to 3.5×10^{17} He$^+$/cm^2 without apparent helium release. Above this fluence, expansion increased further over the "as implanted" room temperature value during the annealing and was accompanied by the formation of surface bubbles and blisters. It has now been observed that the virtual disappearance of expansion on annealing occurs only if the anneal temperature does not significantly exceed 400°C, regardless of dose. Conversely expansion is accentuated at all doses investigated if the implanted film sample is annealed in the 500°C to 600°C range. Surface bubbling does not occur unless the critical dose is present at a given anneal temperature even though accentuated expansion may be noted as a result of annealing. Finally the magnitude of the expansion decreases from its peak value after surface bubbling in the area under consideration is fully mature.

Reduction of the expansion for anneal temperatures below 400°C and for fluences less than 3×10^{17} He$^+$/cm^2 has been ascribed to the movement of implanted helium atoms from interstitial sites (end of range location) to substitutional sites.[8,9] At fluences greater than 3×10^{17} He$^+$/cm^2 and temperatures greater than 500°C, bubble nucleation and growth apparently predominate over atom site redistribution, resulting in a

sharpened onset of the expansion nonlinearity and an accentuation of its magnitude at a given fluence. Expansion reaches a maximum as the critical temperature for surface bubbling occurs for a given fluence and then reduces somewhat after the surface is fully bubbled, perhaps as the natural result of partial release of the implanted gas.

In Figure 2, the variation with fluence of surface bubble size and spacing is exhibited. Zone (a) of the film surface is unimplanted. Zone (c) is implanted with helium to a fluence four times that in zone (b). Though no surface bubbling occurred in the sample at room temperature, vacuum annealing at 600°C produced the observed variation in surface bubble size and spacing. The surface bubbles are about twice as large, more hemispherical, and more widely separated in the lower fluence region, and also are far less likely to be ruptured despite their larger size. Features such as those illustrated in Figure 2 have been observed on each of several samples, some implanted in contiguous areas to 12 discrete (but uniform) fluences on a single film surface.

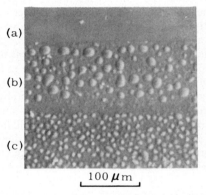

(a)

(b)

(c)

$\underline{100 \, \mu m}$

FIGURE 2   SEM micrograph of (a) an unimplanted erbium film surface (b) an adjacent area implanted to a fluence of 1×10^{17} He$^+$/cm^2 at 160 keV and (c) an area implanted to a fluence of 4×10^{17} He$^+$/cm^2 at 160 keV.

A dependence on fluence of subsurface bubble density and diameter similar to that reported above has been reported previously.[9] In the case of high dose implantation, there seems to be a fluence above which a substantial portion of the implanted atoms begin to interact with existing or radiation-induced defects such as vacancies and vacancy clusters at room temperature and above. It has been shown that interstitial helium atoms will migrate in certain metals at room temperature until a trap of the type described above is encountered.[10] This suggests that higher dose samples will exhibit a finer scale of nucleation, since the damage induced trap density will be proportionately higher.

The temporal character of surface bubble development has been followed in the SEM in real time,[4] which has allowed a better insight into the details of the mechanism involved. The onset of surface bubbling in a given fluence area is quite sensitive to temperature and occurs within 0.5°C of the critical temperature for the given dose; bubbling of the surface is usually complete in 3 to 5 seconds. If the helium fluence in the area under observation is sufficient to produce bubbles which rupture, the rupturing is complete within this time frame. Annealing for hour-long periods at a given temperature does not add materially to the number of bubbles, their size or their probability of being ruptured, but bubble density and maturity are enhanced as higher temperatures are attained, although the rate of new bubble formation is less than that observed when the film surface initially bubbles.

The critical temperature for a given fluence has been difficult to determine exactly and may vary for different films implanted to the same nominal dose. This variation is not surprising considering the high sensitivity of the "critical-dose-to-critical-temperature" relationship, the variation in morphology and intrinsic stress levels normally found in thin films even when prepared under seemingly identical conditions, and the small (but important) errors inherent in counting implanted atom fluence. Each of these parameters affects the "critical" temperature observed at bubbling onset. A plot of the temperature at which surface bubbling occurs versus the corresponding fluence is approximately linear (with a slope of -3.3×10^{15} He$^+$/cm^2 °C) between temperatures of 520°C to 600°C and between fluences of 1×10^{17} He$^+$/cm^2 to 3.5×10^{17} He$^+$/cm^2.

3.2 Subsurface Phenomena

It was clear from previous observations[2] that large ruptured surface blisters were composed (before their rupture) of several subsurface lenticular bubbles 2 to 6 μm in diameter in the implanted subsurface layer. These observations were explained in terms of a "two stage model" of surface bubble development. The term "surface bubble" was applied to circular surface protrusions resulting from lenticular subsurface gas pockets; conversely, irregular shaped surface features resulting from the combining of two or more subsurface lenticular gas pockets were termed "blisters". The same convention is used in this paper.

To formulate ideas on the progress of bubble development in the subsurface layer and to discover the details of the evolution of subsurface bubbles into surface formations, a means was sought to make direct SEM observations of this layer in samples implanted to various fluences and annealed at various temperatures in a manner which would not obscure the desired microscopic details, introduce artifacts, or alter the gas distribution. The "notch and fracture" method described in the experimental section was found to meet the stated criteria and yielded samples of superior quality to those obtainable by microtoming or similar methods: fine structure beneath the surface was not smeared or distorted, there was no need for subsequent etching or electropolishing which can preferentially enlarge[11] or otherwise alter the morphological features, and the possible influence of deformation-induced recrystallization (with concomitant redistribution of gas atoms or bubbles)[12] was minimized or circumvented.

A cross-sectional view obtained by this technique can be seen in Figure 3. The implanted layer is clearly visible by the dispersion of small bubbles most of which are 1000 Å to 3000 Å in diameter. The depth of the layer corresponds to the calculated projected range[3] when measured expansion and sputtering loss at the film surface are taken into account.[8] Visible at the left is a surface bubble which has been split by the line of

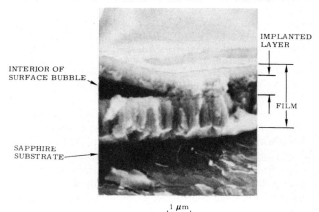

FIGURE 3 Cross-sectional view of the implanted layer and one side of a surface bubble. This film sample was implanted to a nominal fluence of 3×10^{17} He$^+$/cm^2 and isothermally annealed at 500°C before fracturing.

fracture. The material comprising the bubble lid was plastically deformed when the surface feature originally formed as evidenced by its remaining in an inflated position even after the gas has escaped. When the surface bubble formed originally, it separated the implanted layer near its center so that the small subsurface bubbles in the implanted layer can still be seen both in the under part of the bubble lid and beneath the crater base.

In Figure 4 at progressive magnification is shown another surface bubble and its relation to the sub-surface implanted layer. At higher magnification a few of the interior features of the bubble can be seen as well as a transverse view of the crack in the bubble lid. Other surface bubbles can be seen in the background.

From viewing a large number of similar SEM micrographs and by observing real-time bubble formation and rupture in the SEM-TV mode for a variety of doses and anneal temperatures, surface bubble development can be characterized as shown in Figure 5. Initially the implanted gas can be assumed to be atomically dispersed as it comes to end of range during implantation.[9,10] Because of helium atom and/or vacancy migration as a result of high dose and/or annealing, helium

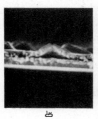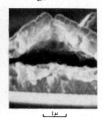

FIGURE 4   Cross-sectional view at progressive magnification of a surface bubble. This film sample was implanted to a fluence of 4×10^{17} He+/cm^2 and annealed by heating at a rate of 5°C/minute from 25°C to 615°C in the SEM while simultaneously observing the surface bubble development by the TV scan mode. The sample was later fractured to obtain view shown here.

atom-vacancy complexes are formed which coarsen at higher temperatures or doses. Continued buildup in helium concentration plus coalescence provides still larger aggregates. Lateral development is favored because implanted gas concentration decreases from its peak value as e^{-ax^2} in thickness direction. When two or more of the larger subsurface bubbles form in adjacent positions, the surface begins to yield locally as the subsurface layer plastically deforms and separates. It is to be expected that the film should rend in the center of the implanted layer because the bubble density is greatest there resulting in a reduced load-supporting cross section. The walls separating adjacent bubbles are stretched plastically and when these fail, they give the appearance of stalactites and stalagmites when viewed in cross section. Remnants of the stalagmites and stalactites seem to disappear as the surface bubble develops further. This is thought to be the result of rather rapid surface diffusion; evidence for a similar effect has been observed in copper samples annealed at the same temperature used in this work.[13]

DEVELOPMENT OF SURFACE BUBBLES

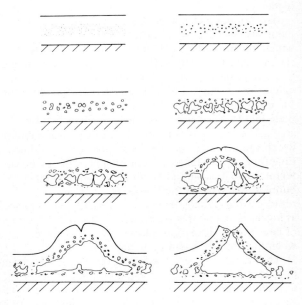

FIGURE 5   Schematic representation of the development of bubbles in the implanted layer into surface bubbles. Initially the implanted helium is atomically dispersed. As higher concentrations and/or temperatures are reached, microscopic bubbles develop in a band centered on the mean implantation range. Ultimately adjacent subsurface bubbles which are 3000 Å or larger combine as the wall separating them is plastically deformed resulting also in a localized film surface deformation. For extremely small (non-columnar) grain films or large grained materials, bubble rupture occurs near the surface bubble periphery. For other films the bubble is more likely to rupture near its apex along a line of grain boundaries. Continued development of bubbles on the underneath side of the surface bubble lid causes a change from convex to concave curvature after rupture.

The surface bubbles tend to rupture in one of two modes:

1) Edge Rupture: If the sample grain size is either very much smaller[2] or much larger[6,14] than the radius of the surface bubble, rupture is likely to occur at the bubble periphery.

2) Apex Rupture: If the film grain size is of the same order as the surface bubble radii (or if the film is small grained but highly columnar), rupture is likely to occur at the bubble apex along a line of grain boundaries (for columnar films) or along a grain boundary which traverses the surface bubble (for bulk material).

For films whose bubble lids are blown away when rupture occurs, after the gas in the interior of the bubble is released, the lid changes from convex to concave curvature, probably as a result of the compressive stress in the lower part of the lid caused by continued development of bubbles in the implanted layer. This outward curling is clearly observable in the real time SEM viewing even after the crack in the bubble surface is large enough to release the gas.

Gas release is observed to occur only after surface features have formed and ruptured, and the fraction of gas released is related to the number and size of the bubbles which have ruptured. This suggests that gas released through bulk or grain boundary diffusion is either negligible or far less important than that released through surface bubble or blister rupture at the temperatures investigated.[2]

The same experiments reported here have also been performed using helium implanted erbium deuteride. Topographical and expansion effects qualitatively similar to those observed in implanted erbium metal were found except that lower critical doses and temperatures were required for surface bubbling. The principal rationale for investigating erbium deuteride was the comparison of its response to high concentrations of implanted helium to corresponding phenomena in erbium tritide[15] where helium is generated in the lattice without causing significant radiation damage. Microscopic bubbles have been observed in tritide films by transmission electron microscopy[16] and surface bubbles form[15] which bear a strong resemblance to those induced by helium implantation in erbium deuteride.[2] In experiments by others,[17] no significant difference has been found to exist in the gas release characteristics of helium implanted silver after introducing more than 10 times as much displacement damage in the lattice (by proton irradiation) as was produced by the helium implantation itself. Results for materials such as these suggest that, though lattice damage is a sufficient condition, it is not a necessary condition for the formation of inert gas bubbles and that helium release is not sensitive to lattice displacement at levels which accompany the implantations cited above.

SUMMARY:

1) Thin films of erbium exhibit no surface bubbling at room temperature after helium implantation at a fluence more than twice as high as that required for surface bubbling in bulk erbium.

2) The reduction toward zero of implantation induced expansion (for fluences below 3.5×10^{17} He^+/cm^2) does not occur if the anneal temperature exceeds 400°C, even for considerably lower fluences;

accentuated expansion occurs for all fluences investigated if the anneal temperature exceeds 500°C.

3) The diameter of surface bubbles appearing on a fluence area of 1×10^{17} He^+/cm^2 after a 600°C anneal is larger, the spacing is greater and the bubbles are much less likely to be ruptured than those which form in an area implanted to four times higher fluence.

4) An inverse relationship exists between critical dose and annealing temperature; between the temperatures of 520°C and 600°C and doses of 1 to 3.5×10^{17} He^+/cm^2, the relationship is approximately linear with a proportionality factor of -3.3×10^{15} He^+/cm^2 °C.

5) The stages of development through which the bubbles in the implanted layer pass have been observed with the SEM using a sample fracture technique. The manner in which surface bubbles evolve from combining subsurface bubbles has been illustrated.

6) Implantation induced displacement damage does not appear to be a prerequisite for inert gas bubble formation and related effects in implanted metals.

ACKNOWLEDGEMENT

It is a pleasure to acknowledge J. K. Maurin for performing the scanning electron microscopy, G. Skinner and H. I. Taylor for video-taping services, and E. P. EerNisse for the use of the Sandia 300 kV Heavy Ion Accelerator.

REFERENCES

1. For a compendium of recent progress in this area, see the Surface Damage and Topography Section of the Proceedings of the Conference on Ion-Surface Interaction published in this issue and the references included there.
2. R. S. Blewer and J. K. Maurin, *J. Nucl. Mater.* **44**, 260 (1972).
3. D. K. Brice, private communication, April, 1970, and *Appl. Phys. Lett.* **16**, 103 (1970).
4. R. S. Blewer and J. K. Maurin, 30th Annual Conference Electron Microscope Soc. Amer., Los Angeles, Calif., August 14–18, 1972.
5. J. M. Pearson, *Th. Sol. Films* **6**, 349 (1970).
6. S. K. Erents and G. M. McCracken, *Rad. Effects* (this issue).
7. A. J. Summers, J. Freeman and N. R. Daly, private communication, (October 1972).
8. R. S. Blewer and W. Beezhold, *Rad. Effects* (to be published).
9. I. J. Hastings and B. Russell, *J. Nucl. Mater.* **23**, 1 (1967); B. Russell and I. J. Hastings, *J. Nucl. Mater.* **17**, 30 (1965).
10. E. V. Kornelsen, *Can. J. Phys.* **48**, 2812 (1970).
11. J. E. Bainbridge and B. Hudson, *J. Nucl. Mater.* **17**, 237 (1965).
12. M. V. Speight and G. W. Greenwood, *Phil. Mag.* **9**, 683 (1964).

13. D. J. Mazey, R. S. Nelson and P. A. Thackery, *J. Mat. Sci.* **3**, 26 (1968).

14. G. J. Thomas and W. Bauer, *Rad. Effects* (to be published).

15. L. C. Beavis and C. J. Miglionico, *J. Less Comm. Met.* **27**, 201 (1972).

16. R. S. Blewer and G. J. Thomas, unpublished data (February 1970).

17. D. H. Garside, Ph.D. Dissertation, Univ. of Sussex (January 1971).

BLISTERING OF POLYCRYSTALLINE AND MONOCRYSTALLINE NIOBIUM †

M. KAMINSKY and S. K. DAS

Argonne National Laboratory, Argonne, Illinois 60439

Radiation blistering of niobium surfaces has been investigated for normally incident He^+ and D^+ projectiles. For the He^+, the bombarding energies were 0.5 and 1.5 MeV, the targets were both monocrystalline and polycrystalline niobium at room temperature and at 900°C, and the total dose was from 0.01 to 1.0 C/cm^2. For cold-worked samples at room temperature, 0.5-MeV He^+ ions produced large blisters (up to 500 μm in diameter), many of which were ruptured; 1.5-MeV He^+ ion implantation produced one big blister covering most of the irradiated area. For the D^+ ion implantation, the bombarding energies were 0.250 and 0.5 MeV and the targets were all polycrystalline. Implanting 0.250-MeV D^+ ions in annealed polycrystalline Nb at 700°C produced small blisters for doses ranging from 1.6 to 3.0 C/cm^2.

Preliminary results on the skin thicknesses of ruptured blisters indicate a close relationship between the thickness and both the projected range of the ions and the peak in the defect distributions.

1 INTRODUCTION

The formation of gas-filled blisters in the surface regions of irradiated solids has been observed by several authors (as reported, for example, in Refs. 1–6 and in the reviews of earlier work in Refs. 7–9). The blisters can rupture in energetically favored regions such as grain boundaries and thereby release bursts of gas. The present authors have recently reported the formation of gas-filled blisters by implantation of 0.5-MeV helium ions into monocrystalline and polycrystalline Nb at room temperature[10-12] and into monocrystalline Nb at 900°C.[13]

Blistering can play an important role in the operation of controlled thermonuclear fusion devices or reactors[14-17] since it can lead (1) to serious damage and erosion of bombarded wall surfaces and (2) to the release of gases which will contaminate the plasma and thereby cool it below the minimum temperature for thermonuclear reaction. In ion-accelerator technology, blistering is important in the erosion of collimators, apertures, targets, and the like.

Since the basic mechanisms underlying the blistering process are still only poorly understood, it is impossible to make reliable theoretical estimates of the size, shape, and surface density of blisters, of the amount of gas released, or of the degree of surface erosion in materials considered for fusion devices and reactors and for accelerator components.

In the studies described in the present paper, the previously reported helium blistering studies[10] of poly-

† Work performed under the auspices of the U. S. Atomic Energy Commission.

crystalline niobium have been extended to helium-ion energies of 1.5 MeV and to target temperatures of 900°C. Furthermore, deuteron blistering of polycrystalline niobium will be reported for deuteron energies of 0.25 MeV and doses ranging from 1.6 to 3.0 C/cm^2 at 700°C, and results obtained for 1.6 C/cm^2 will be compared with earlier reports[18] and with our previous results[12,16,19] on the blistering of polycrystalline niobium at room temperature under irradiation with 0.5-MeV deuterons.

2 EXPERIMENTAL PROCEDURES

Target Preparation

The niobium monocrystals and the cold-worked polycrystalline foils were obtained from the Material Research Corporation (MARZ grade). The niobium monocrystals were spark cut from 1.2-cm-diameter cylindrical ingots with the [111] direction parallel to the long axis of the cylinder. In cutting the ~0.25-mm-thick slices from the ingots, the crystals were oriented with great care to ensure that the flat faces of these disks were normal to the [111] axis. The back-reflection Laue photographs of the final polished disks show the orientation of the surface normal to be within 0.5° of the [111] direction. Both the monocrystalline disks and the polycrystalline foils were mechanically polished and lapped and then subsequently electropolished in an electrolyte containing 92.5% methanol, 5% H_2SO_4, and 2.5% HF at −60°C at a current density of 0.13 A/cm^2. This process removed approximately 35 μm from the thickness of the polished pieces and left them micropolished (more than 90% of the polished area had an average surface roughness of less than 0.02 μm, as determined with the aid of a scanning electron microscope with a resolution limit of about 0.02 μm). Thin films suitable for transmission electron microscopy to characterize the microstructure of both the cold-worked polycrystalline foils and the monocrystalline disks were prepared by electropolishing by use of the window method and the electrolyte described above. Transmission electron micrographs of the cold-worked polycrystalline material indicate a dislocation density greater than 10^{11} lines/cm^2.

Target Irradiation

A mass-analyzed beam of ^{4}He$^+$ ions with energies of 0.5 or 1.5-MeV or of D$^+$ ions with energies of 0.25 or 0.5-MeV from a 2-MeV Van de Graaff accelerator was highly collimated (half-angle of divergence = 0.01°) and was incident within ~0.1° of the normal to the surface plane of the niobium targets. During the He irradiation, the ion flux was kept constant at 1×10^{13} ions cm^{-2} sec^{-1}, with total doses ranging from 0.01 to 1.0 C/cm^2. For the D$^+$ irradiation, the ion flux was kept constant at 4×10^{14} ions cm^{-2} sec^{-1}, with total dosages ranging from 0.1 to 3.0 C/cm^2. A more detailed description of the experimental arrangement has been given previously.[20]

In each case, the irradiated area on each target was a circular spot about 2 mm in diameter. Because of the high beam collimation, there was a sharp boundary between the bombarded and the unbombarded area. During the irradiation the vacuum in the target chamber was maintained at about 1–2 $\times$ 10^{-8} Torr by ion pumping. Each target irradiation was repeated to check the consistency of the results.

Examination of the Irradiated Target

The irradiated surfaces were examined with a scanning electron microscope (Cambridge Stereoscan model Mark II) and micrographs were taken. From the enlarged micrographs, the size distribution of the blisters was measured with a Zeiss particle-size analyzer. Since the blisters were irregular in shape, the average diameter of a blister was approximated as the diameter of a circle having the same area as the blister.

3 RESULTS

Irradiation by ^{4}He$^+$ Ions

The blistering observed in cold-worked polycrystalline niobium bombarded by 1.5-MeV He$^+$ to a total dose of 1.0 C/cm^2 at room temperature is shown in Figure 1a. This scanning electron micrograph, shows the entire

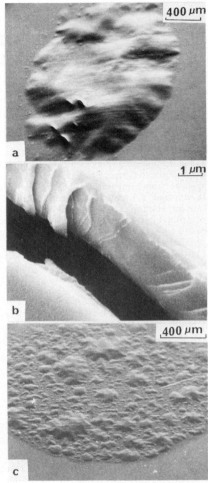

FIGURE 1   Scanning electron micrographs of cold-worked polycrystalline niobium surfaces after irradiation at room temperature to a total charge density of 1.0 C/cm^2. (a) Back-scattered-electron image of the entire area bombarded with 1.5-MeV He$^+$ ions. (b) Secondary-electron image of the skin of the ruptured blister seen in (a). (c) Secondary-electron image of the area irradiated with 0.5-MeV He$^+$ ions.

bombarded area, where most of the irradiated area appears to be covered by one big blister. This big blister appears to have resulted from the coalescence of several smaller blisters. It can be noticed that the big blister has ruptured in several positions and an enlarged view of the ruptured skin is shown in Figure 1b. From this enlarged micrograph one can estimate the skin thickness to be approximately 2 μm. Although no correction has been applied for the angle between the fractured edge and the surface normal, this estimate of the thickness cannot be off by more than 40%. It is of interest to note that while bombardment with 1.5-MeV He$^+$ results in one big blister covering most of the bombarded area (Figure 1a), the area bombarded with 0.5-MeV He$^+$ (Figure 1c) is occupied by many blisters of various sizes.[10] Although only a small fraction of the latter blisters are more than 100 μm in diameter, these larger ones occupy a large fraction ($\sim$55%) of the total bombarded area.

The effect of elevated target temperature is shown in Figure 2, in which the scanning electron micrograph

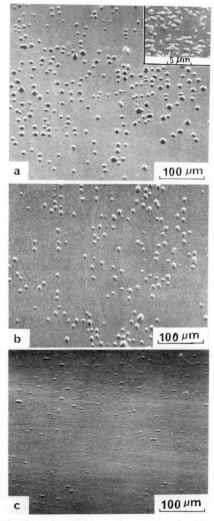

in Figure 2a shows the blisters formed by a total dose of 1.0 C/cm^2 of 0.5-MeV He$^+$ ions incident on a target at 900°C. One notices that the blister density is considerably less than in Figure 1c and that the blisters are considerably smaller. The blister sizes fall into two broad distributions: the dome-shaped blisters that are readily visible in Figure 2a and the very small blisters that can be seen at the higher magnification in the inset. These small blisters are relatively irregular in shape and occupy about 26% of the bombarded area. The helium blister formation for the same irradiation conditions but for a lower total dose (0.1 C/cm^2) is shown in Figure 2b. A comparison of the two micrographs indicates only a slight reduction of blister density and average blister size at the lower dose. The fraction of total irradiated area that is blistered decreases from 32% to 28% as the total dose is decreased from 1.0 C/cm^2 to 0.1 C/cm^2 at this temperature of 900°C. It is of interest to note that even when the total dose is reduced by another order of magnitude (to 0.01 C/cm^2), helium blisters are still readily observable in the bombarded area in Figure 2c. Small blisters comparable to those shown in the inset in Figure 2a were also observed at the lower doses (0.1 and 0.01 C/cm^2).

Irradiation by D$^+$ Ions

For cold-worked polycrystalline niobium irradiated at room temperature to total doses of 0.1 and 1.0 C/cm^2 with normally incident 0.5-MeV D$^+$ ions, no blisters could be resolved with a scanning electron microscope (resolution limit $\approx$ 200 Å), as reported earlier.[12,16,19] Blistering was observed, however, for higher doses and for a target temperature of 700°C, as shown in Figure 3 for the case of irradiation with 0.25-MeV D$^+$ ions. For a total dose of 1.6 C/cm^2 (Figure 3a), the blister density is $\sim$2 x 10^8 blisters/cm^2 and the average blister diameter is approximately 0.4 μm. These results qualitatively confirm those reported by Donhowe and Kulcinski,[18] although their reported average blister diameter (0.72 μm) is larger than ours. The blisters obtained for deuteron irradiation under similar irradiation conditions but for higher doses (2.0 and 3.0 C/cm^2 for Figures 3b and 3c, respectively), appear to be more elongated than those shown in 3a; their average length ranges from 1 to 3 μm. The blister shape appears to depend on the grain orientation, as will be discussed in the next paragraph. In addition to these elongated blisters, many smaller ones (average diameter of the order of 0.15 μm) can be seen as small dots in the background in Figures 3b and 3c. Over the small range studied here, the dose seems to have no significant effect on the size of blisters.

FIGURE 2   Scanning electron micrographs of cold-worked polycrystalline niobium surfaces after irradiation at 900°C with 0.5-MeV He$^+$ ions. The total doses were (a) 1.0 C/cm^2, (b) 0.1 C/cm^2, and (c) 0.01 C/cm^2. Corrections for the 45° tilt of the specimen when these secondary-electron images were taken were applied in (a) and (b) but not in (c).

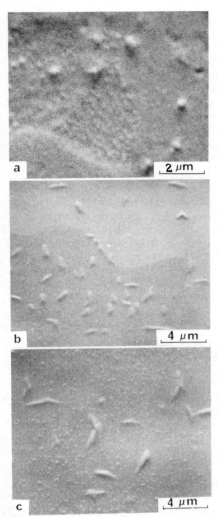

FIGURE 3   Scanning electron micrographs (back-scattered-electron images) of polycrystalline niobium annealed at 1250°C for 2 hours before D$^+$ irradiation at 700°C. The total doses of 0.250-MeV D$^+$ ions were (a) 1.6 C/cm^2, (b) 2.0 C/cm^2, and (c) 3.0 C/cm^2.

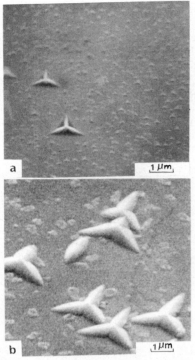

FIGURE 4   Scanning electron micrographs (secondary-electron images) of niobium surfaces: (a) after irradiation of an annealed polycrystalline sample at 700°C to a total dose of 3.0 C/cm^2 of 0.250-MeV D$^+$ ions and (b) after irradiation of a (111) monocrystalline sample at 900°C to a dose of 1.0 C/cm^2 of 0.5-MeV He$^+$ ions.

That the grain orientation affects the shape of the deuteron blisters is shown rather dramatically in Figure 4a. These blisters result from irradiation at 700°C to a total dose of 3.0 C/cm^2 of 0.25-MeV D$^+$ ions. Most of the blisters exhibit threefold symmetry in the plane of the irradiated surface. Since the orientation of this particular grain is unknown (as it is difficult to determine) one can only conjecture that the prongs of the crow-foot shaped blister are aligned along the traces of {110} planes of the niobium lattice, as was observed previously[13] in the scanning electron

micrograph (Figure 4b) of an irradiated monocrystalline niobium target with the [111] direction parallel to its surface normal (0.5-MeV He$^+$ beam, total dose of 1.0 C/cm^2, target temperature of 900°C). The blisters in Figure 4a fall into two different size classes. The larger blisters measure about 0.7 μm from the center to the end of a prong, and the maximum width of a prong ranges from 0.20 to 0.25 μm. For the smaller blisters, the corresponding length of a prong is about 0.15 μm. The blister density is substantially higher for the smaller blisters than for the larger ones. Although the results shown in Figures 4a and 4b are not entirely comparable since the irradiation conditions were different, it is interesting to note that the average length of the prongs of the large blisters produced by helium irradiation (lower dose but higher target temperature) is 2–3 times that for those produced by deuteron irradiation (higher dose but lower temperature).

4 DISCUSSION

In a cold-worked polycrystalline niobium sample, the appearance of one large blister over most of the irradiated surface area when bombarded at room temperature with 1.5-MeV He$^+$ ions (Figure 1a), but of numerous smaller blisters when bombarded with 0.5-MeV ions (Figure 1c) can possibly be understood in the following way. As the projectile energy increases, the peak of the damage distribution and the region of maximum concentration of implanted ions shifts to larger depths below the irradiated surface. This results in the formation of helium bubbles at larger depths, and this in turn leads to an increase in the thickness of the blister skin (our earlier measurements[10] for 0.5-MeV He$^+$ bombardment gave a skin thickness of ~1.2 μm, while the thickness in the present work with 1.5-MeV He$^+$ is ~2.0 μm). Furthermore, a higher gas pressure is required for plastic deformation of the thicker blister skin formed at the higher projectile energy, and hence a larger number of bubbles from a larger irradiated area coalesce to form larger blisters. The formation of smaller blisters by the 0.5-MeV ions may also be favored because the concentration of radiation-induced defects near the surface is higher at this energy than at 1.5 MeV.

The fact that blistering by helium ion irradiation was less at 900° than at room temperature, both for monocrystalline and for polycrystalline niobium, may be due to several important factors. One may be that the more rapid annealing of lattice defects at higher sample temperatures would lead to a reduced rate of bubble nucleation. Another reason is that some of the implanted helium may be lost through the surface by atomistic helium diffusion or migration of small bubbles. Although the size of such bubbles appears to be below the resolution limit of the scanning electron microscope used in these experiments, the evidence for the fact that helium atoms or bubbles can migrate such distances during irradiation has been obtained by our earlier observation[11] of denuded zones near the large crow-foot blisters. In some recent mass-spectrometric measurements of the helium re-emitted from niobium surfaces during helium ion implantation, Bauer and Morse[21] found that the re-emission rate at doses greater than 1.1 x 10^{18} He atoms/cm^2 is higher at 500–700°C than at room temperature. This agrees with our observation that the blistering is less at 900°C than at room temperature.

The difference between the degrees of blistering in the monocrystalline niobium samples (Figure 4b) and in the cold-worked polycrystalline ones (Figure 3) is partly due to the extremely large difference in the initial dislocation density of the two cases: it is about 2 x 10^5 lines/cm^2 in the monocrystal and >10^{11} lines/cm^2 in the cold-worked polycrystalline niobium. The smaller amount of blistering by deuteron irradiation than by helium irradiation under similar conditions can be related to the fact that the solubility and diffusivity in niobium is many orders of magnitude larger for deuterium than for helium. The diffusion coefficient of deuterium in niobium[22,23] is D$_D$ = 1.3 x 10^{-4} cm^2/sec at 800°C, while that for helium in niobium[24] goes from 10^{-19} to 10^{-14} cm^2/sec between the temperatures of 600°C and 1200°C. The dependence of the blister shape on the crystallographic orientation of the irradiated sample has been described in more detail in our earlier papers.[11,13]

ACKNOWLEDGEMENTS

We would like to thank Mr. Thomas Campana and Mr. Thomas Dettweiler for their help in target preparation. We are grateful to Mr. Peter Dusza for his help in irradiating the samples and to Mr. G. T. Chubb for his cooperation in our use of the scanning electron microscope.

REFERENCES

1. R. S. Barnes and D. J. Mazey, *Proc. Roy. Soc. (London)* **275**, 47 (1963).
2. M. Kaminsky, *Adv. Mass Spectrometry* **3**, 69 (1964).
3. W. Primak, *J. Appl. Phys.* **35**, 1342 (1964).
4. W. Primak and J. Luthra, *J. Appl. Phys.* **37**, 2287 (1966).
5. L. H. Milacek, R. D. Daniels and J. A. Cooley, *J. Appl. Phys.* **39**, 2803 (1968).
6. W. Bauer and G. J. Thomas, *J. Nucl. Mater.* **42**, 96 (1972).
7. G. Carter and J. Colligon, *Ion Bombardment of Solids* (American Elsevier Publishing Co., New York, 1968).
8. M. W. Thompson, *Defects and Radiation Damage in Metals* (Cambridge University Press, London, 1969).
9. M. Kaminsky, *Atomic and Ionic Impact Phenomena on Metal Surfaces* (Springer-Verlag, Heidelberg/New York, 1965).
10. S. K. Das and M. Kaminsky, *J. Appl. Phys.* **44**, 25 (1973).
11. S. K. Das and M. Kaminsky, *J. Appl. Phys.* (in press).
12. M. Kaminsky and S. Das, *Bull. Am. Phys. Soc.* **17**, 678 (1972).
13. M. Kaminsky and S. K. Das, *Appl. Phys. Lett.* **21**, 443 (1972).
14. M. Kaminsky, *IEEE Trans. Nucl. Sci.* NS 18, 208 (1971).
15. J. E. Draley, B. R. T. Frost, D. M. Gruen, M. Kaminsky, V. A. Maroni, *Proceedings of the 1971 Intersociety Energy Conversion Engineering Conference* (American Institute of Chemical Engineering, New York, 1971), pp. 1065–1075.
16. M. Kaminsky, *Proceedings of the International Working Sessions on Fusion Reactor Technology,* Oak Ridge National Laboratory, Oak Ridge, Tennessee, June 28– July 2, 1971, Conf-710624, pp. 86ff, 120 ff.
17. D. G. Martin, Culham Laboratory Report No. CLM-R 103, 1970.
18. J. M. Donhowe and G. L. Kulcinski, in *Fusion Reactor First Wall Materials,* edited by L. C. Ianiello, U. S. Atomic Energy Commission Report No. WASH-1206, 1972, p. 75.

19. M. Kaminsky, in *Fusion Reactor First Wall Materials,* edited by L. C. Ianiello, U. S. Atomic Energy Commission Report No. WASH-1206, 1972, p. 63.
20. M. Kaminsky, in *Recent Developments in Mass Spectroscopy*, edited by K. Ogata and T. Hayakawa (University of Tokyo Press, Tokyo, 1970), p. 1167.
21. W. Bauer and D. Morse, **44**, 337 (1972). *J. Nucl. Mater.* **44**, 337 (1972).

22. J. Volkl, *Proceedings of Sixth Symposium on Fusion Technology* (Aachen, Germany, Commission of the European Communities, Luxembourg, 1970), p. 519.
23. G. Schaumann, J. Volkl, and G. Alefeld, *Phys. Stat. Sol.* **42,** 401 (1970).
24. S. Blow, U.K.A.E.A. Report AERE-R 6845, Atomic Energy Research Establishment, Harwell, Berkshire, England, 1971.

BLISTERING OF MOLYBDENUM UNDER HELIUM ION BOMBARDMENT

S. K. ERENTS and G. M. McCRACKEN

U.K.A.E.A. Research Group, Culham Laboratory, Abingdon, Berkshire

Polished molybdenum targets have been bombarded with helium ions of energy from 7 to 80 keV in ultra-high vacuum. During bombardment the release of gas was continuously monitored and after bombardment the targets were examined in a scanning electron microscope.

Blister formation was observed to occur after a critical dose of $\sim 5 \times 10^{17}$ ions/cm^2, and the appearance of blisters coincides with gas release from the surface. The average size of the blisters increases with energy but not with ion dose. In addition to room temperature observations, blisters have also been examined following high temperature bombardment of molybdenum, and room temperature bombardments of W, Pt, Ni, Cu and Zr targets.

1 INTRODUCTION

Radiation blistering of materials due to high energy, α, proton and deuteron bombardment has been known for some time.[1,2,3,4] Initial work on bubble formation[5,6] all indicated that temperatures sufficiently high to allow vacancy migration were required before bubbles (hence blisters) could be observed. However, Nelson,[7] has observed bubbles in copper following 60 keV He$^+$ irradiation at a temperature of only 20°C. In addition, recent work on blister formation in palladium[8] showed 5 μ diameter blisters after 300 keV He$^+$ bombardments at 93 K. However, observations of these blisters were made at room temperature and the thermal release spectra of helium from palladium indicated mobility at temperatures as low as –160°C.

Blisters have also been observed in erbium and erbium deuteride thin films following 160 keV He$^+$ ion irradiation.[9] The size and density of blisters was found to depend on the post implantation annealing temperature. At 400°C, a critical dose (3 x 10^{17} atoms/cm^2) was required before blistering occurred.

In the present work blister formation on molybdenum is examined, both at temperatures near ambient (< 50°C) and at elevated temperatures (up to 500°C) during ion bombardment. Examination was by means of the Surrey University scanning electron microscope (S.E.M.) from which blister diameters could be recorded both as a function of total ion dose, and incident ion energy. The He gas thermal release spectra have also been obtained following bombardment at 50°C.

2 EXPERIMENT

Molybdenum samples 2 cm x 0.5 cm x 0.1 cm were cut from 99.96% purity sheet. Prior to bombardment these were mechanically polished using finally a $\frac{1}{4}$ μ diamond abrasive and then annealed at 2000 K for 5 minutes. Each sample was examined for scratches on the optical microscope with magnification 1500 x.

Bombardments were carried out in the U.H.V. light ion accelerator used previously for the study of gas release from surfaces.[10] The vacuum system was modified for use with helium by adding two mercury diffusion pumps. A background gas pressure of 1 x 10^{-9} torr in the target chamber was obtained after baking. The beam was electrostatically scanned which results in a more uniform ion dose over the bombarded area and hence better resolution in the helium release curves measured as a function of dose.

A turntable allowed up to six targets to be positioned in the beam during an experiment. Each one could be heated from the back face by electron bombardment, up to temperatures in excess of 2000 K. Platinum, platinum-rhodium thermo couples were used for temperature measurement. The helium re-emission both during and after bombardment, was measured using the quadrupole mass spectrometer as in previous work.[11] Care was taken not to disturb the bombarded

239

areas prior to observation on the Stereoscan electron
microscope, with magnifications of 1000-20,000 x.

3 RESULTS AND DISCUSSION

3.1 Gas Release During Bombardment

The helium partial pressure, beam current and target
temperature are monitored continuously during bombardment. With the highest ion dose rate (2×10^{15}
ions/cm^2/sec), the target temperature rises by ~50 K.
Figure 1 shows the helium re-emission as a function
of ion dose during bombardment with 20 keV He$^+$
ions. The curve exhibits features similar to those observed when nickel is bombarded by deuterons at low
temperatures[12] namely a constant ($< 10\%$) re-emission
of the beam, followed by a sharp break point and a
rapid rise in the release rate towards 100% re-emission.

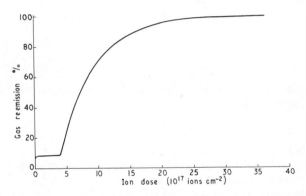

FIGURE 1   Helium re-emision during bombardment of Mo
by 20 keV He$^+$ ions.

The ion dose at which the break point occurs
changes slowly with both ion energy and target temperature. It also appears to be a function of the crystal
face presented to the beam—samples annealed at 2000 K
contain large grains with differing orientations. The
break points ranged from $1-2 \times 10^{17}$ for 7 keV ions,
$4-6 \times 10^{17}$ for 20 keV ions and $5-7 \times 10^{17}$ ions/cm^2
for 36 keV ions. S.E.M. micrographs are shown in
Figure 2 for the different doses. It is seen that the
size distributions are very similar, large blisters being
observed at doses just above the threshold. Samples
examined just below the threshold as determined by
the gas release curve show no sign of any blister formation even at the highest magnification used (25,000 x).
The size distributions obtained by measuring the size
of blisters on the micrographs are shown in Figure 3.
It is apparent that the distributions are approximately
gaussian in shape and that the average size increases

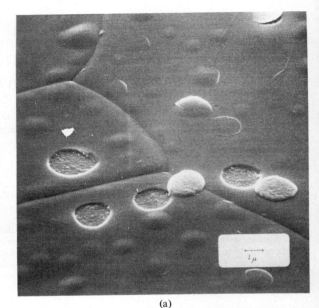

(a)

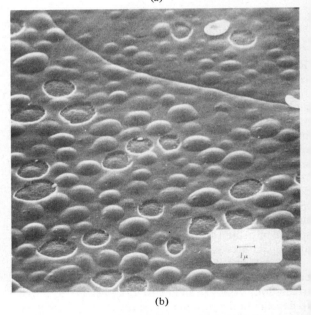

(b)

FIGURE 2   Dose effect on blistering of a molybdenum
surface after bombardment with 36 keV He$^+$ ions. (a) 6×10^{17}
ions/cm^2. (b) 1.2×10^{18} ions/cm^2.

with energy. Micrographs of blisters at different energie
are shown in Figure 4.

3.2 Mechanism of Blister Formation

The results of gas released versus dose indicate that
initially a large fraction of the incident ions are trapped

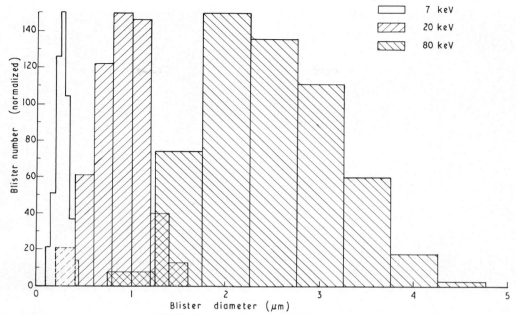

FIGURE 3   Blister size distributions He$^+ \rightarrow$ Mo.

in the surface. The fact that they do not come out is consistent with the low observed rate of diffusion of helium in metals, and contrasts with the behaviour of hydrogen isotopes at room temperature.[11] The concentration of the gas in the surface must build up during this time. Samples examined in a transmission electron microscope (T.E.M.) show that very small bubbles with diameters in the range 20–40 Å are produced.[13] However, the size of these bubbles does not increase with dose, only the number density increases. The fact that blister formation at the surface occurs abruptly suggests that there is a critical concentration above which the metal lattice is disrupted. It is suggested that this is due to the mean distance between bubbles reducing to the point where they interact and there is then a sudden coalescence of bubbles within the range distribution of the incident ions. This coalescence would exert a pressure on the surface from a depth corresponding to the projected range of the incident ions in the metal. From the observed form of the blisters it is clear that there is plastic deformation at the surface with eventual fracture at the periphery of the blister. At the point of plastic deformation we have that

$$\pi r^2 P = 2\pi r\, RY \tag{1}$$

or

$$r = \frac{2RY}{P} \tag{2}$$

where r is the blister radius, R the projected range of ions in the solid, Y the yield strength of the material and P the gas pressure within the blister. Assuming that all the implanted gas is collected in the coalescence process then the pressure will be given approximately by

$$P = D/2.7 \times 10^{19}\ \Delta R \text{ atmospheres} \tag{3}$$

where D is the total ion dose per cm^2 and ΔR is the average height in cm of the void in the metal. It is assumed as a first approximation that ΔR will be equal to the range distribution of ions in the metal. The basis of the coalescence hypothesis is that the bubble density must reach a critical value before the blister can form. If a constant fraction of the incident ion flux goes into bubbles then the critical concentration C will be given approximately by the ion dose divided by the width of the range distribution or $C = D/\Delta R$. Thus

$$r = 5.4 \times 10^{19}\ RY/C \tag{4}$$

By definition both C and Y are constant for a given material at any one temperature and hence we have that to a first approximation the blister radius is proportional to the range of the ions in the metal. Knowing $C \cdot \Delta R \simeq 5 \times 10^{17}$ ions cm^{-2} and $\Delta R \sim 1000$ Å we have $C \simeq 5 \times 10^{22}$ atoms $\cdot$ cm^{-3}. Thus taking the yield strength of molybdenum to be $\sim 5 \times 10^3$ atmospheres[14] we obtain:

$$\frac{r}{R} \sim 5 \tag{5}$$

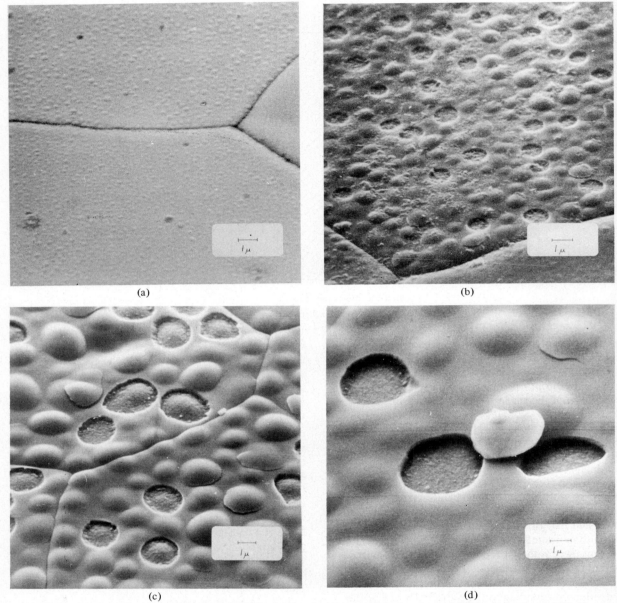

FIGURE 4   Blistering of a molybdenum surface after bombardment with helium ions of different energies. (a) 7 keV. (b) 20 keV. (c) 50 keV. (d) 80 keV.

which is in quite good agreement with the relation between the radius and depth of the craters as observed in the S.E.M. micrographs. The yield modulus Y varies with temperature and material. Moreover it must also be a function of crystal orientation and be higher in single crystals than in polycrystalline material. The experiments carried out are essentially in single crystals since the size of the blisters are much smaller than the grain size in the molybdenum surfaces.

The projected range R increases approximately linearly with energy E in the energy range investigated, and is given approximately by[15]

$$R = \frac{C_i(\mu)100\, A_2(z_1^{2/3} + z_2^{2/3})^{\frac{1}{2}}E}{z_1 z_2 \rho} \text{ Å for } E \text{ in keV} \quad (6$$

where $c_i(\mu)$ is a function of the ratio of the mass of

incident and target atoms ρ is the target density in gms/cc and the other symbols have their usual meanings. Thus

$$R = 57.5 \, E \, \text{Å} \qquad (7)$$

We would therefore expect that the blister radius should be proportional to energy. The experimental results shown in Figure 5 indicate this to be approximately correct.

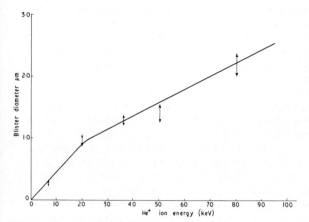

FIGURE 5   Molybdenum blister size as a function of He$^+$ incident ion energy.

Blisters have also been observed after helium irradiation, in W, Pt, Zr, Ni and Cu. Some of these results are shown in Figure 6. The general character of the blisters is very similar to those observed in molybdenum, though only in tungsten are the blisters as clearly seen as in molybdenum. In metals other than W and Mo the number of burst blisters observed is very small, whereas for molybdenum it is ~10%.

3.3 Effect of Temperature

Up to now we have discussed gas release and blister formation only near ambient temperatures. Experiments have also been carried out at higher temperatures and the thermal desorption spectra measured. Desorption spectra are shown in Figure 7 for molybdenum after irradiation at room temperature to a number of specified doses. Although the desorption peaks occur at much higher temperatures the general behaviour is again very similar to that previously observed for deuterium.[12] As the initial dose is increased a greater and greater fraction of the gas is desorbed from the low temperature peaks. It is well known that at very low doses (say < 10^{14} ions/cm^2) no helium is released from molybdenum or tungsten at temperatures below 2000 K following bombardments with

(a)

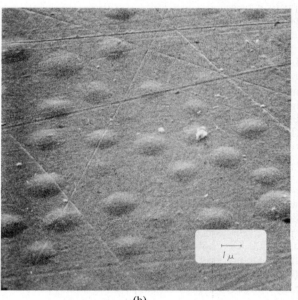

(b)

FIGURE 6   Blistering of (a) platinum and (b) nickel surfaces after bombardment with 36 keV He$^+$ ions.

ions $\gtrsim$ 5 keV. In the present range of doses the desorption peak at 1350 K is rapidly increasing with dose while the peak at 1700 K appears to decrease. Since it is reasonable to attribute the peak at 1700 K (0.59 T_m) to the self diffusion of lattice atoms, its decrease with increasing dose is consistent with the transfer of gas from substitutional sites to the small bubbles observed

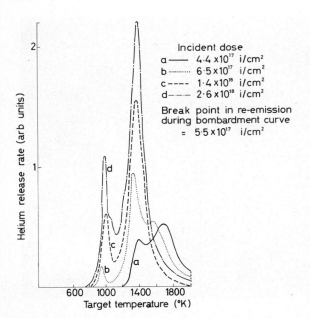

FIGURE 7   Thermal release spectra of He from molybdenum, following 36 keV He$^+$ bombardments.

in the T.E.M. The peak at 1000°K is interesting in that there appears to be a threshold dose below which it is not observed. This dose of 5×10^{17} ions cm^{-2} coincides with that observed for the appearance of surface blisters and for the release of gas at room temperature (Figure 1). It is possible therefore that this peak is due to release of gas from blisters which have not actually ruptured before heating.

Micrographs of surfaces implanted at various temperatures are shown in Figure 8. These temperatures were chosen on the basis of the desorption spectra; 800 K is just below the first peak in the desorption spectra, 1100 K is between the first and second peak and 1600 K is above the second peak. The results obtained are not easy to explain. At 800 K the surface is flaking, at 1100 K it has formed into smooth blisters rather larger than at room temperature and at 1600 K it has an irregular appearance characterized by small pinholes. On the basis of Eq. (4) one would expect the blister radius to decrease, since the yield modulus decreases with increasing temperature. Many additional micrographs have been taken of samples which have been implanted at one temperature and then annealed at a higher temperature. The general conclusion from examining these is that the temperature of implantation is the most important factor in determining surface structure and the subsequent anneal modifies the surface only slightly. It is intended to present these results in more detail in another publication.

4 CONCLUSIONS

The present results confirm the previous evidence[2,3,8] that surface blistering of metals under irradiation results in the release of gas, and indicates that this is in fact the principal mechanism by which the implanted gas is released. The blisters occur after a well defined critical dose of ions has been implanted and this dose agrees quite well with an earlier estimate.[16]

The size of the blisters increases with energy and a simple analysis indicates that this is explained by the increasing depth of implantation. Results are similar for a wide range of metals and the variation in the size from one metal to another is qualitatively in agreement with the variation in range. The appearance of the blisters in practice does not appear to be critically dependent on the properties of the metals although the simple model indicates that it should. Moreover the critical dose for blisters to appear is the same within 50% for all metals so far investigated.

The size distributions of the blisters examined are approximately gaussian within the statistics of the number of blisters counted. There appears to be no change in the mean size of the blisters with the incident ion dose. As the temperature of molybdenum increases the nature of the blistering changes radically and four different characteristic forms of surface blistering have been identified at temperatures of 300 K, 800 K, 1100 K, and 1600 K. The mechanism of the change in blistering is not understood but may be due either to changes in the mechanical properties of the metal or to changes in the size and distribution of the initial bubble formation in the metal before blistering occurs.

The blistering phenomenon must be confined to light ions. With heavier ions, because of the combination of higher sputtering coefficient and lower range, the surface is eroded and gas released before sufficiently large concentrations have built up to produce blistering. With helium the probability of blistering is increased because of the low solubility resulting in no loss of gas by permeation. This contrasts with the case of hydrogen where there is considerable gas release due to permeation,[17] and accounts for the fact that blistering occurs at much lower doses in helium ($\sim 5 \times 10^{17}$ ions cm^{-2}) than in hydrogen ($\sim 10^{19}$ ions cm^{-2}).[17] The bursting of blisters and the cracking of the surface by gas pressure must lead to surface erosion which in practice will be superimposed on the erosion by atomic sputtering processes. Whether erosion due to blistering is measured in any conventional sputtering experiment will depend on whether or not the critical dose has been exceeded. It would be of interest to compare the rate of erosion of a surface at doses lower and higher than the critical dose.

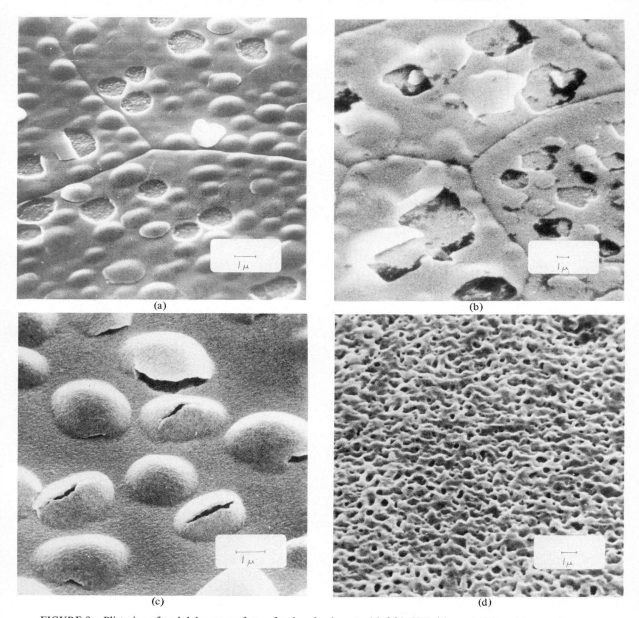

FIGURE 8   Blistering of molybdenum surfaces after bombardment with 36 keV He$^+$ ions at different temperatures. (a) 300 K. (b) 800 K. (c) 1100 K. (d) 1600 K.

ACKNOWLEDGEMENTS

The authors are grateful to B. L. Eyre, (A.E.R.E. Harwell) for his examination of specimens in the T.E.M. and for some valuable discussions, to P. Goldsmith who was responsible for much of the experimental measurements and to A. Summers (A.W.R.E. Aldermaston) for the implantations at 50 and 80 keV.

REFERENCES

1. W. Primak and J. Luthra, *J. Appl. Phys.* **37**, 2287 (1966).
2. M. Kaminsky, *Adv. Mass Spectrometry* **3**, 69 (1964).
3. R. D. Daniels, *J. Appl. Phys.* **42**, 417 (1971).
4. R. Behrisch and W. Heilund, *6th Symposium on fusion technology Aachen 1970*, p. 465.
5. R. S. Barnes and D. J. Mazey, *Proc. Roy. Soc.* **275**, 47 (1963).
6. R. Kelly and E. Ruedl. 3rd European Conference on Electron Microscopy Prague 1964.
7. R. S. Nelson, *Phil. Mag.* **9**, 343 (1964).
8. W. Bauer and G. J. Thomas, *J. Nucl. Mat.* **42**, 96 (1972).
9. R. S. Blewer and J. K. Maurin, Sandia Laboratories, Report No. SC-DC-714060, 1971.

10. G. M. McCracken, J. H. C. Maple and H. H. H. Watson, *Rev. Sci. Inst.* **37,** 860 (1966).
11. S. K. Erents and G. M. McCracken, *Brit. J. Appl. Phys.* **2** (2), 1397 (1969).
12. S. K. Erents and G. M. McCracken, *Rad. Effects,* **3,** 123 (1970).
13. B. L. Eyre, Private Communication 1972.
14. W. H. Kohl, *Materials and techniques for electron tubes* (Reinhold, New York, 1960), p. 315.
15. H. E. Schiott, *Rad. Effects* **6,** 107 (1970).
16. G. M. McCracken and S. K. Erents, B.N.E.S. Nuclear Fusion Reactor Conference, Paper 4.2 p.   Culham September 1969.
17. R. Behrisch. Private Communication, April 1971.

THERMAL EVOLUTION SPECTROMETRY OF LOW ENERGY INERT GAS IONS INJECTED INTO Cu, UO₂ AND GLASS

A. O. R. CAVALERU[†], C. M. MORLEY[‡], D. G. ARMOUR[§], and G. CARTER[§]

Low energy ions were injected into single crystal Cu, sintered UO₂ and glass and the probabilities of entrapment measured, as a function of ion energy from 50 eV to 3 keV, by heating the targets after irradiation and recording the ratio of quantity of gas released to incident ion fluence. Values of entrapment probability increase from zero at low ion energies to about unity at 1 keV. The rate of thermal gas evolution was also recorded as a function of temperature during the post-bombardment target tempering, enabling deduction of activation energies for evolution. These activation energies are correlated with published data on the annealing of point and extended defects in the solids and with self diffusion energies.

INTRODUCTION

The injection of energetic ions into solids usually results in their solution in the matrix, upon coming to rest, and even in the case of inert gases, which are generally quite insoluble, high gas concentrations can be accumulated since the gas is able to occupy native defects in the solid, defects created in the collision cascade generated by the ion and is also able to cluster in bubbles. The ability of an ion to penetrate into a solid depends intimately upon the forces of interaction between the ion and surface atoms of the solid and at low ion energies (⩽1 KeV) one would expect rapid variation of the penetration probability with increasing ion energy as the ion becomes more able to penetrate the surface potential barrier. Measurements with inert gas ions incident upon polycrystalline[1,2] and single crystal[3] W, single crystal Nb,[4] polycrystalline Au[5] and Re,[6] and amorphous glass[7] have indicated the reality of this form of penetration probability variation and attempts have been made[8] to evaluate the magnitude of the interatomic forces from measured data by computer simulation of the injection process. A difficulty in making such evaluations has been recognized,[2,8] that experimental data do not measure penetration probabilities alone, but are in fact the product of this parameter with the subsequent probability of ion solution in the target. Consequently, it is desirable to measure the mechanisms of ion solution and one method of attempting this is to observe the post-injection thermal migration kinetics of dissolved inert gas atoms. In this technique, which has been successfully employed with W,[1,2,3] Nb,[4] Au,[5] Re,[6] Ni[9] and glass targets,[7] the irradiated target is heated so that its temperature rises as a well defined time function and the rate of gas migration to and evolution from the surface as a function of temperature is monitored either by mass spectrometric or radioactive assay methods. Alternatively, the target may be heated in a stepwise manner and similar rate measurements made. This gas evolution rate data may then be analysed to yield activation energies for the migration processes and the orders of the reaction, assuming either single step migration to and from the surface[10,11] or multistep diffusion before evolution.[12,13] As a result of analyses of the above and other ion-solid systems over a broad ion energy range (50 eV to 100 keV) Kelly and his colleagues[14,15] have classified gas release processes into five stages.

With low energy incident ions all five stages have been observed, but in recent, elegant experiments Kornelsen[16] has employed such low incident ion fluences that cooperative effects between injected ions could be totally suppressed (stage III). Kornelsen was then able to find firm evidence of He dissolved in single vacancies in single crystal W, in multiple vacancy clusters and of He associated with clusters of previously injected heavier inert gas atoms. On the other hand, experiments with single crystal Nb,[4] admittedly with considerably higher ion fluence exhibited such a multiplicity of activated migration processes (as evidenced by many peaks in the evolution rate–temperature spectra) as to render interpretation in terms of atomic processes almost impossible. For this reason it was decided to observe low energy ion entrapment and gas release processes in single crystal Cu, a

† Institute of Atomic Physics, Bucharest, Romania.
‡ Associated Electrical Industries Ltd., Manchester.
§ Department of Electrical Engineering, University of Salford, Salford M5 4WT, Lancashire, U.K.

material in which many atomic migration processes have been catalogued,[17] if not fully understood or resolved. At the same time it was decided to make measurements with a polycrystalline material, sintered UO_2 was chosen because of its technological importance in reactor applications and because gas migration data are required for swelling rate predictions, and an amorphous solid, glass in order to be able to observe differences and make comparisons. The preliminary results of these studies are presented in this communication.

EXPERIMENTAL

A detailed description of the apparatus and the experimental technique has been presented recently[18] by the authors and only a brief resume will be given here for clarification of the form of the data.

Monoenergetic (to within 5 eV) inert gas ions were generated in an ion source adapted from a design by Kornelsen,[1] focused into an almost parallel beam and manipulated onto a target, of the desired material. The target could be heated during bombardment, or cycled thermally according to a programmed temperature rise schedule following bombardment, by means of controlled electron bombardment on the opposite face of the target to the ion beam. An evaporated Au layer on the glass target prohibited charge build up during electron bombardment and allowed uniform heating. In addition an electron emitting filament was positioned opposite the ion bombarded surface of glass and used to bathe the surface with electrons to prohibit positive charging during ion irradiation. It was not possible in these experiments to suppress secondary electrons emitted from the targets so that errors estimated at less than 10%, which vary with ion energy, in ion dosimetry were present.

All of this apparatus was contained within a stainless steel ultra high vacuum system in which pressures in the low 10^{-10} torr region were attained routinely. During ion bombardment the inert gas pressure rose to about 10^{-5} torr but this would be ineffective in contaminating target surfaces. Included in the U.H.V. system, a 5 cm radius sector field mass spectrometer was used to monitor, by electronic differentiation of its collector current, the rate of gas evolution during the post bombardment tempering cycle during which period the target volume and mass spectrometer were isolated from any pumping action.

Sensitivity limitations of the mass spectrometer unfortunately rendered impossible the detection of evolved quantities of gas less than 10^{12} atoms/cm^2 of target area and so we could not expect, nor did we observe the well resolved gas evolution structures

obtained by Kornelsen. At low ion energies, with concomitantly low entrapment probabilities, we do believe however, that cooperative effects between injected atoms were minimized as the data will subsequently reveal.

The Cu, UO_2 and glass targets were disc shaped and were vibratorily polished and chemically etched before insertion into the vacuum system where they were spring mounted with their flat faces perpendicular to the ion beam (the ⟨100⟩ face for single crystal Cu). The spring mounting ensured that the targets did not deform (UO_2 showed a marked tendency to crack) during heating. Temperature measurement was effected by thermocouples inserted radially into the discs at a point mid-way between the front and rear surfaces. Subsidiary experiments showed that temperatures recorded here were no more than 10°K different from those of the ion bombarded surfaces. Before irradiation all targets were subjected to many cycles of heat treatment to temperatures above those subsequently employed in the evolution studies, to ensure, together with the U.H.V. capabilities, rather clean surfaces for bombardment.

It is not claimed that the surfaces were atomically clean and gas free, since the time required to execute experiments was usually of the order of several minutes. The presence of adsorbed gas layers is undoubtedly influential in determining ion entrapment probabilities and location of embedded gas, particularly for low mass (He) ions and at low incident energy. It is believed, however, that using the heavier ions (the results for which are reported here) and particularly at higher energies (1–3 keV) the influence of adsorbed impurity gases would be small, largely because of the gas sputter cleaning action of the incident ions themselves.

As noted above, the mass spectrometer was used to monitor instantaneous rates of gas evolution during post bombardment tempering and this required precalibration of the system with known rates of inert gas influx. The total quantity of gas evolved into the system up to any temperature was determined directly from the undifferentiated mass spectrometer signal, and in this way the total quantity of gas evolved over the full tempering cycle could be deduced. Using such a tempering technique there is always the uncertainty that all injected gas may not be evolved but we believe that the order of this error is about the 10% determined by dosimetry uncertainties. Measurements of the entrapment probability for the highest energy ions used here ($\simeq 3$ keV) for all materials showed that this approached to within a few per cent of unity (for all three materials) indicating that all entrapped gas was evolved (to within the dosimetry 10% error). Further checks were made by injecting somewhat higher energy (10 keV) radioactive Kr^{85} ions into UO_2 and upon subjecting these targets to the same tempering schedules used in the lower energy work, only 20% of the gas trapped at room temperature remained in the target. One would expect the retained percentage to be lower for lower energies.

Following the procedures outlined above, evolution rate–temperature data were recorded for several different inert gas ion species (He, A, Kr and Xe) injected into Cu, UO_2 and glass over the ion energy range 50 eV to 3 keV and for incident ion fluences between 10^{12} ions/cm^2 and 5×10^{14} ions/cm^2. Unfortunately, it was not possible to vary the ion flux over a significant range. Simultaneously the total number of gas atoms entrapped (or evolved) was measured. For all ion species and energies and for each target material the number of ions entrapped followed an initially linear dependence upon ion flux, but then the rate of entrapment declined at higher fluence until saturation quantities were reached as has been observed earlier with W,[1,2] Ni,[4] Re,[6] and glass[7] and is the result of dynamic equilibrium between gas entrapment and removal by a sputtering process. Consequently, entrapment probabilities were deduced from the initial low fluence regime and Figure 1 shows entrapment probabilities for all three targets as a function of ion energy for A$^+$ ions. For all three materials the entrapment probability, η, approaches zero at energies below 100 eV (the threshold energies for Cu, UO_2 and glass are 190 eV, 12 eV and 20 eV respectively) then rises rapidly with ion energy up to about 1 keV and then saturates at a near unity value (remembering the 10% error possibility) at higher energies. This is precisely the form of behaviour expected theoretically and already observed experimentally for W, Re and Au.

EXPERIMENTAL RESULTS AND DISCUSSION

As noted above evolution spectra were recorded for four ion species, as were entrapment probabilities. A general result of the latter measurement was that for each target material, at any energy below the value at which η approached unity, then the entrapment probabilities were in the order $\eta_{He} > \eta_A > \eta_{Kr} \simeq \eta_{Xe}$ indicating that, as expected for any reasonable interatomic potential law the surface repulsive potential barrier decreases with decreasing incident ion atomic number. It should be noted that whereas the threshold energies for glass (a borosilicate "Pyrex") of mean atomic mass of about 30 a.m.u. and sintered UO_2 of mean atomic mass of about 90 a.m.u. are similarly

low (<20 eV) the threshold energy for Cu of atomic mass mid-way between glass and UO_2 is very much higher (190 eV) indicating that structure and interatomic spacings may be parameters of similar or greater importance than atomic number in dictating penetration into the solid phase. Andersen and Sigmund[19] have indicated how it is theoretically possible to estimate the energy of an ion necessary to carry it through (or penetrate in the current context) the central axis of a "ring" or symmetric atoms if the interatomic potential law is known. In the present case of (100)Cu such a ring consists of two surface plane atoms and then behind these a further ring of two atoms with their axis at right angles to that joining the two surfaces atoms and with the same spacings. If it is assumed that ions must penetrate the second layer to entrap in the solid, then as shown computationally by Smith and Carter,[20] this requires about twice the energy to penetrate a single two atom ring, which assuming a Born-Mayer interatomic constant with appropriate constant suggested by Abrahamson[21] leads to a threshold energy of 51 eV. This is in poor agreement with the experimental result and may result from:

1) a poor choice of Born-Mayer constants;

2) ions which do just penetrate the solid may not create defects which may be required to ensure their retention so that higher values of energy are required before experimental measurements of η are meaningful;

3) little experimental data was accumulated at low incident energies and reproducibility was relatively poor because of detection sensitivity limitations thus rendering the extrapolated value for the threshold energy inaccurate.

However, the results for threshold energy for He penetration are 7 eV (theoretical) and 15 eV ± 3 eV (experimental) which are in rather better accord. Despite the present lack of quantitative agreement for A however it is believed that further more careful measurements should be made to improve the observation of the threshold energy. Similar calculations are not possible with glass and UO_2 but since these materials both possess rather "open" structures it is readily understood why penetration thresholds are low and entrapment probabilities high.

It has already been noted that the η values in Figure 1 were derived at low ion fluence levels because of "saturation" effects at higher fluences. It is not intended to comment much further on the sputtering mechanism responsible for this saturation since more detailed analysis of this effect is to be published elsewhere. However, it should be noted that if the reported

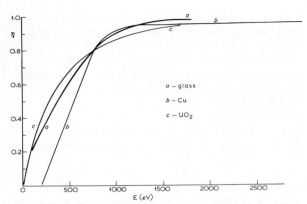

FIGURE 1   Entrapment probability (η) as a function of incident A^+ ion energy for Cu, UO_2 and glass targets.

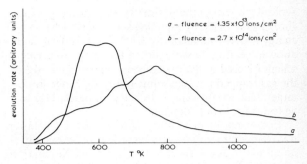

FIGURE 2a   Evolution rate spectra of 1 keV A^+ injected into Cu at two different incident ion fluences. (Evolution rate scale is arbitrary and is different for each fluence.)

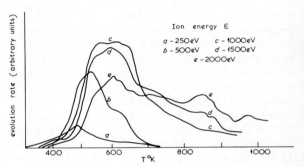

FIGURE 2b   Evolution rate spectra of 1.35×10^{13} ions cm^{-2} of A^+ injected into Cu at various energies.

value[22] for the sputtering coefficient of Cu by 1 keV A^+ ions ($\simeq 3$ atoms/ion) is adopted then an incident fluence of 10^{15} ions/cm^2 is sufficient to remove about 1 surface atomic layer of Cu. Since the mean projected range of 1 keV A^+ in Cu is expected to be equivalent to only about two atomic spacings[23] then the removal of one plane of atoms would cause considerable gas evolution and the importance of target sputtering in a relatively high sputtering coefficient material in determining saturation effects is evident.

In presenting and analysing the gas evolution rate data, only the data for A^+ ions are considered. A more comprehensive treatment for the several ion species will be presented for each target material in subsequent reports but for the present it is important to recognize that A^+ may be regarded as an archetype for the three heavier ions, A, Kr and Xe. Thus, although there are minor differences in the energies of migration evaluated for each ion for a given target and even major differences in the populations of the different migrating specie densities as ion energy or incident fluence is varied, nevertheless the evolution peak structures are typified, on temperature scales by the behaviour of A. The same cannot be said for He ions and these form a special case, again to be treated in detail elsewhere, since as has been observed in W,[2,16] Re[6] He ions, because of their low mass and inefficient energy transfer to target atoms, create few defects and this grossly influences the entrapment process.

The form of evolution rate–temperature spectra are influences, for each ion species and target material by both ion fluence and energy. Figures 2a and 2b indicate the variation of such spectra for Cu as a function of ion fluence at one constant energy and as a function of ion energy at a constant fluence respectively. Figure 3a and 3b and 4a and 4b illustrate analogous

data for UO_2 and glass targets respectively although it should be noted that Figure 4a records data for Kr evolution, rather than A, in order to reinforce the point made above regarding the similarity of data for different species.

It is necessary to consider each target separately but there are also some important generalizations which apply to all three target materials. Firstly, each spectrum has a multiple peak appearance which is nowhere as well defined as the distinctly resolved, almost line spectra reported for W[16] and for Nb.[4] This cannot be attributed to lack of resolution in the experimental system since this was capable of the theoretical resolution for *any* such system,[11] (e.g. that two evolution peaks the difference in activation energies for which were less than 5% of the activation energy for either were well resolvable) and this indicates that the peaks therefore resulted from either a continuum of activation energies (e.g. in glass, Figures 4a and 4b) or closely spaced discrete activation energy processes (e.g. in Cu and UO_2 as seen in Figures 2a and 2b and 3a and 3b). Secondly, in all materials one can

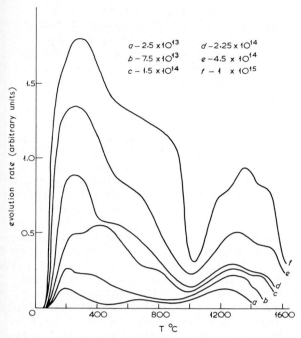

FIGURE 3a   Evolution rate spectra of 1 keV A$^+$ injected into UO$_2$ at various incident fluences (ions cm^{-2}).

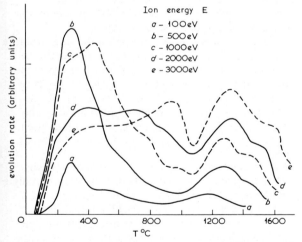

FIGURE 3b   Evolution rate spectra of 1.5 x 10^{14} ions cm^{-2} of A$^+$ injected into UO$_2$ at various ion energies. (Evolution rate scales are arbitrary and are different for each energy.)

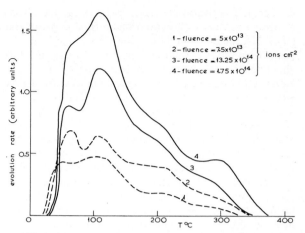

FIGURE 4a   Evolution rate spectra of 1 keV Kr$^+$ injected into glass at various incident fluences.

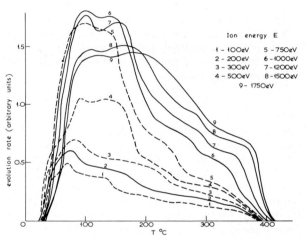

FIGURE 4b   Evolution rate spectra of 1.75 x 10^{14} cm^{-2} of Ar$^+$ injected into glass at various ion energies.

broadly group the evolution peaks into lower temperature and higher temperature regions and with both increasing ion fluence and energy the increase in population in the higher temperature peaks is greater than in the lower temperature regime. Thus, at constant ion energy the population of the higher temperature regime increases more rapidly with ion fluence than the population of the lower temperature regime, although the former does not increase population at the expense of a decrease in population of the latter. A similar statement applies at constant ion fluence with increasing ion energy, since the total population increases with increasing energy (i.e. η increase) as well as individual population increases in both regimes. A ready interpretation of the latter behaviour is that as ion energy increases, ions are injected deeper into the solid so that a larger fraction diffuse from more deeply located entrapment centres before evolution. This effectively transfers an increasing proportion of gas from near surface, low activation energy centres (stage IA release) to deeper centres with paranormal diffusion parameters (stage IIA release) and indeed the higher temperature peaks themselves exhibit an upwards temperature shift characteristic of a longer path diffusion process. The enhanced population of the higher temperature regime with increasing ion fluence reflects the fact that the low temperature centres, presumably close to the surface saturate early (since there are a limited number available) whilst the more deeply lying centres (of high density) saturate later. Additionally, at high ion doses one must expect that collaborative effects will become important and that gas clusters and gas precipitated upon defects will increase in importance leading to a stage IA-IB transition behaviour. Indeed, in some of our more recent experiments with single crystal UO_2, similar spectra to those reported here have been measured where the ion fluences were similar, but at much higher fluence ($>10^{16}$/cm^2) a substantial fraction of the gas was not evolved up to the highest temperatures used indicating the precipitation of gas into immobile bubbles (stage III).

We now turn to a more detailed consideration of the three target materials.

Cu

As already noted, and much to our disappointment, discrete well resolved spectra as measured in W and Nb were not observed, although as shown in Figures 2a and 2b there is evidence of about six relatively well defined peaks which overlap and may be superimposed on a general continuum. In such a case it is difficult to estimate activation energies precisely but approximate values for these six peaks were determined from first order single jump evolution rate theory to be 1.3 eV, 1.4 eV, 1.7 eV, 1.8 eV, 2.1 eV and 2.3 eV. It is possible to identify the lowest of these (1.3 eV) with the activation energy (1.26 eV) reported by Wynblatt[24] for the surface migration of Cu atoms indicating that a fraction of trapped A may be held just below the surface and is released as the surface is reordering during heating. Alternatively this lowest energy peak (or indeed higher energy peaks) may be due to the thermally induced motion of bulk vacancies formed during the A$^+$ irradiation but requiring an energy of migration modified by the presence of the gas atom to above the reported[25] activation energy for vacancy migration in Cu (1.1 eV).

A consistently major peak in the evolution spectrum is that centred around 2.1 eV (800°K) peak and it is very tempting to associate this with the migration of vacancies thermally formed during tempering and

with an activation energy close to the measured activation energy[26] for self diffusion in Cu. Indeed the evidence of increasing importance of this peak with increasing ion injection energy and ion fluence, where vacancies will be increasingly formed and available for gas atom occupancy favours this interpretation. Similar associations with self diffusion (stage IIA) have been made in W[2,16] and Re.[16] One cannot specify, with any confidence, the nature of the processes responsible for gas release with energies of 1.4 eV, 1.7 eV and 1.8 eV (i.e. between 535°K and 700°K). However, electron microscopic evidence of low energy ion bombardment induced black spot and dislocation formation in Au[27] and Cu (R. S. Nelson, Private communication) shows that this disorder generally begins to anneal at temperatures below the self diffusion temperature via slip and climb processes and the influence of the large defects upon the energy of point defect migration. Thus, the intermediate peak may well be associated with dislocation motion and sweeping of gas atoms, or even piping of gas via dislocations to the surface (i.e. stage IB type of release).

Finally, the highest temperature processes, which at high ion energies and high ion doses ($\gtrsim 10^{15}$/cm^2), show evidence of continuing beyond 1000°K can readily be associated with gas cluster and bubble formation (stage III) since here the gas atom concentrations will be in the order of 10% where bubble precipitation can become influential.

Thus, although Cu does not reveal the fine structure reported for W, and one reason for this may well be the rather higher than desirable fluences it was necessary to employ which lead to increasing cooperative effects and spectral complexity, it does appear possible to make some reasonable suggestions as to the processes responsible for gas solution and migration.

UO_2

Electron microscopic examination of the polished sinters revealed these to consist of large polycrystalline grains with much evidence of superficial voids. In short, the surface was a mess looking rather like Gruyere cheese. It was therefore not surprising that fine line shaped evolution spectra were not observed but it was gratifying that, as with Cu some dominant peak characteristics could be deduced. Moreover, entrapment and gas release data were highly reproducible indicating that ions sampled all of the surface topography. Again it was possible to confidently identify six resolvable evolution peak energies at 1.5, 1.7, 2.6, 3.3, 4.1 and 4.4 eV. The lowest energy processes (at 1.5 and 1.7 eV) can be ascribed to stage IA mechanisms in which the gas atom is trapped in ome (unspecified) site or is so close to the surface that strain fields lower its migration energy. The evolution at 2.6 and 3.3 eV are ascribed to stage IB processes requiring disorder annealing and indeed Cornell and Bannister[28] have reported on the annealing of bombarded induced dislocations in UO_2 in this energy (or temperature) regime. Additionally the entrapment of A into these centres exhibited a superlinear dependence of population upon ion fluence, indicative of either cooperative gas-gas or gas-disorder effects.

It should be noted that previous studies of UO_2[29,30] have mentioned the occurrence of these lower temperature gas release processes but assigned minor significance to them, whereas in the present studies gas release below 1000°K is seen to be a major factor. The clear reason for this difference is that earlier studies have employed higher ion energies and fluences than in the present work and such irradiation leads naturally to deeply located gas, dissolved in the lattice and in bubbles so that stages IIA, IIB and III evolution assume greatest importance. As indicated previously stage III behaviour has recently been observed in our own continuing studies with high fluence injected single crystal UO_2, but no real evidence of this process was observed in the present work at lower fluences. However, the present studies do reveal evidence of normal diffusion (stage IIA at an activation energy of 4.12 eV) and retarded diffusion (stage IIB at an activation energy 4.45 eV). The mechanism of self diffusion in UO_2 is believed to be[34] via migration of the less mobile U vacancy and to require an activation energy of 3.8–4.4 eV in agreement with the measured 4.1 eV. Retarded diffusion should occur with higher activation energy and increase in importance with increasing ion fluence and energy as gas and defect concentrations increase as indeed is the case for the 4.4 eV peak. Matzke[30] suggests that the activation energy for this stage IIB process may be between 0.3 and 0.9 eV higher than the self diffusion energy and this is in accord with the observed difference of 0.33 eV.

Thus, as in single crystal Cu, sintered polycrystalline UO_2 exhibits identifiable gas evolution processes which can be correlated with observed lattice defect migration processes. It must certainly be true that at the higher dose levels used in the present studies both materials will possess a rather complex point and extended defect structure near the surface where the injected ions come to rest and it is immediately apparent why the release processes should be dictated by the migration and annealing of the structural defects in the solid. It is for this reason that only small differences are observed in the gas evolution

behaviour of the three heavy ions studied. It must be stated, however, that in order to obtain the more unequivocal assignation of gas location studied by Kornelsen for W, much lower ion fluences than used in the present studies are required.

GLASS

One would not expect ion fluence to exert such an influence on evolution spectra in such a material since even before irradiation it is generally considered amorphous, and indeed increase of ion fluence does not appear to materially change the relative populations of the two broad peaks of Figure 4a whilst increase in ion energy appears only to shift these peaks to slightly higher temperature, commensurate with an increase in injected atom location depth. The existence of broad peaks also indicates the existence of a continuum of activation energy for migration processes and this could certainly be anticipated for a solid with randomised atomic array and concommitant variety of capture and binding forces. Because of this continuum character for the evolution spectra it is possible only to define most probably migration activation energies from Figure 4 and these are evaluated as 0.9, 0.9, 1.1, 1.3 and 1.5 eV. In an earlier study of A injected Pyrex glass James and Carter[32] also observed the broad continuum over the same energy and temperature range shown in the present data but, probably because of inadequate system resolution, could not identify the existence of any finer structure as seen by the five peaks here.

It is very difficult even to attempt assignment of these five energies to inert gas migration processes in glass since alternative measurements by conventional diffusion processes are scarce. In fact, only one relevant measurement,[33] for A in fused silica, of about 1.3 eV is known to us and this might indicate that either the 1.3 eV or the 1.5 eV measured value is to be associated with normal (but undefined) diffusion. Since glass is amorphous, however, the other activation energies may correspond to preferential trapping centres within the glass matrix (again of undefined character) but one cannot exclude the possibility that even with an initially amorphous matrix, ion injection can create reordered zones due to the development of the collision cascade and local heating (or thermal spike) effects. Certainly the refractive index of glass, quartz and vitreous silica[34,35] can be changed by inert gas injection and this indicates some form of atomic reordering in the matrix. Thus, all that one can say in this case is that gas release processes appear to occur in temperature regimes associated with stages

I and II migration but whether ion irradiation provokes the existence of centres associated with stage IB and IIB migration is uncertain.

CONCLUSION

Inert gas ion entrapment and subsequent thermal evolution from three diverse materials, single crystal Cu, polycrystalline UO_2 and amorphous glass, show surprisingly similar features of broad ill resolved gas release behaviour and similar entrapment probabilities. In all materials the entrapment probability rises to unity (to within 10%) at ion energies of about 1 keV, whilst the evolution spectra consist of overlapping, non-discrete activated processes. In Cu and UO_2 it is possible to suggest correlations of measured migration energies with defect annealing and self diffusion mechanisms, whilst in glass all gas release appears to occur at and below the "normal" diffusion temperature.

ACKNOWLEDGEMENTS

A. O. R. Cavaleru gratefully acknowledges receipt of a maintenance grant from U.K.A.E.A. (Harwell) for supporting his studies with UO_2 and glass. The authors are pleased to acknowledge permission from that establishment to publish the data accumulated during the period of this grant and particularly Dr. R. S. Nelson for stimulating discussions on the data observed in UO_2.

REFERENCES

1. E. V. Kornelsen, *Can. J. Phys.* **42**, 364 (1964).
2. K. Erents and G. Carter, *J. Phys. D.* **1**, 1323 (1968).
3. E. V. Kornelsen and M. K. Sinha, *J. Appl. Phys.* **39**, 4546 (1968).
4. R. O. Rantanen and E. E. Donaldson, *J. Vac. Sci. Technol.* **8**, 23 (1971).
5. K. Erents, R. P. W. Lawson and G. Carter, *J. Vac. Sci. Technol.* **4**, 252 (1961).
6. A. L. M. Davies and G. Carter, *Rad. Effects* **10**, 227 (1971).
7. W. A. Grant and G. Carter, Proc. 7th International Conf. on Phenomena in Ionised Gases, Belgrade Yogoslavia (1965).
8. M. Pryde, A. G. Smith and G. Carter, *Atomic Collision Phenomena in Solids* edited by D. G. Palmer, P. A. Townsend and M. W. Thompson (North Holland Publishing Co. Amsterdam, 1970), p. 573.
9. R. O. Rantanen, A. L. Moon and E. E. Donaldson, *J. Vac. Sci. Technol.* **7**, 18 (1970).
10. P. Redhead, *Vacuum* **12**, 203 (1962).
11. G. Carter, *Vacuum* **12**, 245 (1962).
12. G. Farrell, W. A. Grant, K. Erents and G. Carter, *Vacuum* **16**, 295 (1966).
13. G. Farrell and G. Carter, *Vacuum* **17**, 15 (1967).
14. F. Brown and R. Kelly, *Acta. Met.* **13**, 169 (1965).
15. C. Jech and R. Kelly, *Proc. Brit. Ceram. Soc.* **9**, 259 (1967).
16. E. V. Kornelsen, *J. Vac. Sci. Technol.* **9**, 624 (1972).
17. See for example A. C. Damask and G. J. Dienes, *Point defects in metals* (Gordon and Breach Ltd, New York, 1963).
18. A. O. R. Cavaleru, C. Morley, D. G. Armour and W. A. Grant, *Vacuum* (To be published) (1972).
19. H. H. Andersen and P. Sigmund, *Mat. Fys. Medd. Dan. Vid. Selsk.* **34**, No. 15 (1966).
20. A. G. Smith and G. Carter, *Rad. Effects* **12**, p. 63–71 (1972).
21. A. A. Abrahamson, *Phys. Rev.* **133**, A990 (1964).
22. G. Carter and J. S. Colligon, *Ion Bombardment of Solids* (Heinemann Educational Books Ltd., London, 1968).
23. J. Lindhard, M. E. Scharff and H. Schiøtt, *Mat. Fys. Medd. Don. Vid. Selsk.* **33**, (1963).
24. P. Wynblatt, *Phys. Stat. Sol.* **36**, 797 (1969).
25. A. Seeger and D. Schumaker, *Lattice Defects in Quenched Metals* edited by Cotterill *et al.* (Academic Press, New York, 1965), p. 15.
26. A. Kuper, H. Letaw, L. Slifkin, E. Sonder and C. T. Tomizuka, *Phys. Rev.* **98**, 1870 (1955).
27. G. J. Ogilvie, J. V. Sanders and A. A. Thompson, *J. Phys. Chem. Solids* **24**, 247 (1963).
28. R. M. Cornell and G. H. Bannister, *Proc. Brit. Ceram. Soc.* **7**, 355 (1967).
29. O. Gautsch, C. Mustacchi, A. Scharenkamper and H. Wahl, Euratom Report. EUR-3267 E (1967).
30. Hj. Matzke, *Proc. Int. Summer School on the Physics of Ionized Gases* (Herceg Novi, Yugoslavia, 1970), p. 326.
31. R. Kelly and C. Jech, *J. Nucl. Mat.* **30**, 122 (1969).
32. L. H. James and G. Carter, *Br. J. Appl. Phys.* **13**, 3 (1962).
33. R. M. Barrer, *Diffusion in and through solids* (Cambridge University Press, 1951).
34. R. L. Hines, *J. Appl. Phys.* **28**, 587 (1957).
35. R. L. Hines, *Phys. Rev.* **120**, 1626 (1960).

OPTICAL PROPERTIES OF ION BOMBARDED SILICA GLASS

A. R. BAYLY

*University of Sussex, School of Mathematical and Physical Sciences,
Falmer, Brighton BN1 9QH, Sussex, England*

Refractive index profiles, produced by ion irradiation in silica glass, have been measured, using ellipsometry, by sectioning the implanted region, obtaining a series of measurements at progressively deeper levels and analysing the profile as a stack of thin homogeneous layers.

Results have been obtained for He^+, Ar^+ and Kr^+ implantations using hydrofluoric etching as the sectioning technique. It is found that the profile consists essentially of a shallow peak near the surface of $1.0 \to 1.5\%$ change in index, and a deeper peak which seems to saturate at a change of $\sim 1.0\%$. The deep peak shows good agreement with theoretical damage distributions for low dose He^+ but shows signs of saturation after a dose of $5 \cdot (10)^{15}/cm^2$ 20 kV He^+. For Ar^+ and more so for Kr^+ at doses of $1 \cdot (10)^{16}$ ions/cm^2 at 300 kV there appears significant broadening of the profiles compared to theory. Although this might be explained by damage saturation the lack of any obvious implantation effect suggests a possible explanation in terms of a redistribution of the implanted species by the compaction effect which occurs in silica. Alternatively the implanted species may oppose compaction. Back-scattering studies are proposed to resolve these questions.

INTRODUCTION

Interest in the use of ion beams for polishing or machining optical surfaces and also for modifying the optical constants of the sub-surface layers is increasing rapidly as the concept of "integrated optical devices" approaches reality. Already various demonstrations of fabrication by ion beam of light-guiding channels[1,2,3] and other optical components[4] have appeared in the literature. However, the nature of the mechanisms responsible for the changes in optical properties are not well understood. A question of major importance is what are the respective contributions of radiation damage and implanted ions? Ideally, for implantation to realize its full potential it is desirable that the ions used should provide the dominant effect. However, current opinion generally is that most of the effects observed have been the result of damage.[1,2,5,6,7] The methods used for these studies have required various assumptions about the relationship between ion implant and refractive index change depth profiles in order to evaluate the changes in terms of a single homogeneous layer, but these procedures are open to question.

The need for such assumptions would disappear if direct measurements of the refractive index profiles were taken. A method using ellipsometry was described by Schroeder and Dieselman[6] whereby theoretical profiles were adjusted by trial and error to match the variation, with angle of incidence of the ellipsometer beam, of the ellipsometric parameters of the original implanted surface. Although the feasibility of this was demonstrated, the matching procedure could be difficult if the initial theoretical profile were drastically wrong. For the present work a more direct approach has been adopted. This will be described[8] in detail with a discussion of the accuracy and merits elsewhere, and is outlined below. The results presented below show the potential of the technique for identifying various damage and implant processes. In fact three separate effects may be present, two due to damage and one to implantation, with considerable deviation of the implant profile from that predicted theoretically.

EXPERIMENTAL PROCEDURE AND ANALYSIS

A brief outline only is given here of the experimental and analytical techniques used. For a full description see Ref. 8.

Samples

"Spectrosil" Synthetic vitreous silica discs, 25 mm x 1 mm were supplied optically polished by Thermal Syndicate Ltd. Before implantation samples were degreased followed by etching, for $\sim$10 mins in a solution of HF 5%, HNO_3 30%, "Teepol" 2%, and distilled water, and finally washing in distilled water to remove any polish-damage layer.

Implantation

Implantations of 300 kV inert gases at room temperature were performed using a Dynagen 300 keV accelerator with magnetic separation and a slow electro-static scan to improve implantation uniformity. Implantation of 20 kV helium was performed on a smaller machine using a Nelson–Hill ion source. All implantations were performed under high vacuum conditions ($\lesssim 5 \cdot 10^{-6}$ torr) with liquid nitrogen trapping and also a liquid nitrogen cooled tube in front of the target.

Beam current densities were $\leqslant 5$ $\mu A/cm^2$ and surface neutralization was ensured by a free supply of low energy electrons from a shielded tungsten filament.

Profile Measurements

Ellipsometric measurements were made on the implanted specimens at fixed angles of incidence and polarizer as shown in Figure 1, with the 5461 Å light

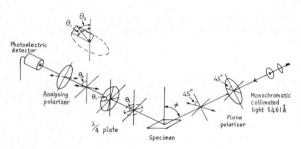

FIGURE 1   The basic ellipsometer arrangement: the automatic faraday-modulation nulling system is not included for clarity.

from a high pressure mercury lamp. Thin layers were stripped from the surface by etching in a 0.3% HF acid bath stirred continuously at room temperature. Successive measurements and strippings were made until the whole of the affected depth of silica had been removed.

Etch Rate Measurements

Etch rates were measured for each profile on samples which had been implanted under identical conditions to those used in the ellipsometry measurements.

Etch steps were formed using petroleum jelly as a mask. The jelly was removed and the samples were coated with silver, by evaporation, prior to step height measurements on a "Veeco" multiple beam interferometer.

Analysis

The ellipsometric data, a set of readings (θ_1, θ_2) as a function of etch time (see Figure 1), was translated into a set (θ_1, θ_2) as a function of depth below the original surface using the etch rate data obtained

from an identically treated sample. This set of data was then analysed by a computer program using exact reflection equations to construct a complex refractive index profile composed of a stack of thin homogeneous films superimposed on a silica glass substrate where the film interfaces corresponded to the depths at which the ellipsometric data had been obtained. The condition for a solution is that the ellipsometric data that would be obtained at each film interface when all the shallower layers had been removed should match the observed data to within experimental error.

RESULTS AND DISCUSSION

Etch Rates

Figure 2 shows the etch rates for implants of 300 keV He^+, Ne^+, Ar^+ and Kr^+ all to a dose of $1 \cdot (10)^{16}$ ions/cm² and for 20 keV He^+ doses of $2 \cdot 5(10)^{14}$ ions/cm² and $5.0(10)^{15}$ ions/cm². Also shown is the etch rate for the unirradiated silica which did not vary appreciably between samples. The curves for 300 kV He^+ and Ne^+ are included to show the relation between ion mass and etch rate, at fixed energy and dose.

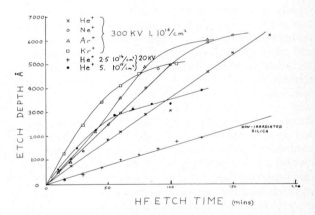

FIGURE 2   Hydrofluoric acid etching rates for inert gas implantation.

Experimental and Theoretical Profiles

a) *Refractive index profiles*   Figs. (3a, 4a and 5a) show the real parts of the complex refractive index profiles for 20 keV He^+ to doses $2.5(10)^{14}$ ions/cm² and $5(10)^{15}$ ions/cm², and 300 kV Ar^+ and Kr^+, both to doses of $1 \cdot (10)^{16}$ ions/cm², respectively. The smooth gaussian-like curves are calculated from Winterbon *et al.*'s[9] damage and range distribution theory for heavy

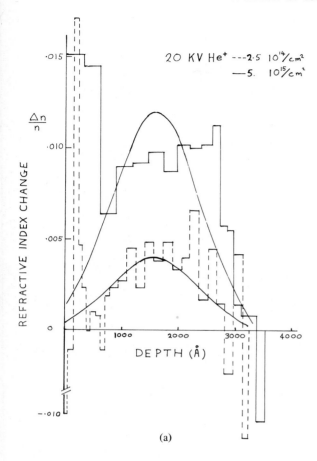

(a)

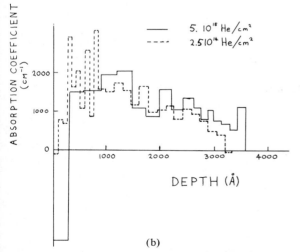

(b)

FIGURE 3 (a) Refractive index profiles and (b) Absorption profiles for 20 kV He+ 2.5(10)[14] ions/cm[2] and 5(10)[15] ions/cm[2].
The smooth curves are theoretical gaussian damage profiles with arbitrary peak heights.

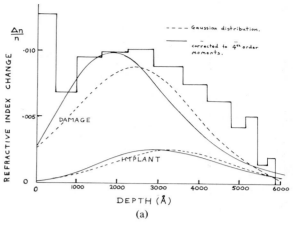

(a)

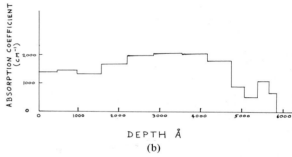

(b)

FIGURE 4 (a) Refractive index profile and (b) Absorption profile for 300 kV Ar+, 1(10)[16] ions/cm[2].
The smooth curves are theoretical damage and implant profiles, both simple gaussian and corrected to 4th order moments. The peak height of the damage profile is arbitrary but that for the implant is absolute.

ions using L.S.S.[10] path lengths calculated from Winterbon's tabulations.[11] In general a mono-atomic approximation of $Z = 10$ and $A = 20$ has been used for SiO₂ although the path lengths for Ar and Kr were calculated for the binary atomic system in the manner described in Ref. 10. However, the difference in path length between a mono and a binary system is small. In the case of Ar the distributions corrected to 4th order moments are included. This is seen to shift the peaks towards the surface. All other distributions are simple gaussian. In the case of He the straggling was calculated for 12 keV He+ to allow for the calculated 40% energy loss due to electronic stopping for 20 keV He+ (see below) whereas the mean damage depth was calculated for 20 keV He+.

The peak heights of the theoretical damage distributions have been chosen arbitrarily to give reasonable fitting over some region whereas the Ar and Kr range distributions are calculated absolutely as described below.

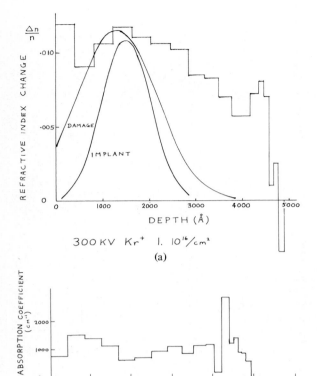

300 KV Kr⁺ 1. 10¹⁶/cm²

(a)

(b)

FIGURE 5 (a) Refractive index profile and (b) Absorption
profile for 300 kV Kr⁺, $1(10)^{16}$ ions/cm².

The comments of Figure (4) apply to the smooth curves
except that only the simple gaussian approximations are shown.

b) *Absorption profiles* Figures (3b, 4b and 5b) show
the imaginary refractive index, or absorption coefficient
profiles for the respective refractive index profiles.

c) *Effects of annealing* Figures (6a and 6b) show
the Ar profiles for identically implanted samples, one
having been annealed for 90 minutes at 425°C in air.

DISCUSSION

Refractive Index Profiles

It is apparent that refractive index, n, profiles are
formed from at least two separate peaks: a large but
shallow surface peak within the first 500 Å or so
followed by a slightly smaller deep peak extending
several thousand angstroms below the surface.

Considering the He profiles: at $2.5(10)^{14}$ ions/cm²
the deep peak is fitted very well by the theoretical

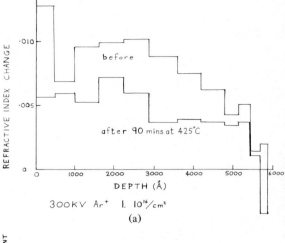

300KV Ar⁺ 1. 10¹⁶/cm²

(a)

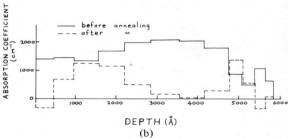

(b)

FIGURE 6 (a) Refractive index profiles and (b) Absorption
profiles for 300 kV Ar⁺, $1(10)^{16}$ ions/cm² before and after a
90 minute anneal in air at 425°C.

damage profile, while the surface peak is already far
developed. At $5(10)^{15}$ ions/cm² there are signs of
saturation as the deep peak has broadened out with a
maximum change of ∼0.010 in $\Delta n/n$. Comparison with
the Ar and Kr profiles shows close agreement in the
deep peak maximum changes. A theoretical estimate
of the refractive index change can be made from the
Lorentz–Lorenz relationship:

$$\frac{4}{3}\pi N\alpha = \frac{n^2-1}{n^2+2} \tag{1}$$

where N = molecular density, α = molecular polarizability, n = refractive index. From this $N\alpha = 0.0863$ for
vitreous silica at 5461 Å ($n = 1.4601$, density = 2.20
gm/cm³). Differentiation of (1) leads to:

$$\frac{dn}{n} = 5.7\, d(N\alpha) \tag{2}$$

for vitreous silica at 5461 Å.

A phenomena well known to occur in vitreous silica
under various forms of radiation, and also thermal,
pressure, or shock treatment, is compaction which can

only be reversed by high temperature annealing.[12] In neutron irradiation a saturation density increase of 2.6–2.8% is found which is identical to the saturation damage state of quartz crystal. If it is assumed that α is not altered and therefore $d(N\alpha) = 0.027\ N\alpha$ the resulting dn/n is 0.013. This is remarkably close to the $\Delta n/n$ value of $0.010 \leftrightarrow 0.011$ seen in these profiles.

Using Eq. (2) it is also possible to calculate the expected change due to the implanted ions if the following assumptions are made:

i) the range distribution is as predicted[9]

ii) the ions rest as neutral atoms interstitially and are not chemically bonded in the lattice. This then justifies the assumption that . . .

iii) the neutral atom polarizabilities are essentially the same as for the atoms in gaseous state i.e. $\alpha(\text{He}) = 0.20$, $\alpha(\text{Ar}) = 1.63$ and $\alpha(\text{Kr}) = 2.46$ in units of $(10)^{-24}\ \text{cm}^3$.[13]

This is probably true since the polarizabilities of molecules in liquid and vapour form are found to be virtually equal (e.g. H_2O).[14]

iv) there is negligible volume expansion due to the inclusion of the interstitial atoms. This may also be well justified, ignoring the well established fact of lattice compaction, since in vitreous silica the basic building block is the (SiO_4) tetrahedra, as in quartz crystal, but the density is only 2.20 gm/cc compared to 2.65 gm/cc for quartz. Thus, there is the concept of "free volume" which accounts for 13% of the total glass volume although only about 3% appears to be available for dissolving gas under thermal "implantation".[15] The total volume per SiO_2 chemical unit (molecule) is $(2.2(10)^{22})^{-1}\ \text{cm}^3$ giving a usable free volume of $1.36(10)^{-24}\ \text{cm}^3$ per unit or a total free volume per unit of $5.9(10)^{-24}\ \text{cm}^3$. In the doses used here the highest concentration is for Kr where the peak would be theoretically 0.035 Kr per SiO_2. The volume of a Kr atom is $4\pi/3\ (2.01(10)^{-8})^3\ \text{cm}^3$, $= 34(10)^{-24}$ cm^3,[16] thus using $1.2(10)^{-24}\ \text{cm}^3$ per SiO_2 i.e. less than the normally available free volume.

With these assumptions the implant curves for Kr and Ar in Figures (5(a) and 6(a)) were calculated (the changes for He are negligible). If it is assumed that the deep peaks, being similar in magnitude for He, Ar and Kr, are mainly due to damage and have approached saturation in the high dose cases, there is little evidence of the implant distributions, particularly the Kr peak. Since damage is the only explanation for the magnitudes of the observed He and Ar index profiles it is assumed that this is so for Kr also. Some justification for this may be obtained from the average energy deposition densities by nuclear stopping for the three gases. These are shown in Table I with the proportion of energy loss by electronic stopping calculated from:

$$\int_0^E k\ \epsilon^{1/2}\ dp \qquad (3)$$

where the electronic stopping formula and symbols are as in Ref. 10 and the calculations were made from Winterbon's tabulations.[11]

The mean energy deposition is calculated crudely by assuming uniform deposition by nuclear stopping over a depth equal to the theoretical full-width-half-peak (FWHP) damage distribution depth.[9]

Table I shows that if the densification effect is closely related to energy deposition by nuclear stopping and is already saturating for the high dose of He then saturation should be reached in both the Ar and Kr implants, and the profiles should be considerably broadened. This is seen to be so although the broadening appears to be confined to the deep end of the distribution. However, this does not explain the absence of the Ar and Kr implant peaks. This requires either (i) that the implanted atoms have suffered considerable redistribution to leave only small concentrations in the theoretical distribution region, or (ii) that there is a volume expansion due to the implanted ions which opposes the radiation compaction effect but has a similar polarizability per unit volume, i.e. the $d(N\alpha)$ terms are similar to that for the compacted silica.

TABLE I

Ion	Dose ions/cm^2	Energy kV	Total energy loss by electronic stopping eV/cm^2	FWHP (theoretical) Å	Mean energy deposition density nuclear stopping eV/cm^3
He	$2.5(10)^{14}$	20	$2.0(10)^{18}$	1850	$1.7(10)^{23}$
He	$5(10)^{15}$	20	$4.1(10)^{19}$	1850	$3.2(10)^{24}$
Ar	$1(10)^{16}$	300	$2.7(10)^{20}$	3600	$7.5(10)^{25}$
Kr	$1(10)^{16}$	300	$6.0(10)^{19}$	2050	$14.6(10)^{25}$

It was shown above that the implant concentrations were not sufficient to fill the available free volume in undamaged silica. However, this free volume ($\sim 3\%$) is equal to the change in volume produced during radiation compaction so that in the damaged silica the implant may in fact oppose compaction. The possibility of normal diffusion decreasing the implant concentrations is unlikely since significant movement of low doses of Xe in silica or quartz does not occur below $800°C$, whereas for high doses, when amorphization (compaction) has occurred, the implant exhibits a very rapid desorption associated with the annealing-out of the amorphization. This rapid desorption is not a thermal diffusion process[17] and the phenomena suggests that the gas is swept out by the rearrangement of the lattice. If this is so it may also happen that the compaction process alters the implant distribution at lower temperatures by sweeping the implant atoms away from the more heavily damaged regions towards regions of greater free volume. This might account for the deep tail of the Kr index profile. It is also noted that after annealing to $425°C$ the Ar index profile shows less annealing towards the tail end which might be due to the presence of Ar. It is expected that back-scattering studies will help to resolve these questions.

The Surface Peak

Although contamination could be responsible to some extent for the surface peak it may not be the major cause. Hines and Arndt reported[5] trouble due to contamination and found it difficult to remove by polishing or etching. This does not appear to be the case here. Also the peak heights are similar for the three ions whereas Hines and Arndt found severe contamination for He irradiation but little for Kr irradiation.

Either of two phenomena confined mainly to the first few hundred angstroms beneath the surface may be relevant here:

1) The surface layers are depleted of electrons by secondary electron emission and also by neutralizing the incoming ions. These are replaced by the neutralizing electrons (100 eV) supplied during implantation to prevent a net surface charge accumulating.

2) The energy lost in electronic stopping is deposited mainly in this region. Table I shows the magnitudes of this energy deposition, which if assumed to be deposited uniformly over 500 Å to within a cylindrical region about the ion path of ~ 10 Å radius (corresponding to the dimensions of fission fragment tracks[18]) the mean energy deposition per Si–O bond per ion is $\gtrsim 1$ eV, while the bond energy is ~ 6 eV.[12] Thus, in close encounters significant lattice ionization and bond breakage may occur.

In either case there will be an increase in the polarizability of the total defect e.g. ion + separated and trapped electron. An idea of the magnitude of change possible is gained from Table II compiled from Ref. 14. This shows the changes in polarizability of oxygen in various states of organic bonding.

TABLE II

Bond	Symbol	Polarizability (D line) $(\times (10)^{-24}$ cm$^3)$
Hydroxyl	—OH	0.60
Ether	—O—	0.65
Carbonyl	=O	0.88
Ionized	O^{2-}	3.90

The changes are quite large especially from the bonded state to the ionized state. To account for the surface peak, assuming an increase in polarizability per defect of $1 (10)^{-24}$ cm^3, would require $2.6(10)^{21}$ defects/cm^3, or a total number in the 500 Å surface layer (corresponding approximately to the peak width) of $1.3(10)^{16}$ defects/cm^2, i.e. ~ 1 defect/ion or a production efficiency of $\lesssim (10)^4$ eV/defect. All these figures seem plausible including the high defect concentration, ($\sim 5\%$ of all bonds) assuming that such a defect is responsible, since silica is capable of recording fission fragment tracks[19] which requires a continuous (i.e. every atomic layer) path of defects for detection to be possible.[20] The number of defects is also of the right magnitude to make the concept of polarizable trapped neutralizing electrons plausible.

The broken Si–O bond is in fact one model[21] for the well known "C-band" colour centre produced by x-rays, γ-rays, electrons, and neutrons in fused silica.[22] The supposed existence of strained bonds in silica glass, as opposed to quartz, overcomes the requirement for atomic displacement, which is necessary in quartz[23] for C-band production, since lattice relaxation prevents reformation of the broken bond. The other model of this colour centre involving a displaced oxygen ion[24] has been refuted on theoretical grounds.[25] However, which ever model is more correct broken bonds and displaced atoms presumably exist in the nuclear stopping region; indeed this is probably necessary to explain the enhanced etching rates and their apparent relation to the energy deposition density suggested by the mass dependence shown in Figure 2. Associated with these defects must be an increased polarizability and therefore there should be an associated change in refractive index. However, the observed change appears to be accounted for mainly by density changes which implies either a low polarizability or a low density of the defects. Therefore, if the surface peak is associated with the type of point defects which may exist in the nuclear stopping region there must be significant annealing during irradiation within the latter region. Furthermore, if the etch rate is closely related to the density of point defects then the surface etch rate should be very high. Present measurements have not been accurate enough to detect this but the absorption coefficient profiles for He irradiation show negative coefficients at the surface which can only be explained by measurement errors, and therefore suggest that the surface etch rate is considerably different from the deeper layers. An attempt to measure the C-band absorption coefficient profile by spectrophotometry for 300 kV H^+ irradiation, where electronic stopping losses are dominant and the nuclear stopping is confined to the deep end of the range, showed a decrease in C-band absorption from the surface to the end of the range, although the small optical density meant rather poor accuracy.[26] Finally, the peak has annealed after the 425°C treatment as shown in Figure 6a, in agreement with C-band annealing in electron irradiated fused silica.[22]

Absorption Profiles

The absorption profiles have been included for completeness, and to demonstrate the surface effects. At present, little comment can be made except that they appear to be rather large. The closest known absorption bands are the A bands associated with an aluminium impurity[24] but the absorption coefficient did not exceed 10 cm^{-1}, although these experiments were stopped well below saturation level. Meek *et al.*[26] have shown apparent absorption in GaAs to be caused by scattering from defect zones. The absorption profiles in Figure 6b show considerable annealing or bleaching as did Mitchell and Paige in their neutron irradiation studies.[24]

CONCLUSIONS

Irradiation with inert gas ions leads to an increase in refractive index which can be explained mainly in terms of density increases. Agreement with theoretical damage distributions is good except for broadening at high doses due to saturation or implanted atoms. The contribution of the latter is not obvious and alternative experiments for checking range distributions will be attempted.

The surface peak is not well understood but might be related to energy deposition in electronic processes giving rise to stable defects.[†]

† *Note added*: It would appear from the work of R. Kelly that non-stoichiometric sputtering could help to account for some effects within the first few layers. However, from a practical applications view-point damage by electronic processes is the most important possibility since it is not fundamentally only a surface effect.

Future work will aim at distinguishing positively between damage and implant by parallel studies in index profiling and back-scattering. The effects of active implants are to be studied shortly.

ACKNOWLEDGEMENTS

The author wishes to thank Dr. P. D. Townsend for his encouragement, advice and dexterity in acquiring financial support.

The assistance of Mr. E. Turpin with some of the implantations was welcome.

Thanks also to Dr. H. G. Jerrard and Mr. D. Henty of Southampton University for the use of, and assistance with, their ellipsometer.

The financial support from the S.R.C. and previously A.E.R.E., Harwell, has not passed unnoticed.

REFERENCES

1. E. R. Schineller, R. P. Flam and D. W. Wilmot, *J. Opt. Soc. Amer.* **58**, 1171 (1968).
2. J. E. Goell, R. D. Standley and W. M. Gibson, *App. Phys. Letts.* **21**, 72 (1972).
3. E. Gamire, H. Stoll, A. Yariv and R. G. Hunsperger, *App. Phys. Letts.* **21**, 87 (1972).
4. R. V. Pole, S. E. Miller, J. H. Harris and P. D. Tien, *App. Opt.* **11**, 1675 (1972).
5. R. L. Hines and R. Arndt, *Phys. Rev.* **119**, 623 (1960).
6. J. B. Schroeder and H. D. Dieselman, *J. App. Phys.* **40**, 2559 (1969).
7. R. D. Standley, W. M. Gibson and J. W. Rodgers, *App. Opt.* **11**, 1313 (1972).
8. A. R. Bayly and P. D. Townsend (paper in preparation).
9. K. B. Winterbon, P. Sigmund and J. B. Sanders, *Mat. Fys. Medd.* **37**, 14 (1970).
10. J. Lindhard, M. Scharff and H. E. Schiott, *Mat. Fys. Medd.* **33**, 14 (1963).
11. K. B. Winterbon, AECL-3194 Chalk River, Ontario (1968).
12. W. Primak, *J. App. Phys.* **43**, 2745 (1972).
13. Polarizabilities taken from M. Born *Atomic Physics* (Blackie, London, 1962) p. 251.
14. V. N. Kondratyev, *The Structure of Atoms and Molecules* (Foreign Languages Publication House, Moscow, 1963), pp. 429–431.
15. R. H. Doremus, *J. Amer. Ceram. Soc.* **49**, 461 (1966).
16. Atomic radii taken from: H. Ebert, *Physics Pocket Book* (Oliver and Boyd, Edinburgh and London, 1967).
17. Hj. Matzke, *Phys. Stat. Sol.* **18**, 285 (1966).
18. D. V. Morgan and D. Van Vliet, A.E.R.E.-R6278, Harwell (1969).
19. R. L. Fleischer and P. B. Price, *J. App. Phys.* **34**, 2903 (1963).
20. R. L. Fleischer, P. B. Price and R. M. Walker, *J. App. Phys.* **36**, 3645 (1965).
21. C. M. Nelson and J. H. Crawford, *J. Phys. Chem. Solids* **13**, 296 (1960).
22. G. W. Arnold and W. Dale Compton, *Phys. Rev.* **166**, 802 (1959).

23. G. W. Arnold, *Phys. Rev.* **139**, A1234 (1965).
24. E. W. J. Mitchell and E. G. S. Paige, *Phil. Mag.* **1**, 1085 (1956).
25. A. R. Ruffa, *Phys. Rev. Letts.* **25**, 650 (1970).
26. R. L. Meek, W. M. Gibson, C. G. Maclennan and D. M. Maker, *Appl. Phys. Lett.* **20**, 400 (1972).

Dose Regularities Under Ion Bombardment of Solids

E. S. MASHKOVA and V. A. MOLCHANOV

Institute of Nuclear Physics, Moscow State University, U.S.S.R.

Among other parameters which influence various processes associated with ion bombardment of solids (such as sputtering, secondary electron emission, ion scattering and so on) there is "ion dose". As the ion dose the product of ion current density (or total ion current) and time of irradiation is usually accepted. However, this definition is valid in such cases only when the time interval required for the actual experiment (or for the actual measurement) is small as compared with a certain time interval (relaxation time) which may be approximately determined as the bombarded ion penetration depth divided by the velocity of the irradiated surface motion due to target sputtering. The examination of the situations which take place in typical ion bombardment experiments (ion current densities of about 0.01–1.00 ma/cm^2, sputtering ratios of about 1–10 at/ion) shows that the relaxation time turns out to be of the order of some minutes to some seconds depending strongly, in particular, on the crystalline target orientation with respect to the ion beam direction. When the time interval required for the performing of the experiment exceeds considerably the relaxation time the critical ion dose must be determined as the product of ion current density and the relaxation time. In fact, the damaged layer of the irradiated target is continuously sputtered and this process prevents the accumulation of damage. Because the relaxation time is inversely proportional to the bombarding ion current density in this case the critical ion dose proves to be independent of ion current density. This peculiar fact must be taken into account in particular when the dependence of various characteristics of the ion–solid interaction process upon bombarded ion current density are analysed. When the time interval during which the measurements are performed is comparable with the relaxation time one can expect that transient characteristics must be observed. In particular they must be observed when an abrupt change of irradiated crystalline target orientation with respect to the bombarded ion beam is performed.

The aim of these remarks is to direct the attention of experimentalists to the problem.

REFERENCES

1. E. S. Mashkova and V. A. Molchanov, *Canad. J. Phys.* **46,** 713 (1968).
2. E. S. Mashkova and V. A. Molchanov, *Izvest. Acad. Sci. ser. Phys.* **33,** 757 (1969).

PHOTON EMISSION FROM LOW-ENERGY ION AND NEUTRAL BOMBARDMENT OF SOLIDS

N. H. TOLK, D. L. SIMMS, E. B. FOLEY and C. W. WHITE

Bell Laboratories, Murray Hill, New Jersey

Low-energy heavy particle bombardment of solid surfaces is observed to be accompanied by the emission of infrared, visible and ultraviolet radiation. Line radiation arising from transitions between discrete atomic or molecular levels may be attributed to the decay of sputtered or backscattered excited particles which have escaped the surface. Broadband continuum radiation which is also observed in low-energy heavy particle collisions with surfaces arises from the surface and appears to be a strictly solid state phenomenon. Measurement of collision induced optical radiation constitutes a powerful tool for studies of the fundamental outershell electronic processes which result from the interaction of low-energy atomic particles with solids.

1 INTRODUCTION

Recent experimental investigations of low-energy (10 eV to 20 keV) heavy particle bombardment of solid surfaces[1-7] have found that infrared, visible and ultraviolet radiation is emitted as a result of the bombardment. The radiation characteristics as well as efficiencies for producing the radiation may vary substantially depending upon the identity of the solid surface and upon the energy and identity of the bombarding particle.

In the experimental observations to be discussed below, three distinct kinds of collision induced optical radiation have been identified. The first of these is line radiation due to sputtering with simultaneous excitation. The interaction of the impinging ion or neutral beam with the surface results in the sputtering of neutral atoms, neutral molecules, and ions from the surface. A significant portion of the sputtered particles leave the surface in excited electronic states which then decay and give rise to optical line radiation characteristic of surface constituents. The second kind of radiation observed has a similar nature but arises from backscattered excited beam particles which have escaped the surface after having experienced violent collisions in in the bulk material. To the extent that the violent momentum changing collisions do not effect the final excited state configuration of the emerging particles, this class of phenomena is identical to the beam foil radiation phenomena. Studies of backscatter radiation provide more complete information on final states than conventional charge state measurements and give important insights into the nature of the interaction of the emerging particle with the surface. The final and perhaps least understood type of radiation is the broad continuum of radiation that is observed from many of the solid materials studied. In the case of metals, for a given beam projectile, at a given energy, the shape of the continuum is observed by Kerkdijk and Thomas not to change for a variety of metal surfaces.[7] For insulators, the shape of the continuum as a function of wavelength is entirely characteristic of the material and independent of the beam particle species. An important conclusion from these studies is that the collision participants, both the solid itself as well as the escaping atomic and molecular fragments, in general are in excited states following the collision interaction. The collision participants may then decay by radiationless de-excitation processes, Auger de-excitation for example, or decay radiatively thereby producing the observed photons.

2 EXPERIMENTAL APPARATUS AND TECHNIQUE

Low energy experimental investigations of heavy particle impact induced optical radiations require an apparatus which can produce monoenergetic beams of ions and neutrals in the general range of 10 eV to 10 keV. An apparatus which satisfies this criteria has been previously described.[1,2] Beams of both ions and neutrals are created. Neutral beams are useful both for comparison with corresponding ion impact measurements as well as to avoid problems associated with charge accumulation on nonconducting surfaces.

In this work, two different kinds of measurements are taken. The first of these is spectral analysis of the emitted radiation obtained by scanning the monochrometer through the accessible wavelength range which is for this work 2000 Å to 8500 Å. The second type of measurement involves observation of the intensity of a single line or a limited segment of the spectrum as a function of energy using either a monochrometer or an interference filter. As will be described below, a wide variety of beam species and target materials have been used in these measurements. For all measurements, the pressure in the target chamber was approximately 1×10^{-7} torr.

3 METALLIC TARGETS

3.1 Radiation from Sputtered Particles
Recent experimental studies have shown that optical radiation arises from excited sputtered surface atoms or molecules produced when beams of low-energy (30 eV and above) ions or neutrals impact on surfaces.[1] In this work, we have obtained emission functions taken as a function of energy showing structure and energy thresholds. This data when analyzed in terms of excited-state sputtering efficiencies and radiationless de-excitation processes contributes to a more fundamental understanding of the sputtering process.

The spectra of radiation observed when Ar^+ ions at 4000 eV impact on Ni and Cu targets are shown in Figure 1. Most of the prominent lines in these spectral scans have been identified as arising from low-lying energy levels of neutral Ni and Cu, sputtered off the surface in excited states by the incident ion beam. In addition to spectral lines from the surface target materials, radiation is also often observed which is characteristic of surface contaminants.[1,2] The molecular radiation centered at 4300 Å and 3900 Å has been identified as arising from the $A^2\Delta \rightarrow X^2\Pi$ and $B^2\Sigma - X^2\Pi$ electronic transitions of the CH molecule. The origin of this radiation is believed to be collisional excitation of adsorbed hydrocarbon surface contamination. The prominent line which is often observed at approximately 5900 Å is the NaD line doublet which is assumed to arise from sodium contamination deposited on the surface. The surface can be cleaned to such an extent that the contaminant radiation is negligible by prolonged exposure (approximately 20 minutes) to a 3-keV argon ion beam or by heating the target. Typical relative emission functions (photons detected per incident ion) are shown in Figure 2. In most cases the emission functions exhibit a completely undefined threshold and a gradual rise with increasing energy although there are exceptions which show definite structure (for example N_2^+ on Ni, NiI 3369 Å and 3374 Å; and N_2^+ on Cu, CuI 3247 Å). The gradual monotonic rise with energy in the emission functions can be understood by assuming that radiationless de-excitation processes, Auger de-excitation or resonance

ionization, compete with the radiative de-excitation of the excited states of the sputtered atoms and molecules. Figure 3 shows schematically the possible radiationless de-excitation processes in the ion-surface energy level system. The transition rate $P(s)$ for radiationless de-excitation processes has been approximated by

$$P(s) = A\,e^{-as}$$

where s is the distance from the ion or atom to the surface and A is the radiationless transition rate at the surface.[1] It can be easily shown that the probability, R, which is the probability that a sputtered atom in an excited state can escape (to "infinity") without undergoing radiationless de-excitation, is given by

$$R(v_\perp) = e^{-A/av_\perp}$$

where $v_\perp$ is the velocity component of the sputtered atom normal to the surface. Assuming this expression, it is clear that the fast excited sputtered particles will contribute most efficiently to the observed radiation.

Using the above expression to account for radiationless transition processes, we can approximate the emission function, $F(E_0)$, in the following way

$$F(E_0) = k\eta(E_0) \int_0^{v_{\perp max}(E_0)} f(v_\perp)R(v_\perp)\,dv_\perp$$

where $\eta(E_0)$ is the excited state excitation efficiency (the number of sputtered atoms initially created in a given excited state per incident ion), k is a constant dependent on geometry but independent of the incident ion energy E_0, and $f(v_\perp)$ is the perpendicular velocity distribution of sputtered excited atoms. The quantity $F(E_0)$ has been calculated over the appropriate range of energies and fit to measured emission function of CuI 3247 Å observed in the impact of Ar^+ on Cu as shown in Figure 2. The values of the parameters used in this calculation have been reported previously.[1] It should be noted that the value for $v_{\perp max}$ was obtained assuming elastic biparticle collisions. The calculated values of $F(E_0)$ for various values of A/a are shown in Figure 4a where both experimental and calculated curves are normalized at 3 keV. The best fit over a broad range of higher energies is for $A/a = 2 \times 10^6$ cm/sec. This value for A/a agrees very well with the value (also 2×10^6 cm/sec) obtained by Van der Weg and Bierman who studied the Doppler broadened line profile of CuI 3247 Å observed in the impact of Ar^+ on Cu at much higher energies (80 keV).[8] This gratifying agreement between widely different experiments appears to reinforce a basic assumption made in the present calculation that fast excited neutrals which arise from

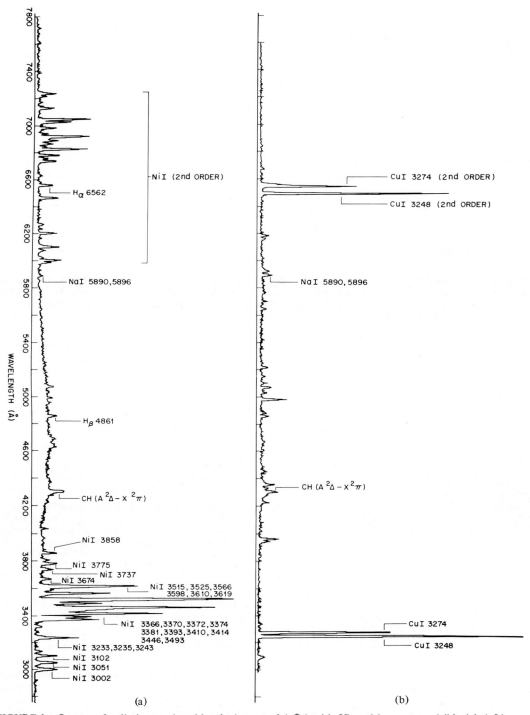

FIGURE 1   Spectra of radiation produced by the impact of Ar$^+$ (at 4 keV) on (a) copper, and (b) nickel. Lines arising from excited states of sputtered copper, nickel and various contaminants are observed in this wavelength interval.

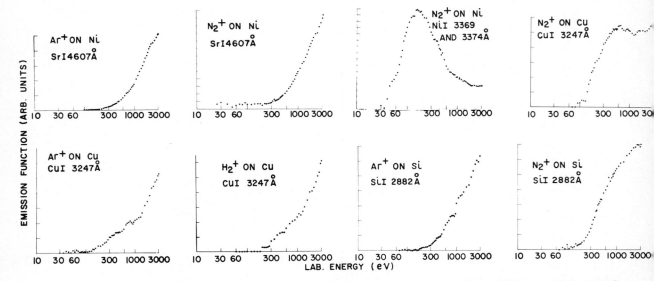

FIGURE 2   Emission functions for optical lines produced in ion-surface collisions. The collision combination and the observed line are indicated with each emission function. In two of the graphs the observed radiation at 4607 Å arises from excited strontium atoms which are sputtered from a strontium-impregnated nickel surface. Each graph is plotted independently in arbitrary units.

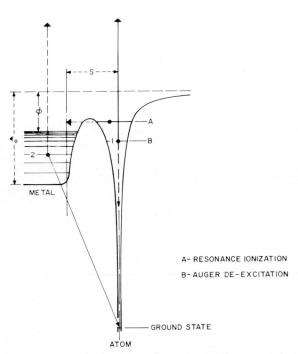

A- RESONANCE IONIZATION

B- AUGER DE-EXCITATION

FIGURE 3   Schematic representation of the ion-metal energy level system showing two possible radiationless de-excitation processes, Auger de-excitation and resonance ionization.

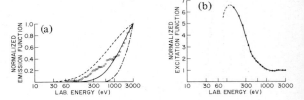

FIGURE 4a   Comparison of three calculated emission functions computed using A/a values of 7.5×10^5 cm/sec (dashed line), 2×10^6 cm/sec (solid line), and 5×10^6 cm/sec (dot-dashed line) with the measured emission function (open circles) for CuI 3247 Å in the case of Ar^+ on Cu.

FIGURE 4b   Postulated relative excitation efficiency (number of sputtered atoms created in a given excited state per incident ion) which produces agreement between calculated and measured emission functions as shown in (a) at low energies for a value of A/a of 2×10^6 cm/sec. The dashed line represents the expected low energy behavior.

one or more nearly elastic biparticle collisions contribute most to the observed radiation.

In Figure 4(a) a substantial discrepancy exists at the lower energies between the calculated and measured emission functions. It is likely that this disagreement is due to the existence of a non-uniform excited state excitation efficiency as well as to the fact that at low energies the inelastic energy transfer becomes comparable to the maximum value of energy transfer for the elastic collision. If we assume that over a significant range of energies the principle contributor to the discrepancy is the variation in the excited state excitation efficiency $\eta(E_0)$, then the excitation efficiency which is needed to cause agreement with the measured values in the range 150 eV to 1.5 keV can be determined and is shown in Figure 4(b). This kind of structure is common in emission cross sections measured in ion-atom and neutral-atom collisional excitation experiments. Experimental studies of this type appear to show great promise, both with regard to understanding the biparticle collision processes associated with sputtering as well as giving insight into interaction of excited particles near surfaces.

3.2 Technique for Measuring Neutral Flux

Our measurements show that for the case of a metallic target, beams of neutrals and ions (of the same species) produce photons due to sputtering with equal efficiency.[2] Low velocity ions with sufficiently large ionization potential impacting on a metal surface are neutralized by non-radiative processes several angstroms in front of the surface, well before the sputtering interaction occurs.[8] It is reasonable then to assume that the radiation which results from the sputtering interaction should be the same for beams of ions or neutrals at the same velocity. This assumption has been verified experimentally using a bolometer to independently measure the neutral "current". These results consequently suggest a means of directly measuring neutral beam flux in the low energy region. It follows from the above discussion that the ratio of the intensity of the optical line radiation produced when a metallic target is bombarded with neutrals to that produced when the same target is bombarded with ions is equal to the ratio of the neutral to ion fluxes thus leading to a direct measurement of the neutral "current". In principle, this technique could also be extended to non-metallic surfaces using a more complicated calibration scheme.

3.3 Radiation from Particles Backscattered from Metals

Low-energy incident ion or neutral particles which are backscattered from bulk or surface metal atoms may also emerge at the surface in excited states. The processes which create excited states in the backscattered particles are likely to be different from those responsible for excited sputtered particles. However, both classes of particles may be treated identically upon leaving the surface in terms of radiationless de-excitation processes which compete with radiative decay. Pioneering experiments using low $Z(H^+, H_2^+$ and $He^+)$ projectiles on a variety of metal surfaces have been performed by Gritsyna et al.,[10] McCracken and

Erents[11] and more recently by Kerkdijk and Thomas.[7] In each case the principal source of optical line radiation was from neutralized backscattered projectiles. In each case also the lines were significantly Doppler broadened with line shapes determined by the wide range of energies and directions of the scattered projectiles as well as radiationless de-excitation effects. Future studies of this kind should provide important information on the relative role of the bulk and the surface in determining excited state distributions of the backscattered particles.

3.4 Continua Observed from Bombarded Metals

Kerkdijk and Thomas have recently reported that they have observed broad emission bands in the wavelength region 3000 Å to 5000 Å for H^+ and He^+ bombardment of gold, nickel, and copper surfaces.[7] This has also been independently observed by the present authors. In each case, the shape of the emission band was found to be independent of the target metal species. The interpretation of this phenomenon is complicated by the facts that similar band structure has been found to arise from bombarded quartz (quartz windows were used in our work) and band emission also arises in this wavelength range from CH contaminants sputtered from the surface.

4 INSULATOR TARGETS

4.1 Radiation from Sputtered Particles

The spectral distribution of radiation arising from excited sputtered particles due to the impact of nitrogen molecules (N_2^0) at a beam energy of 3.5 keV on aluminum oxide (Al_2O_3), lithium fluoride (LiF), and quartz (SiO_2) are shown in Figure 5. Optical scans taken using neutral beams of neon, argon and other heavy particles give similar results. Neutral beams rather than ion beams are used for the bombardment of insulators in order to avoid ion beam energy decrease and defocusing due to charge build-up on the insulator surface. Ion beams may also be used, however, if the insulator surface is bathed in electrons emitted from a nearby heated filament. In Figure 5, the more intense lines are identified as arising from the decay of excited states of neutral aluminum, lithium, silicon, and hydrogen. Because the line widths are found in these experiments to be equal to the instrumental resolution (about 1 angstrom), we may assume that the radiation originates from individual atoms and molecules which have been sputtered off the surface in excited states which subsequently decay by photon emission. The Balmer lines of neutral hydrogen are believed to arise from the sputtering of surface contaminants.

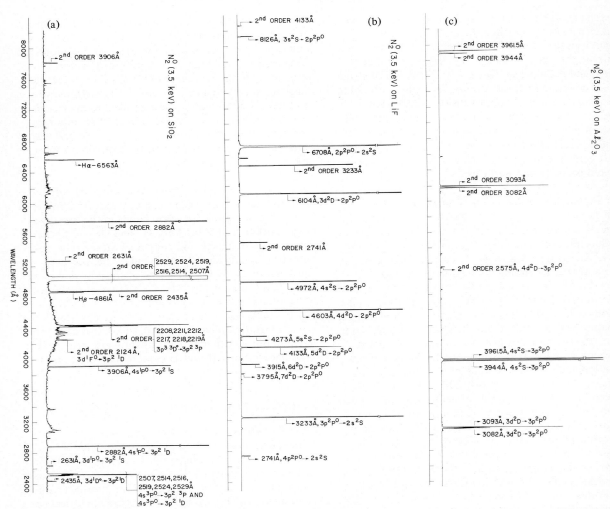

FIGURE 5   Spectrum of radiation produced by the impact of N_2^0 (3.5 kiloelectron volts) on (a) Al_2O_3, (b) LiF, and (c) SiO_2. Lines arising from excited states of neutral aluminum, lithium, and silicon are observed in this wavelength interval. The wavelength and electronic transition are indicated beside each line. Two Balmer lines of neutral hydrogen, H_α and H_β, are also observed on the SiO_2 scan.

For all cases studied, metals semiconductors and insulators, optical radiation has been observed due to sputtering with simultaneous excitation. However, in the insulator case, the excitation efficiency is observed to be much larger than for metals. For the case of metals, typical prominent lines are estimated to have excitation efficiencies of the order of 10^{-4} (10^{-4} photons per incident ion) while in the insulator case, the excitation efficiencies are measured to be two or three orders of magnitude higher. A plausible explanation that the excitation efficiencies are found to be uniformly greater for at least the metal constituents of insulators targets than for metallic targets is found in

an examination of the band structure of insulators and metals. As indicated above (Section 3.1) for metals, the Fermi level is sufficiently high in energy and there exists sufficient unoccupied levels that radiationless processes, resonance ionization or Auger de-excitation, may compete effectively with photon emission as the primary decay mechanism. Consequently the lower energy component of the energy distribution of sputtered particles (which are the greater part of the total sputtered particles over a wide range of incident energies) is de-excited by non-radiative processes in the metal target case. For insulator targets with large forbidden band gaps as shown in Figure 6, it is not

possible energetically for Auger processes to take place
nor is it probable for resonance ionization to occur due
to the paucity of available vacant states. Thus, all the
sputtered electropositive particles created in excited
states (which is believed to be a substantial fraction of
the total number) decay via photon emission.

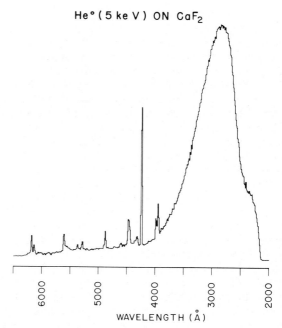

He° (5 keV) ON CaF₂

WAVELENGTH (Å)

FIGURE 7   Spectra of radiation produced in the impact of
He⁰ (5 keV) on a crystal of CaF₂.

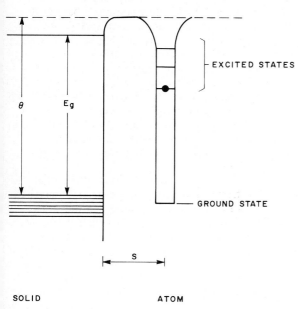

SOLID   ATOM

FIGURE 6   Schematic representation of the ion-insulator
energy level system.

4.2 Broadband Continuum Arising from Insulator Targets

When insulating targets are bombarded with low and
intermediate energy ions and neutrals, in many cases
intense broad continuum of radiation are produced in
addition to line radiation arising from excited states of
sputtered and backscattered particles. An example of
this phenomena is given in Figure 7 which shows the
spectral distribution of radiation produced in the
impact of He⁰ (5 keV) on a crystal of CaF_2. The
impact of neutral helium results in the production of
an intense continuum of radiation extending over
~1800 Å with a single pronounced peak at ~2800 Å.
Prominent line radiation observed in Figure 7 arises
from excited states of Ca and Ca^+. We find the same
continuum radiation is produced when low energy
electrons (200 eV–3 keV) impact on this crystal but
the lines arising from excited states of sputtered atoms
and ions are not observed in electron bombardment.
This continuum of radiation is observed when neutral

He, H, and H_2 are used as projectiles at an energy of
5 keV, but when heavier projectiles (Ne, Ar, and N_2)
are used at the same energy, the intensity of the
continuum is decreased by at least three orders of
magnitude and the spectral distribution is dominated
entirely by radiation from excited states of sputtered
atoms and ions.

In the case of CaF_2, this radiative continuum has
been observed in cathode luminescence experiments at
significantly higher energies (20 keV electron energy).[12]
Hayes et al.[13,14] have provided convincing evidence that
the continuum arises from radiative recombination of
electrons with the self-trapped v_k center (the F_2^-
molecule) of the CaF_2 crystal. We believe that in the
case of low Z neutral particle bombardment, the production of the v_k centers results from the inelastic
energy transfer of beam projectile energy to bound
electrons in the solid (inelastic energy loss in the solid).
The continuum then results from the radiative recombination of the self-trapped hole and mobile
electrons created pair wise with holes in the inelastic
interaction of the projectile with electrons in the
crystal. The absence of a significant continuum in high
Z neutral particle bombardment at the same energy is
consistent with this model since in that case elastic (or
nuclear) collisions with target nuclei is the dominant
energy loss mechanism for the particle in the solid.

The same continuum produced by low Z neutral particle bombardment is also produced by high Z ion bombardment at energies $\geqslant 0.5$ keV. However, the production of the continuum in this case appears to be related to energetic electron bombardment of the crystal which accompanies ion bombardment. When energetic ions bombard an insulator, the surface accumulates positive charge and the bombarded region will have a nonuniform surface potential. Secondary electron emission from the bombarded region and surrounding structure has a finite value, and regions of the bombarded surface may be subjected to electron bombardment simultaneous to ion bombardment at an energy determined by the surface potential. In experiments involving high Z ion bombardment of insulators at energies of the order of a few kilovolts, we find that the radiative continuum can be quenched by neutralizing the insulator surface with electrons from a heated filament during ion bombardment. However, the intensity of the continuum produced by low Z neutral particle bombardment at these energies is not affected by electrons from a filament during bombardment. This supports the conclusion that the continuum produced by low Z neutral particle bombardment results from radiative recombination of mobile electrons and self-trapped holes created pairwise by the inelastic interaction of the neutral projectile with bound electrons in the solid.

4.3 Composition Analysis

Identification of optical line radiation produced in collisions of low energy ions and neutrals with solids can be used as a sensitive technique for surface and bulk composition analysis.[2] This technique, which we have called SCANIIR (Surface Composition by Analysis of Neutral and Ion Impact Radiation) is a very sensitive probe of the surface region since the range of the incident projectiles in the solid is limited to a few monolayers at these low energies. Surface constituents can be determined by identifying prominent optical lines and bands, and the intensities of the individual lines are proportional to absolute concentrations of specific constituents within the sampling depth of the projectile. Damage to the sample can be minimized by using low bombarding energies and low equivalent current densities. High Z neutral atoms and molecules (Ne^0, N_2^0, Ar^0 etc.) are found to be the more useful projectiles in order to maximize the excited state sputtering efficiency and to minimize surface charge accumulation and the generation of luminescent continuum on insulating surfaces.

Using silicate melt targets containing oxide impurities of Al, Fe, Ca, K, Mg, and Na at known con-

centrations, we have recorded the collision induced spectrum of radiation and we have obtained detection sensitivities under present experimental conditions. These measurements show that detection limits range from 6 parts in 10^7 (for sodium) to 8 parts in 10^4 (for potassium) using a neutral beam intensity of 3×10^{-7} amps (of neutral N_2) and a 10 second photon integration time. Details of these measurements have been published previously.[2] In addition to surface analysis, preliminary experiments using ion implanted targets show that depth profile analysis is also possible using this technique. If the sputtering rate of the matrix material is known, then monitoring the intensity of specific optical lines as a function of the bombardment time can be used to determine the depth distribution of selected impurities. Thus observation of radiation in low energy particle solid collisions can be used as a sensitive technique for obtaining surface and depth profile analysis.

In certain cases we have also observed that the spectral distribution of radiation produced in particle-solid collisions reflects the chemical bonding in the solid. This is illustrated in Figure 8 which compares a portion of the spectral distribution of radiation produced in the impact of Ar^0 (3 keV) on MgF_2 and MgO. In the more ionically bonded MgF_2, the intensity of optical lines arising from excited states of Mg^+ is significantly greater (relative to the neutral Mg reson-

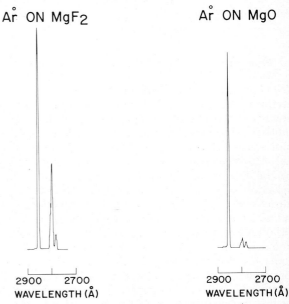

FIGURE 8   Spectra of radiation produced by the impact of Ar neutrals (3 keV) on MgF_2 and MgO illustrating possible chemical bonding effects in the solid.

ance line) than in the case of the more covalently bonded MgO. Comparing these two scans reveals that the Mg$^+$ lines are at least an order of magnitude more intense relative to the neutral Mg line in the case of MgF$_2$. Further work is needed in order to ascertain whether this change in the spectral distribution reflects the nature of the bonding in the solid or is more closely related to the process of electron pickup as the sputtered particle leaves the surface.

5 CONCLUSION

The picture that emerges from studies of photon emission from low-energy ion and neutral bombardment of solid surfaces is that the participants in low-energy collisions, ions, neutrals and the solid itself, may be left following the collision interaction in highly excited states. Since a significant portion of these excited states decay with the emission of radiation, much can be learned both about the detailed processes of the collision as well as about the nature of the coupling between the surface and the excited states of nearby atoms and molecules by studying the emitted radiation. In the case of sputtering, we may conclude that for both metal and insulator targets a substantial fraction of sputtered particles leave the surface in a multitude of excited states. The observed large difference in the photon emission efficiencies due to sputtering between metallic and insulator surfaces may be attributed to the existence of the radiationless de-excitation processes which competes with radiative decay and operate in the metal case and not in general for the electropositive constituents of insulators. It is reasonable to

assume that the sputtered particles in both the metal and the insulator cases are originally created in excited states with more or less comparable efficiencies. These same considerations also apply to backscattered excited particles.

REFERENCES

1. C. W. White and N. H. Tolk, *Phys. Rev. Lett.* **26**, 486 (1971).
2. C. W. White, D. L. Simms and N. H. Tolk, *Science* **177**, 481 (1972).
3. J. M. Fluint, L. Friedman, J. Van Eck, C. Snoek and J. Kistemaker, *Proceedings of the Fifth International Conference on Ionization Phenomena in Gases, Munich, Germany, 1961,* edited by H. Maecker (North-Holland, Amsterdam, 1962).
4. C. Snoek, W. F. van der Weg and P. K. Rol, *Physica Utrecht* **30**, 341 (1964).
5. J. P. Meriaux, J. M. Guttierrez, Ch. Schneider, R. Goutte and Cl. Guilland, *Nouv. Rev. d'Optique appliquee,* **2**, 81 (1971).
6. I. S. T. Tsong, *Phys. Status Solidi,* **A7**, 451 (1971).
7. C. B. Kerkdijk and E. W. Thomas, Manuscript in preparation.
8. H. D. Hagstrum, *Phys. Rev.* **123**, 758 (1961).
9. W. F. van der Weg and D. J. Bierman, *Physica (Utrecht)* **44**, 206 (1969).
10. V. V. Gritsyna, T. S. Kujan, A. G. Koval and Ya. M. Fogel, *Soviet Physics, JETP* **31**, 796 (1970).
11. G. M. McCracken and S. K. Erents, *Physics Letters* **31A**, 429 (1970).
12. D. R. Rao and H. N. Bose, *Physica* **52**, 371 (1971).
13. W. Hayes, D. L. Kirk and G. P. Summers, *Solid State Commun.* **7**, 1061 (1969).
14. J. H. Beaumont, W. Hayes, D. L. Kirk and G. P. Summers, *Proc. Roy. Soc.* **315**, 69 (1970).

PHOTON-EMISSION INDUCED BY IMPACT OF FAST IONS ON METAL SURFACES

C. B. KERKDIJK and E. W. THOMAS†

*FOM-Instituut voor Atoom- en Molecuulfysica, Kruislaan 407, Amsterdam/Wgm.,
The Netherlands*

A study has been made of photon-emission induced by impact of H⁺ and He⁺ ions on gold and nickel metallic surfaces. An energy selected ion beam (3 to 10 keV) is directed onto a metal surface and the emitted light is analyzed by a monochromator (3000–6000 Å). The target is located in a vacuum chamber with base pressure of 1×10^{-9} torr, the target is cleaned by heating and by Ar⁺-sputtering. Lines from the He-spectrum (3^3D, 4^3D, 5^3D $\rightarrow 2^3$P and 3^3P $\rightarrow 2^3$S) and H-spectrum (Balmer α, β, γ) are observed for impact of He⁺ and H⁺ respectively on both targets.

No singlet lines of helium were observed and also no line emissions from the target material.

The line-shape of the He-lines has been studied in detail. The shape can be qualitatively explained by Doppler shifts of emission from projectiles scattered by bi-particle collisions in the surface.

The total intensity of the lines increases as a function of beam energy and of incidence angle on the surface.

1 INTRODUCTION

A study has been made of the light emission induced by 2 to 10 keV H⁺ and He⁺ ions on gold and nickel surfaces. The techniques and experimental arrangements were identical to those used previously[1] for a similar study with a copper surface.

We will give here only a brief summary of the experimental techniques; the reader is referred to the earlier report[1] for full details.

The ion beam is provided by a unoplasmatron source, accelerated and mass analyzed to select the species of interest. The beam then falls on the target, making some angle ϕ with respect to the surface normal. A scanning optical monochromator with photomultiplier detection is used to analyze the emission induced at the surface. The axis of the monochromator is perpendicular to the ion beam and lies in the same plane as the beam and the surface normal. The effective spectral range of the monochromator was 3000 to 6000 Å.

The gold target was a single crystal arranged with a (100) surface orientation. The nickel was a single crystal arranged with a (111) surface orientation. Prior to insertion in the apparatus the crystals were electropolished. After insertion in the apparatus the crystals were outgassed by heating to 300°C and cleaned by bombarding

† Present address: Georgia Institute of Technology, Atlanta, Georgia, U.S.A.

with a 10 keV Ar⁺ ion beam. Background pressures in the target area were typically 2×10^{-9} torr. During bombardment with a He⁺ beam the background pressure would rise to above 10^{-8} torr due to the poor pumping speed of the ion getter pumps for helium gas; since the pressure rise is entirely due to helium this should not cause any contamination of the surface.

For this experiment ion beam currents of 2 to 10 μA were used. The flux of incoming particles should be sufficient to sputter away contaminant layers as rapidly as they are produced. Thus the surface remains clean as long as the ion beam is kept running.

The first part of the experimental program was a general survey of the spectrum induced by particle impact on the surface. The next stage was an attempt to study the coefficients for emission of photons in certain radiative transitions.

2 EMISSION SPECTRA

The general features of the spectra were similar to those previously observed for impact of the same projectiles on a copper target.[1] Proton impact gave emission of the Balmer series. He⁺ impact gave emissions from ³P and ³D levels of helium. No emissions were observed from the singlet series of helium; also there were no line emissions that could be associated with excited atoms of the target materials.

Line emissions from the projectile species must arise through the scattering of the projectile ion with a simultaneous pick-up of an electron into the excited state. The lines are observed to be broadened. In our previous report[1] such broadening has been explained in terms of the Doppler shifts of emission from fast projectiles scattered by bi-particle collisions at the surface. The only significant differences between the case of gold and the case of nickel was that the line emissions for the gold target were somewhat weaker than those for the nickel.

In our previous paper we also reported the observation of a broad band emission in the wavelength region 3000-4000 Å. However, from recent measurements we must conclude that this band is not emitted by the target but by the window through which the observations are made. This radiation is probably due to an electron bombardment of the window. The electron bombardment is induced by bombardment of this window by reflected protons from the target.[3,4]

3 LINE SHAPE OF EMISSION FROM SCATTERED PROJECTILES

The line emission from the scattered projectiles exhibits broadening of as much as 30 Å; this width decreases with decreasing impact energy of the projectile. Such broadened lines have been previously observed by us in the study of He^+ and H^+ impact on a copper target[1]; there have also been observations of such effects in the work of McCracken and Erents[2] which studied the Lyman-alpha line induced by H^+ impact on Molybdenum. In our previous work[1] we ascribe such line broadening effects to the Doppler shifts of emission from scattered projectiles. We have also shown how the shape of such lines may be predicted in a simple manner.

Figure 1 shows the form of some helium lines induced by He^+ on a nickel target. The display is facilitated by plotting the line shape as a function of the fractional shift in wavelength from the true wavelength emitted by a stationary atom, $\Delta\lambda/\lambda_0$; thus the intensity $I(\lambda)$ measured at some wavelength λ is plotted as a function of the quantity $(\lambda-\lambda_0)/\lambda_0$ where λ_0 is the wavelength emitted by a stationary helium atom. The wavelength resolution of the monochromator was 1.7 Å. All four lines have been normalized together in intensity. The resulting display in Figure 1 shows that all four lines have essentially the same shape.

In Figure 1 we show the individual data points; no attempt has been made to draw a line of best fit. Uncertainty in intensity may be estimated by the repro-

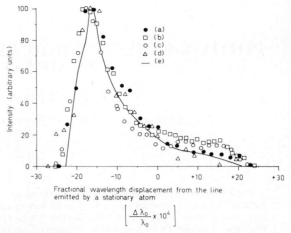

FIGURE 1   Line shapes shown as a function of the fractional displacement of wavelength from the wavelength emitted by a stationary helium atom. Data are for 10 keV He^+ impact on Ni at an incidence angle of 45°. (a) 3889 Å line ($3^3P \to 2^3S$); (b) 5875 Å line ($3^3D \to 2^3P$); (c) 4471 Å line ($4^3D \to 2^3P$); (d) 4026 Å line ($5^3D \to 2^3P$); (e) Predicted line shape. All curves are normalized together in intensity at the maximum.

ducibility of the individual point. The weakest emission is the 4026 Å line and the uncertainty in intensity at any point is ±15 units; for the other three lines the uncertainty is five units. The uncertainty in wavelength scale is less than two scale divisions (2×10^{-4}).

We have also carried out a prediction of the line shape; the method of doing the calculation has already been outlined[1] and we will here confine ourselves to stating the three simplifying assumptions made in the calculation.

The assumptions are†:

a) That the projectiles are scattered by bi-particle collisions on surface atoms with a Rutherford cross section.

b) That the probability of a scattered particle being in an excited state is independent of the angle through which it is scattered.

c) That particles scattered into the surface are lost and only particles scattered out of the surface contribute to the emission.

With these basic assumptions the line form was calculated following the procedures we have already described.[1] In Figure 3 we also show the calculated

---

† In calculating the lineshape one should have a knowledge of the energy-distribution $N(E, \alpha, \beta)$ of the backscattered particles, the probability $P(E, \alpha, \beta)$ to be in a particular excited state, and the probability $D(E, \alpha, \beta)$ for emitting a photon (excited atoms can be involved in a radiationless deexcitation near the surface). Since these functions are not known we performed a calculation using the given assumptions.

line shape; the agreement with experiment is quite good considering the many assumptions made. Discrepancies in detail between the calculated and measured values are probably due to the assumption that the probability of scattered particles being excited is independent of scattering angle; there is no justification for that assumption.

Shown in Figure 2 is the behaviour of the 5875 Å line ($3^3D \rightarrow 2^3P$) as the incidence angle ϕ is varied; all three curves have been normalized together. The figure shows best fit lines drawn through the individual data points. Random uncertainty in intensity is ±5 scale

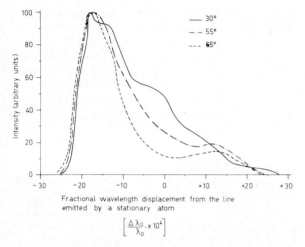

FIGURE 2   Shape of the 5875 Å ($3^3D \rightarrow 2^3P$) line induced by 10 keV He$^+$ impact on nickel for incidence angles of 30°, 55° and 65°. All three curves have been normalized in intensity at the maximum.

divisions; the $\Delta\lambda/\lambda$ axis is accurate to two scale divisions (2×10^{-4}). The most prominent feature is that the signal exhibiting zero Doppler shift ($\Delta\lambda/\lambda_0 = 0$) becomes relatively more important as the incidence angle is decreased. The calculations of the line shape (see Figure 9 of Ref. 1) indicate that this is to be expected.

4   BEHAVIOUR OF THE LEVEL EXCITATION COEFFICIENT

In our previous work[1] we have defined a quantity, γ_i, called the "Level excitation coefficient"; this represents the probability of forming a particular excited state. We measure the total number of photons J_{ij}^T emitted per second in the transition from a state i to a state j; this is the integral over the whole Doppler broadened line. Then with A_{ij} as the transition probability for decay of state i to state j, I_B as the projectile beam current (measured in particles per second) and $C(\lambda_0)$

as the detection efficiency at the centre wavelength of the broad line we define γ_i as follows

$$\gamma_i = S_{ij}^T \sum_j A_{ij}/I_B \, C(\lambda_0) \, A_{ij}$$

In the present work we measure only the relative values of γ_i. Our previous report[1] has described how one may measure S_{ij}^T directly by setting the exit slit of the monochromator sufficiently wide to accept the whole of the emitted spectral line.

In Figure 3 we show the variation of the level excitation coefficients for certain helium levels induced

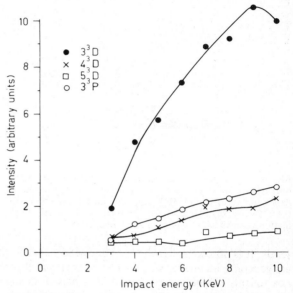

FIGURE 3   Relative variation of the excited state formation coefficients for various helium levels, shown as a function of He$^+$ impact energy. The relative values of the data for different levels are valid. The angle of incidence, ϕ, is 45°.

by He$^+$ impact on a nickel target. All levels studied show a generally increasing excitation coefficient with projectile energy. For one level (3^3D) there is a suggestion that the coefficient has reached a maximum at 10 keV impact energy. The random errors in these excitation coefficients may be established simply by the reproducibility of the data; for all four levels the error should not exceed ±5%. Systematic errors in the ratios of coefficients for different lines may be caused by uncertainties in the calibration of detection sensitivity; these should not exceed ±5%.

In Figure 4 we show the same level coefficients as a function of the projectile incidence angle. The uncertainties in this data are essentially the same as for Figure 5; no more than ±5% random uncertainty for individual data points and no more than ±5% systematic error in the ratio of coefficients for different

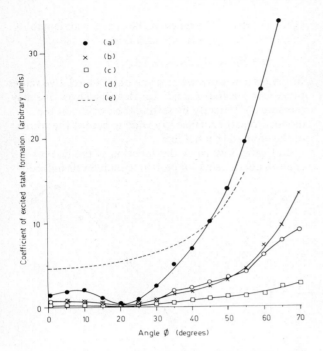

FIGURE 4   Relative variation of the excited state formation coefficients for various helium levels, shown at various incidence angles for 10 keV He$^+$ impact on nickel. (a) 3^3D level; (b) 4^3D level; (c) 5^3D level; (d) 3^3P level; (e) Theoretical prediction normalized to curve (a) at an angle of $45°$.

levels. It is observed that the level excitation coefficients are generally increasing with incidence angle; all three coefficients exhibit a minimum at $20°$ angle of incidence. Also shown in this figure is a predicted value of the excitation coefficient; this is essentially derived by integrating our predictions of line shape over the whole line width and then repeating this at different impact angles. The major assumption of the prediction is that the angular distribution of excited atoms is proportional to $\sin^{-4}(\theta/2)$ where θ is the angle through which the projectile has been scattered. This predicted quantity will give the relative variation of excitation coefficient with incidence angle ϕ; clearly it will be the same for each level. The prediction is shown here normalized to the coefficient for the 3^3D level. While the predicted values show the correct qualitative behaviour it is obvious that it does not correspond in detail to the experimental values. In particular the minimum exhibited by all the data curves at an angle of $20°$ is certainly not predicted. Clearly the cross section behaviour assumed for the predictions is not correct.

5 CONCLUSIONS

The results of this work are very similar to those observed previously with a copper target.[1] He$^+$ impact on Au and Ni gives emission from ^{3}P and ^{3}D states; the line shape of the emission can be predicted satisfactorily by the Doppler shift of emission from projectiles that have suffered bi-particle collisions on the surface. H$^+$ impact on these two targets gives the Balmer series emission; also with a Doppler broadened shape.

ACKNOWLEDGEMENTS

The authors thank Professor J. Kistemaker for his continued interest in this work and for many valuable discussions.

This work is part of the research program of the Stichting voor Fundamenteel Onderzoek der Materie (Foundation for Fundamental Research on Matter) and was made possible by financial support from the Nederlandse Organisatie voor Zuiver-Wetenschappelijk Onderzoek (Netherlands Organization for the Advancement of Pure Research).

The work of one of us (E.W.T.) was partially supported by the Controlled Thermonuclear Research Program of the U.S. Atomic Energy Commission.

REFERENCES

1. C. Kerkdijk and E. W. Thomas, *Physica* (to be published).
2. G. M. McCracken and S. K. Erents, *Physics Letters* **31A**, 429 (1970).
3. N. Tolk, private communication.
4. J. Schutten and Th. van Dijk, *Nature* **211**, 470 (1966).

TIME DECAY OF ION BOMBARDMENT
INDUCED PHOTON EMISSION

JACQUES G. MARTEL† and N. THOMAS OLSON‡

Department of Nuclear Engineering,
Massachusetts Institute of Technology, Cambridge, Massachusetts

Spectral line emission from sodium ion bombardment of aluminum polycrystalline targets has been observed to be dependent upon the target ion bombardment dose. Al and Na emission levels from an unbombarded target decay with ion dose to a saturation level. Qualitative calculations show that the photon emission decay time is consistent with implanted ion saturation times. Cesium ion bombardment, because of its shorter range and smaller saturation time did not yield any transient results.

INTRODUCTION

Optical wavelength photon emission resulting from energetic ion bombardment of solid targets is a well established phenomenon observed by many investigators.[1-4] In all cases, the emission is line spectra characteristic of the target lattice atoms and/or incident ion species. The origin of the photon emission is inelastic collisions between the incident ions and the target atoms, but emission from sputtered metastable atoms[3] and excited ions and atoms within the target lattice[1] have been reported.

The results reported here are from a series of experiments in which a transient in the spectral emission level was measured when an ion beam was switched onto a previously unbombarded target. The transient resulted from embedded ion buildup in the target surface which diminished the characteristic target atom and the incident ion line spectra. The temporal dependence of the photon emission is consistent with range and implanted ion concentrations which are to be expected from the experimental conditions of alkali metal ion bombardment of aluminum poly and single crystal targets.

Time dependent spectral emission measurements are reported for 12.5 to 20 keV sodium and cesium ion bombardment of polycrystalline aluminum.

EXPERIMENTAL TECHNIQUE

The experimental technique, described in more detail in a previous paper,[1] consisted of a 10^{-8} torr all metal ion pumped ion source-target vacuum chamber. The

† Present Address: Centre de l'Energie, INRS, Université du Québec, Varennes, Canada.
‡ Present Address: Northrup Corporation, Northrop Research and Technology Center, Hawthorne, California.

ion source was a surface contact ionization type and it could be operated in the target chamber without contaminating the vacuum. The experimental configuration is illustrated in Figure 1.

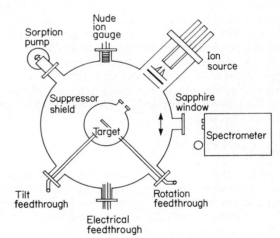

FIGURE 1   Vacuum chamber.

The target material investigated, polycrystalline aluminum, was held in a triple axis goniometer for ion beam-target alignment. All measurements were made at normal incidence. The ion beam-spectrometer axis angle was fixed at 45°. The ion bombardment induced photon emission was monitored with a $\frac{1}{2}$ meter scanning spectrometer which had a photomultiplier output. The ion beam energy was variable from 1 to 20 keV and the ion beam current on the target was 0.5 μa, which corresponded to 5 μa/cm^2.

To measure the time dependent nature of the photon emission from a previously bombarded surface, the ion beam was prevented from striking the target by an electrostatic deflection plate. An unbombarded portion of the target was positioned in the beam area by the goniometer and the ion beam was focused back onto the target.

The spectrometer photomultiplier output and the target ion current were monitored simultaneously with a dual channel strip chart recorder. The normalized photon emission could be measured as a function of time to determine the transient nature of photon emission during ion bombardment buildup.

EXPERIMENTAL RESULTS

During the ion bombardment experiments, only characteristic line spectra were observed and the results reported here were for the 3961.5 Å line of Al and the 5892.5 Å line of Na. These lines correspond to the neutral spectra of both elements.

Aluminum and Sodium Photon Time Decay
During sodium ion bombardment of aluminum poly-crystalline targets, a time dependent decay of both the aluminum and sodium photon emission was observed. The intensity of emission started at a value characteristic of an unbombarded surface and decayed to a constant level as sodium ions were implanted into the target surface. The time dependent portion of the photon emission appeared to decay exponentially so that the following analytical function was fitted to the experimental data,

$$I_{PM}(t) = I_A \, e^{-\alpha t} + I_B$$

From the strip chart recorded time dependent data it was a straightforward matter to plot the time dependent data and obtain the decay constant and the half-lives ($T_{1/2} = 0.693/\alpha$) which characterize the decay.

Figure 2 is a typical plot of the time dependent portion ($I_{PM} - I_B$) of the spectral line emission measured for both aluminum and sodium. The experimental conditions were for ion energies of 12.5 keV and 20 keV, a current density of 5 μa/cm^2, and the ion beam was incident normal to the target surface, $\theta = 0°$. At each energy, the time dependent decay of the aluminum and sodium spectral lines was identical so that the lines at each energy in Figure 2 correspond to both aluminum and sodium.

The decay constants and half-lives of the aluminum and sodium spectral emission decay are tabulated, for various ion beam energies, in Table I. Although cesium ion bombardment experiments were performed, no transients were observed.

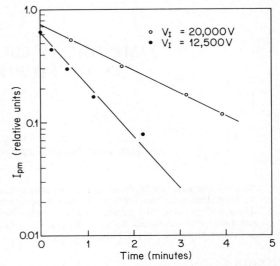

FIGURE 2   Photomultiplier current as a function of time for Na on Al.

TABLE I
Decay constants and half-lives of aluminum and sodium photon emission.

	Ion Energy–keV		
	12.5	18	20
α min^{-1}	1.0	0.59	0.43
$T_{1/2}$ min	0.693	1.17	1.61

DISCUSSION OF RESULTS

The experimentally measured time dependent character of the aluminum and sodium spectral lines was such that they both decayed exponentially with increasing ion bombardment. Additionally, both decays had the same decay constant.

The origin of the time dependent portion of the photon emission is not firmly established, but qualitative calculations indicate that it results from the implanted ion buildup in previously unbombarded targets. The level of photon emission from a previously bombarded surface is assumed to be characteristic of only aluminum and sodium inelastic collisions. As the implanted sodium concentration builds up, sodium–sodium collisions reduce both the aluminum and sodium emission levels.

The most probable penetration depth of 20 keV sodium ions incident upon polycrystalline aluminum is approximately 250 Å for ion dose levels of 10^{14} ions/cm^2. The time constant characteristic of implanted ion buildup should thus be related to the time required to saturate a surface layer 250 Å deep. The target ion beam current density for the data of Figure 2 or the results tabulated in Table I was 5 $\mu a/cm^2$ which corresponds to an ion flux at the target surface of 3×10^{13} ions/cm^2/sec. If these ions are distributed uniformly throughout a volume of 250 Å the rate of implanted ion buildup is 1.2×10^{19} ions/cm^3/sec. The half-life of the photon emission decay is 1.6 minutes, or 100 seconds, which corresponds to an implanted ion density of 1.2×10^{21} ions/cm^3. This latter quantity is one tenth the aluminum atom density and approximates the saturation density of implanted sodium ions.

The qualitative agreement of the photon emission half-life and the calculated time for implanted ion saturation are based upon the reported penetration distance of 250 Å. Davies, et al.[6] have reported that the ion penetration ranges in single crystals are dependent upon the total ion dose per cm^2 and that the range decreases with increasing ion dose. The reason appears to be ion damage to the surface crystal structure. If polycrystalline targets experience a similar reduction in ion penetration ranges, the 250 Å volume utilized in the saturation time calculation was too large. Although no data for this effect is available at dose levels of 10^{15} ions/cm^2 and above, a reduced saturation volume would reduce the saturation time constant. This would bring the measured and calculated time constants into closer agreement.

The energy dependence of the photon decay time constant is also qualitatively correct. The ion penetration range at these energies is proportional to the ion energy. At 18 keV and 12.5 keV, the ion ranges are proportionately shorter as are the saturation half-lives in Table I.

Ion dose dependence measurements of photon emission during cesium ion bombardment of aluminum did not yield measurable results. Within the sensitivity of the experimental technique, which was approximately 0.1 minutes minimum detectable period, no dose dependence of the emission level was observed. The range of cesium in aluminum is less than that of sodium by a factor of 2.2. Additionally, the sputtering yield is greater by a factor of 5 so that cesium implant buildup should occur at a rate 10 times that of sodium. This puts the transient measurement at the limit of the instrumentation sensitivity and probably accounts for the lack of a measured dose dependence.

CONCLUSIONS

The level of ion bombardment induced photon emission has been demonstrated to be dependent upon the target ion bombardment dose. Aluminum and sodium spectral emission intensities decay with increased sodium ion buildup within the bombarded target. Reduced Na–Al and increased Na–Na collisions with increased sodium ion bombardment appear to account for the decreased emission. The level of photon emission measured correlates well with calculated rates of ion saturation buildup within the bombarded target.

REFERENCES

1. J. G. Martel and N. T. Olson, *Nuc. Inst. and Methods,* **105**, 269–275 (1972).
2. J. M. Fluit, L. Friedman, J. vanEck, C. Snoek and J. Kistemaker, *5th Int. Conf. on Ionization Phenomena in Gases,* 131–149 (1961).
3. C. Snoek, W. F. van der Weg and P. K. Rol, *Physica* **30**, 341–344 (1964).
4. I. Terzic and B. Perovic, *8th Int. Conf. on Phenomena in Ionized Gases* 2.4.10 (1967).
5. G. R. Piercy, M. McCargo, F. Brown and J. A. Davies, *Can. J. of Phys.* **42**, 1116–1134 (1964).
6. J. A. Davies, G. C. Ball, F. Brown and B. Domeij, *Can. J. Phys.* **42**, 1070–1115 (1964).

ANGULAR AND ENERGETIC DISTRIBUTIONS OF SECONDARY ELECTRONS EMITTED BY SOLID TARGETS UNDER IONIC BOMBARDMENT

N. COLOMBIE, C. BENAZETH, J. MISCHLER and L. VIEL

Laboratoire de Physique des Solides, Associé au C.N.R.S., Université Paul Sabatier,
118, Route de Narbonne, 31 077 Toulouse Cedex, France

The energy distributions of secondary electrons emitted by targets bombarded with ions in the range 10–100 keV, are studied.

The energy spectra show an intensive peak at 2 eV. Beyond, in the experiments performed on light elements ($Z \leqslant 30$) we observe an Auger peak relating to the creation of a hole on the first atomic energy level situated below the valence band (KVV, L_{23}VV and M_{23}VV transitions) accompanied by two or three peaks corresponding to excitation of surface plasmons.

Energy spectra obtained from phosphorus and sulphur compounds reveal Auger electrons corresponding to L_{23}(P) VV (or L_{23}(S) VV) and L_{23}(P) M_{IV-V}(Ga) V (or L_{23}(P) N_{IV-V}(In) V) transitions; moreover peaks due to plasmon excitations and d band excitations can be distinguished.

The first results obtained from some alloys show Auger spectra, characteristic of the different elements which compose the alloy.

An anisotropy in the energy angular distribution of secondary electrons emitted from copper single crystal is observed.

1 INTRODUCTION

Few studies relating to secondary electron spectroscopy from targets bombarded by ions (in the energy range 10–100 keV) have been made.

In this domain two problems, in particular, can be studied: On the one hand, the energy distribution of emitted electrons, on the other hand, the energy angular distribution of these secondaries.

To carry out such researches, we used two experimental arrangements. We first adapted a retarding potential method and an electronic derivation which give the energy distribution $N(E)$ (or the derivative $dN(E)/dE$) of the totality of electrons emitted from the target with an energy between 0 and 200 eV. Some of results obtained from this method already have been published.[1-3] An electrostatic energy selector (127°) travelling into vacuum has been so realized. It allows us to analyse the electrons ejected in a given direction. This apparatus is principally intended for the study of the energy angular distribution of electrons emitted from single crystal targets. The first results obtained in this domain will be presented and analysed.

2 ENERGY SPECTRA

The energy spectra obtained with these two apparatus show, in all cases, one intensive peak centred at 2 or 3 eV. Its very steep slope on the low energy side allows an accurate determination of the zero of the energy scale.

a Energy Spectra from Elements

For the studied elements ($Z \leqslant 30$) we observe, beyond the intense low energy peak, a spectrum on the high energy tail of the energy distribution curve $N(E)$.

The high energy spectra present a particularly fine structure for elements belonging to the 3rd line of the periodic table. These results are illustrated by the Figures 1 and 2. The first shows the Si energy spectrum drawn with retarding field analyser and Figure 2 gives the Mg energy distribution $N(E)$ realized with electrostatic energy selector.

In the high energy spectrum, the main peak (part II) corresponds to L_{23}VV electrons. In this Auger transition the last two electrons are valence electrons. Discontinuities in the surface density of states may explain the sharpness of the observed Auger peak. The measured

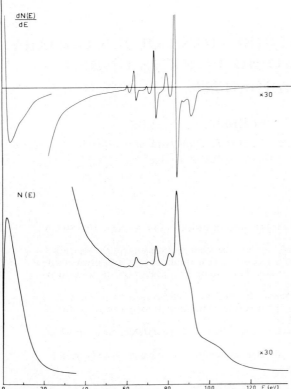

FIGURE 1   Energy spectrum $N(E)$ and its first derivative $dN(E)/dE$ for secondary electrons emitted by Si under 40 keV Ar$^+$ ion bombardment.

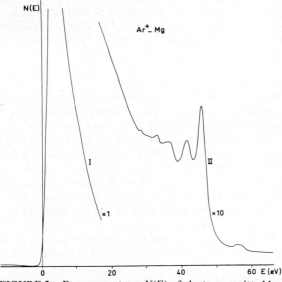

FIGURE 2   Energy spectrum $N(E)$ of electrons emitted by Mg under 40 keV Ar$^+$ ion bombardment.

value of the high energy edge is consistent with the calculated value ($L_{23} - 2\varphi$).

The equidistant peaks observed on the low energy side of $L_{23}VV$ peak can be attributed to surface plasmon excitation losses of Auger electrons. The experimental values of the energy losses (10 eV for Si) denote a "modified" surface† because the value for Si "ideal" surface is 11.9 eV.

For the other elements studied (2nd and 4th lines of periodic table) the Auger electron rate is less intense so, the observed structure is less rich and the first derivative of the distribution $dN(E)/dE$ is often needed to analyse the high energy spectrum.

The mechanism of the hole formation can be deduced from the analysis of diabatic correlation diagrams for the symmetric and asymmetric collisions.[4] The observation of Auger transitions for each of the elements studied agrees with this analysis. So, Koster–Kronig transitions were observed with aluminum and magnesium targets. The experimental position of all the observed peaks is in good agreement with correlation diagrams. The relative intensity of Auger peaks can be explained by taking into account both the probability of hole formation and the fluorescence rate of the considered electronic shell.

Table I summarizes our experimental results.

b Energy Spectra from Binary Compounds[5] and Alloy
By using the retarding potential method we have undertaken the study of some binary compounds, in particular phosphorus and sulphur compounds.

The spectra obtained from GaP and InP show an intensive peak at 2 eV and, with a suitable amplification, a detailed spectrum between 80 and 140 eV. Figure 3 gives the spectrum obtained by bombarding an InP sample with 40 KeV Ar$^+$ ions. The analysis of the curve II (amplified 100 times with respect to the part I) reveals:

—Auger electrons corresponding to a transition between the phosphorus L_{23} level (134 eV) and two levels belonging to InP valence band (peak 1:120 eV).

—L_{23}(P) VV electrons having lost 8.2 eV by excitation of a surface plasmon (peak 2:111.8 eV).

—Electrons due to an Auger process between a phosphorus L_{23} hole, an indium N_{IV-V} electron and an indium phosphide band electron (peak 3:106.3 eV).

—L_{23}(P) VV Auger electrons having excited in $4d$ band electrons (shoulder 4:98.5 eV). L_{23}(P) N_{IV-V}(In V electrons having undergone plasmon surface excitation losses may also contribute to the peak 4.

† It appears unlikely that this change in the observed energy losses can be attributed to surface oxidation.

TABLE I

		Li[a]	Be	Mg	Al	Si	P[a]	S[a]	Ti	V	Cr	Mn	Fe	Ni	Cu
Atomic energy levels (eV)[b]	K	58	115												
	L$_{23}$			54	77	104	134	168							
	M$_{23}$								38	43	48	54	60	74	81
Measured energy of Auger peak maximum (eV)		47.5	104	45.5	63.5	85	120	146	26	30	33.5	40	44	50	58
Highest calculated Auger energy (eV)[c]	K-2φ		107												
	L$_{23}$-2φ			46.5	69	96									
	M$_{23}$-2φ								30	35	39	46.5	52	64	73
Measured surface plasmon losses (eV)	1st			4.0	7.0	10.0									
	2nd			8.5	14.0	20.0									
	3rd				21.0											

[a] Measured values in following compounds: MgLi (10%), InP, CdS.
[b] M. O. Krause[15]—Zero energy is taken at vacuum.
[c] Calculated values, in agreement with measured high energy edges of Auger spectra. (φ: work function of the element).

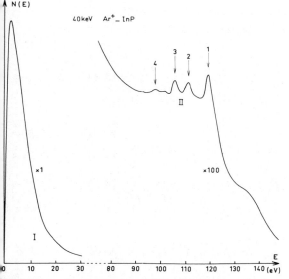

FIGURE 3   Energy spectrum $N(E)$ of electrons emitted from a InP target bombarded by 40 keV Ar$^+$ ions.

The energy distribution curves $N(E)$ obtained by bombarding a CdS sample with 40 KeV Ar$^+$ ions show a peak centred at 146 eV attributed to the L$_{23}$ (sulphur) VV Auger transition. The shoulders observed on the low energy side of this peak can be attributed to energy losses due to surface plasmon excitation and d band electron excitation.

The first results obtained from some alloys (Fe$_3$Al, AlMg (15%), Cu$_2$MnAl) show Auger spectra characteristic of the different elements which compose the studied alloy (Figure 4). The intensity of each Auger peak depend on the concentration and the nature of the element in the alloy.

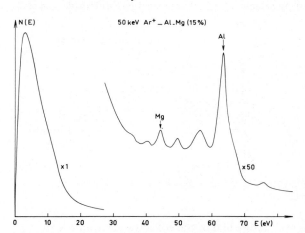

FIGURE 4   Energy spectrum $N(E)$ of electrons emitted from Al–Mg (15%) alloy bombarded by 50 keV Ar$^+$ ions.

These experiments will be continued but we can already notice that these first observations are in good agreement with the studies concerning the asymmetric collisions in solid compounds.[6,7]

3 ENERGY ANGULAR DISTRIBUTIONS

A second point we are studying is the angular dependence, at a fixed energy, of the secondary electronic emission. Such investigations have first been carried out by Burns[8] and Appelt[9] but under electronic bombardment. Some results concerning the angular and energetic dependence of ion induced electron emission were obtained by Yurasova,[10] Hliscs[11] and Begrambekov;[12–14] these authors observe an anisotropic angular distribution of low energy electrons (6 eV Yurasova, 10–20 eV Hliscs, 5–35 eV Begrambekov).

Burns and Appelt showed that maxima of emission inside the crystal correspond to reciprocal lattice vectors of low indices whatever the energy of the electrons may be. From the position of maxima outside the crystal, the authors deduced the refractive index and its variation as a function of secondary electron energy.

Our first results concerning angular distributions at a fixed energy were obtained by bombarding, under normal incidence, a (111) copper surface with 40 KeV argon ions. The electrons ejected 35° away from normal incidence are analysed with the selector and collected in a Faraday cup (Figure 5). This method also allows us to obtain well-defined energy spectra ($\Delta E < 0.5$ eV) between 0 and 200 eV (Figure 2). Angular distributions were measured by rotating the single crystal around the normal to the surface.† In these experimental conditions, we observed an anisotropy which is both characteristic of the target crystal-

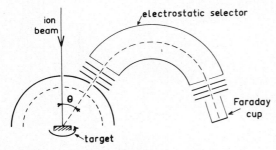

FIGURE 5   Schematic drawing of electrostatic energy selector.

line structure and of the collected electron energy. Figure 6 shows the angular dependence of electron emission as a function of azimuthal angle for different energies of the electrons. The relative anisotropy is about 10% at 13 eV.

All these curves present threefold symmetry (or a multiple of three), which is the (111) copper face

† From a polycrystal the number of collected electrons remains constant.

symmetry, but the modulation of these curves varies with the electron energy. Let us notice the important variation of the modulation between 22.5 eV and 23.5 eV.

This angular dependence can result from:

—an initial orientation of the electron momentum;[8]

—diffraction effects.

Let us notice that this latter assumption is rejected by Burns mainly because, in his experiments, the widths of the observed peaks are too narrow in relation to the wide range of energies (10 eV) covered by each group of secondaries observed.

In our experiments these considerations do not apply ($\Delta E \sim 0.5$ eV).

Furthermore, this resolution allows us to observe important variations of anisotropy as a function of the electron energy (for instance between 22.5 and 23.5 eV Figure 6). This variation cannot be explained

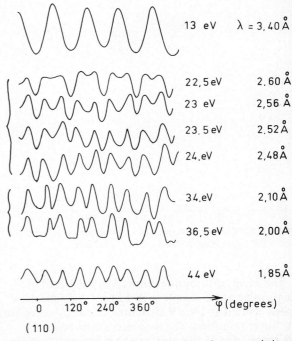

FIGURE 6   Anisotropy of the secondary electron emission as a function of azimuth φ for different energies. Corresponding de Broglie wavelengths are also indicated.

by the assumption proposed by Burns because in this theory the variation of anisotropy as a function of the electron energy is only due to the variation of the refractive index. But this index remains constant in their experimental results between 20 and 25 eV.

Without rejecting this assumption, we are trying to explain the angular dependence in terms of bulk diffraction effects.† Maxima of electron emission can be expected if the wavelength of electrons satisfied the Bragg conditions for reflection from various planes. Localisation of peaks in our first experiments seems in good agreement with Kossel diagrams established for various wavelengths. For instance three maxima are expected for $\lambda = 3.40$ Å ($E = 13$ eV) in the direction which is nearest to the (100) direction. Kossel diagrams also explain the evolution of the shape of the modulation when the electron wavelength decreases. For $\lambda = 2.56$ Å the six maxima can be explained by the presence of a new Kossel cone centred on (110) direction. When the wavelength decreases new cones appear and the Kossel diagram becomes more and more complex.

Further experiments are planned to allow us to determine the contribution of diffraction effects to the anisotropy of the angular dependence.

CONCLUSION

Some results concerning two aspects of the spectroscopy of secondary electrons emitted under ionic bombardment were presented: the energetic distribution of electrons and the angular distribution of these secondaries.

These studies can be connected both with the problems of atomic collisions and with the problems of solid state. Effectively, the initial impact between an incident ion (or a recoiling atom) and a lattice atom

† Our experiments allow us to exclude the possibility of diffraction by the two-dimensional surface lattice because the (111) copper plan symmetry is 6 and the observe symmetry is 3.

can be treated, in first approximation, as a collision between two free particles. The diabatic correlation diagrams for symmetric and asymmetric collisions agree with our experimental results. The target crystalline structure must be taken into account in the elementary excitation processes creating losses by plasmon excitations and interband transition excitations. This crystalline structure intervenes in the anisotropy of angular distribution of secondary electrons whatever the origin of this anisotropy may be: either a diffraction effect of the internal secondaries by Bragg planes in the crystal or an initial orientation of the electron momentum. In this latter case some information on the ion-electron emission process and on valence electron wave functions can be obtained.

REFERENCES

1. L. Viel, B. Fagot and N. Colombie, *C. R. Acad. Sciences, Série B* 272, 623 (1971).
2. L. Viel, C. Benazeth, B. Fagot, F. Louchet and N. Colombie, *C. R. Acad. Sciences, Série B* 273, 30 (1971).
3. F. Louchet, L. Viel, C. Benazeth, B. Fagot and N. Colombie, *Rad. Effects* 14, pp. 123–130 (1972).
4. U. Fano and W. Lichten, *Phys. Rev. Letters* 14, p. 627 (1965).
5. C. Benazeth, L. Viel and N. Colombie, *Surface Science* 32, 618 (1972).
6. H. Barat and W. Lichten, *Phys. Rev. A.* 6, 211 (1972).
7. F. Joyes, *J. Phys. C.: Solid State Phys.* 5, pp. 2192–2199. (1972).
8. J. Burns, *Phys. Rev.* 119, n°1, pp. 102–114 (1960).
9. G. Appelt, *Phys. Stat. Sol.* 27, 657, pp. 657–669 (1968).
10. V. E. Yurasova, V. M. Buchanov and M. Gold, *Phys. Stat. Sol.* 17, K187 (1966).
11. R. Hliscs and H. J. Binder, *Phys. Stat. Sol.* 38, K27 (1970).
12. L. Begrambekov, V. A. Kurnaev, V. M. Sotnikov and V. G. Tel'kovskii, *Soviet Physics, Solid State* 12, n°7, pp. 1738–1739 (1971).
13. L. Begrambekov, V. A. Kurnaev, V. M. Sotnikov and V. G. Tel'kovskii, *Soviet Physics Solid State* 13, n°2, pp. 379–380 (1971).
14. L. Begrambekov, V. A. Kurnaev, V. M. Sotnikov and V. G. Tel'kovskii, *Soviet Physics Solid State* 13, n°6, pp. 1365–1368 (1971).
15. O. Krause, O.R.N.L. Internal Report (1970).

ESTIMATION OF ELECTRON POTENTIAL EMISSION YIELD DEPENDENCE ON METAL AND ION PARAMETERS

L. M. KISHINEVSKY

Institute for Electronics, Uzbek Academy of Sciences, Tashkent, USSR

The paper discusses the dependence of γ—the coefficient of potential emission of electrons from metals on the work function φ and Fermi energy ϵ_F for metals and the ionization energy of ions E_i.

It has been found that

$$\gamma \approx \frac{0.2}{\epsilon_F}(0.8\,E_i - 2\varphi)$$

in the range $3\varphi < E_i < 2(\epsilon_F + \varphi)$.

Deviations from this dependence exceeding 10% have been observed at $E_i < 3\varphi$ and at $E_i > 2(\epsilon_F + \varphi)$.

Theoretical estimates are in satisfactory agreement with experimental results.

1) Potential emission of electrons from metal targets belongs to those physical phenomena which have been relatively well studied both experimentally[1,3] and theoretically.[2,4] The metal-ion system is an excited one. If the ionization energy of the incident particle exceeds the doubled work function ($E_i > 2\varphi$), then an Auger-neutralization of the ion is possible[4] resulting in the formation of a free electron whose emission is classified as "potential". The electron release may also occur another way—with a preliminary resonance neutralization and subsequent Auger-relaxation. Both of these competing processes make a similar contribution to emission,[4] hence in examining the complete coefficient of emission we shall not dwell at length on Auger-relaxation.

Experimental studies of potential emission from a whole series of metals and metal film have shown that the coefficient of potential ion–electron emission (PIEE) γ depends to a great extent on the work function of the target.[5] However, there have been no special theoretical studies of the dependence of γ on individual parameters of the target and the ion, including the work function. The purpose of this paper is to examine these dependences.

Fundamental research in PIEE falls into two groups of studies—consistently theoretic studies and semi-phenomenological studies. The former demonstrates the possibility in principle of describing a phenomenon and also the practical impossibility of achieving good agreement with the experiment while remaining consistently within the framework of theory; these studies also show the ways and means for selecting the required parameters and functional dependences in the theories of the second kind. The latter strive to ensure agreement with the experiment through the utilization of a sufficient number of physically substantiated parameters.

2) The most detailed description of PIEE on the basis of the semi-phenomenological approach is provided by Hagstrum.[4] The method he had developed is applied in this paper.

Having assumed that the probability of a single Auger transition is representable as co-factors which separately take into consideration the angular and energy dependences and having substituted the co-factor which takes into account the dependence of the transition probability on the energy of the primary state level by a constant, which is equivalent to using the effective density of electron states $N_c(\epsilon)$ within the conductivity zone, Hagstrum[4] derived the following ratio for the PIEE coefficient:

$$\gamma = \int_0^\infty N_0(E_k)\,dE_k = \int_{\epsilon_0}^\infty N_i(\epsilon_K)P_e(\epsilon_K)\,d\epsilon_K \qquad (1)$$

Here $N_0(E_k)$, $N_i(\epsilon_K)$—energy distributions of Auger electrons outside and within the metal;

$$N_i(\epsilon_K) =$$

$$\begin{cases} \dfrac{N(\epsilon_K)T\left[\dfrac{\epsilon_K + \epsilon_0 - E_i(S_m)}{2}\right]}{\displaystyle\int_{\epsilon_F}^\infty N(\epsilon_K)T\left[\dfrac{\epsilon_K + \epsilon_0 - E_i(S_m)}{2}\right]\cdot d\epsilon_K}; & \epsilon_K > \epsilon_F \\[4pt] 0; & \epsilon_K < \epsilon_F. \end{cases} \qquad (2)$$

289

ϵ_K, E_k—energy of electrons, read-odd correspondingly from the bottom of the conductivity zone and the vacuum level, $\epsilon_0 = \epsilon_F + \varphi$.

$E_i(S_m)$—ionization energy of the incident particle at distance S_m from the metal where Auger neutralization at the given velocity is most effective.

If the complete probability of Auger neutralization $R_t(S)$ changes with distance to the surface according to the law $R_t(S) = A\, e^{-as}$, then

$$S_m = \frac{1}{a} \ln\left(A/av\right)$$

$N(\epsilon_K) \sim \sqrt{\epsilon_K}$—state density over the Fermi level; $T(\epsilon)$—is connected with the state density $N_c(\epsilon)$ below the Fermi level by the ratio:

$$T(\epsilon) = \begin{cases} \int\limits_0^\epsilon N_c(\epsilon - \Delta)N_c(\epsilon + \Delta)\, d\Delta; & 0 < \epsilon < \epsilon_F/2, \\ \int\limits_0^{\epsilon_F - \epsilon} N_c(\epsilon - \Delta)N_c(\epsilon + \Delta)\, d\Delta; & \epsilon_F/2 < \epsilon < \epsilon_F, \\ 0; & \epsilon < 0, \epsilon > \epsilon_F \end{cases}$$

and turns out to be an Auger transformant.

$$P_e(\epsilon_K) = 2\pi \int\limits_0^{\theta c} P_\Omega(\epsilon_K, \theta) \sin\theta\, d\theta \text{—probability}$$

that the electron at energy ϵ_K will overcome the potential barrier on the surface of the metal.

$P_\Omega(\epsilon_K, \theta)\, d\Omega$—probability that the momentum of the electron at energy ϵ_K is directed into the solid angle $d\Omega$.

Here and further we examine only the complete PIEE coefficient while the broadening of the energy spectrum of electrons is not taken into account.

If we do not limit ourselves to the assumption that the probability of Auger transition is independent of the energy of primary state level, then the Auger transformant $T(\epsilon)$ includes into the sub-integral expression a weight factor which is characteristic of this probability. P. M. Propst,[6] in particular, applying the quasi-classical investigation has demonstrated that this factor diminishes exponentially with the growing distance between the energy level of the electron and the Fermi level. Works[7,8] arrive at similar conclusions.

3) According to Hagstrum's reasoning,[4] the angular distribution of electrons produced in the course of Auger neutralization is asymmetric. This is due to the fact that the matrix element for free electrons, i.e. those which are capable of escaping from the metal is larger than for bound electrons.

From a comparison with experiment Hagstrum established that

$$P_e(\epsilon_K) = \frac{1}{2} \cdot \frac{1 - \sqrt{\epsilon_0/\epsilon_K}}{1 - \lambda\sqrt{\epsilon_0/\epsilon_K}} \equiv P_{\text{Hag}}, \qquad (4)$$

where $\lambda = 1 - 1/f^2$; $f = 2.2$—ratio of matrix elements for free and bound electron.

At $f = 1$ the distribution of $P_\Omega(\epsilon_K, \theta)$ is spherically symmetric and

$$P_e(\epsilon_K) = \tfrac{1}{2}(1 - \sqrt{\epsilon_0/\epsilon_K}) \equiv P_{\text{sym}}. \qquad (5)$$

Contrary to[4] Propst[6] believes that P_Ω is a spherically symmetric function of angle θ. In his view the change in the form of $P_e(\epsilon_K)$ is due to the contribution of secondary electrons.

The share of primary electrons which fail to be emitted from the metal makes up $N_i(\epsilon_K)[1 - P_e(\epsilon_K)]$.

One electron at energy ϵ_K produces true secondary electrons whose number comprises the coefficient of true secondary emission

$$\delta = \frac{\sigma - r}{1 - r}$$

Here σ—coefficient of secondary electron emission; r—coefficient of elastic reflection of electrons.

These electrons have such an energy distribution $N_{\text{sec}}(\epsilon_K, E)$, that $\int N_{\text{sec}}(\epsilon_K, E)\, dE = \delta$.

Hence the complete number of electrons emitted at energy ϵ_K, i.e. the energy spectrum is derived from the following:

$$\begin{aligned} N_0(\epsilon_K) &= N_i(\epsilon_K)P_e(e_K) + \int N_i(\epsilon_K') \\ &\quad \times [1 - P_e(\epsilon_K')N_{\text{sec}}(\epsilon_K', \epsilon_K)]\, d\epsilon_K' \\ &= N_i(\epsilon_K)P_e(\epsilon_K) + N_{\text{sec}}(\epsilon_K) \\ &= N_i\left(P_e + \frac{N_{\text{sec}}}{N_i}\right) = N_i P_e^*(\epsilon_K) \end{aligned} \qquad (6)$$

On the other hand the complete number of emitted electrons during the formation of one primary electron at energy ϵ_K comprises

$$P_{\text{sym}}(\epsilon_K) + \delta[1 - P_{\text{sym}}(\epsilon_K)] = P_e^{**}(\epsilon_K) \qquad (7)$$

That is why the coefficient of potential emission can be written down as

$$\begin{aligned} \gamma &= \int N_i(\epsilon_K)P_e^*(\epsilon_K)\, d\epsilon_K \\ &= \int N_i(\epsilon_K)P_e^{**}(\epsilon_K)\, d\epsilon_K \end{aligned} \qquad (8)$$

In the second case the sub-integral expression, although it does not describe the energy spectrum of electrons, nevertheless takes the same form as given by Hagstrum.[4]

In order to describe the difference between P_e^* and P_e^{**} let us examine the form of the energy spectrum for true secondary electrons. When the energy of primary electrons is 15–20 eV, then the maximum energy of true secondary electrons does not exceed 4 eV.[9] Hence $P_e^* > P_e^{**}$ within the range of low energies while in the other energy ranges $P_e^* < P_e^{**}$.

The comparison used by Hagstrum for achieving better agreement with experiment for the dependence $P_{Hag}(\epsilon_K)$ with curve $P_e^{**}(\epsilon_K)$ of tungsten (Figure 1) shows that within the low energy range $P_e^{**} < P_{Hag}$,

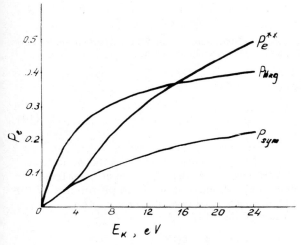

FIGURE 1   Various forms of dependence $P_e(\epsilon_K)$.

while in the range of $E_k > 15$ eV $P_e^{**} > P_{Hag}$. Taking into account the ratio between P_e^{**} and P_e^* we arrive at the conclusion that P_{Hag} and P_e^* are close to each other, which apparently explains the good agreement of Hagstrum's calculations with the experiment.

The coefficient of true secondary emission δ has been taken from monograph[9] where it comprises $\delta/(1-r)$.

By comparing the coefficients of true secondary electron emission for various metals[9] we arrive at the conclusion that they are very close to each other. Hence $P_e^{**}(\epsilon_K)$ are also close. Taking this into account let us in examining PIEE take only the dependence $P_e(\epsilon_K)$ which is used by Hagstrum.

It will be noted that in examining P_e^{**} we did not take into account the asymmetry of the angular distribution of primary electrons, the contribution to the elastically reflected secondary electrons and the

reverse reflection on the surface barrier of true secondary electrons. However, the last factor compensates the former two.

4) A good approximation to the function of electron density within the filled part of the conduction band in the case of tungsten is either $N_c(\epsilon) = K_1$ or $N_c(\epsilon) = K_2\sqrt{\epsilon}$.[4] Further we shall restrict ourselves to the first of these approximations for other metals too. With this:

$$T(\epsilon) = \begin{cases} K_1^2 \epsilon; & 0 < \epsilon < \epsilon_F/2, \\ K_1^2(\epsilon_F - \epsilon); & \epsilon_F/2 < \epsilon < \epsilon_F, \\ 0; & \epsilon < 0, \epsilon > \epsilon_F \end{cases} \quad (9)$$

Taking into account (1) and (2) we derive

$$\gamma = \frac{\int_{\epsilon_0}^{\infty} \sqrt{\epsilon_K}\, P_e(\epsilon_K) T\left(\frac{\epsilon_K + \epsilon_0 - E_i}{2}\right) d\epsilon_K}{\int_{\epsilon_F}^{\infty} \sqrt{\epsilon_K}\, T\left(\frac{\epsilon_K + \epsilon_0 - E_i}{2}\right) d\epsilon_K}$$

$$= \frac{\mathscr{T}_2}{\mathscr{T}_1} \quad (10)$$

The ratio (10) at the above-mentioned assumptions can be taken down in a clearly analytical but extremely cumbersome form. An attempt is therefore made to obtain simple dependences γ on individual parameters.

5) Coefficient γ depends on three parameters—the work function φ, the Fermi energy ϵ_F and the position of the vacant term E_i. However, if $T(\epsilon)$ is a homogeneous function of its argument of the nth order ($n = 1$ at $N_c = K_1$; $n = 2$ at $N_c = K_2\sqrt{\epsilon}$), then

$$\mathscr{T}_j(\alpha\varphi, \alpha\epsilon_F, \alpha E_i) = \alpha^{3/2+n} \mathscr{T}_j(\varphi, \epsilon_F, E_i);$$
$$j = 1, 2$$

Hence

$$\gamma(\alpha\varphi, \alpha\epsilon_F, \alpha E_i) = \gamma(\varphi, \epsilon_F, E_i) \quad (11)$$

Suffice it to determine the behaviour of γ at any arbitrary change of any two parameters to be able to judge the behaviour of γ when the third parameter changes too.

6) Let us begin by examining the dependence of coefficient γ on the Fermi energy ϵ_F at fixed values of φ and E_i. Curve γ_{I} in Figure 2 illustrates $\gamma_{\mathrm{I}} = \gamma(\epsilon_F)$ at $\varphi = 4$ eV, $E_i = 20$ eV in the interval of $4 < \epsilon_F < 12$ eV, which with the increase of ϵ_F diminishes as $\mathrm{const}/\epsilon_F \cdot \gamma_{\varphi,E_i}(\epsilon_F)$ behaves analogically at other fixed values of φ, E_i (see curve γ_{II}, Figure 2 which corresponds to $\gamma_{\mathrm{II}} = \gamma(\epsilon_F)$ at $\varphi = 4.5$ eV; $E_i = 14$ eV). For comparison sake Figure 2 also illustrates sections of the hyperbolic line $1.6/\epsilon_F$; $0.33/\epsilon_F$. They coincide with the rated

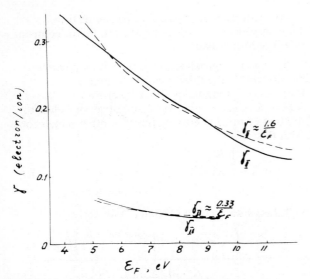

FIGURE 2   Dependence of the PIEE coefficient on Fermi energy ϵ_F at fixed values of φ and E_i.

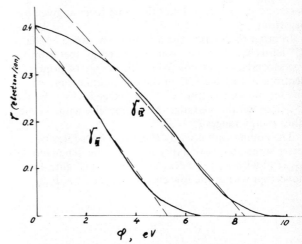

FIGURE 3   Dependence of the PIEE coefficient on the work function φ at fixed values of ϵ_F and E_i.

curves γ_I, γ_{II} with a precision of ±10%. Subsequently

$$\gamma_{\varphi,E_i}(\epsilon_F) \approx \frac{const}{\epsilon_F}, \qquad (12)$$

where const depends on φ, E_i.

At lower values of ϵ_F outside of the mentioned interval the growth of γ is slowed down with the reduction of ϵ_F ($\lim_{\epsilon_F \to 0} \gamma_I = 0.42$; $\lim_{\epsilon_F \to 0} \gamma_{II} = 0.345$) although this carries no practical interest since in the case of metals $\epsilon_F \sim 5$ eV.

Let us now examine the dependence of coefficient γ on the work function φ at fixed values of ϵ_F and E_i. As an example Figure 3 illustrates $\gamma_{III} = \gamma(\varphi)$ at $\epsilon_F = 6$ eV, $E_i = 14$ eV and $\gamma_{IV} = \gamma(\varphi)$ at $\epsilon_F = 6$ eV, $E_i = 20$ eV at a change of φ from zero to $E_i/2$.

As it appears from Figure 3 the central sections of these curves provide a satisfactory approximation by linear dependences:

$$\gamma_{III} \approx 0.4\text{–}0.076\,\varphi; \ \gamma_{IV} \approx 0.5\text{–}0.060\,\varphi \qquad (13)$$

With the reduction of the work function within the range of small φ the growth of $\gamma(\varphi)$ slows down.

7) Since we have derived the dependence of γ on φ and ϵ_F let us now derive the dependence $\gamma(E_i)$. From (11) and (12) it follows that

$$\gamma(\varphi, \epsilon_F, E_i) = \frac{1}{\alpha}\gamma\left(\varphi, \frac{\epsilon_F}{\alpha}, E_i\right) = \frac{1}{\alpha}\gamma(\alpha\varphi, \epsilon_F, \alpha E_i),$$

i.e. at fixed values of ϵ_F:

$$\gamma_{\epsilon_F}(\varphi, E_i) = \frac{1}{\alpha}\gamma_{\epsilon_F}(\alpha\varphi, \alpha E_i) \qquad (14)$$

This ratio may be used in two ways. In the first place by fixing α we can derive from the given curve $\gamma_{\epsilon_F, E_i}(\varphi)$ new and analogical curves at other fixed values of E_i. Secondly having fixed the product $\alpha\varphi = \Phi$ and knowing $\gamma_{\epsilon_F, E_i}(\varphi)$ we derive $\gamma_{\epsilon_F, \Phi}(E_i)$.

Taking down (13) in the form of

$$\gamma_{\epsilon_F, E_i}(\varphi) = a - b\varphi$$

we obtain

$$\gamma_\Phi(\alpha E_i) = \alpha(a - b\varphi) = \frac{(\alpha E_i)a}{E_i} - b\Phi$$

Changing the symbols of the component values

$$\Phi \to \varphi, \qquad \alpha E_i \to E_i, \qquad E_i \to E_i^*$$

and taking into account the dependence of γ on ϵ_F we derive the final:

$$\gamma(\varphi, \epsilon_F, E_i) = \frac{6}{\epsilon_F}\left(\frac{aE_i}{E_i^*} - b\varphi\right) = \frac{3b}{\epsilon_F}\left(\frac{2aE_i}{bE_i^*} - 2\varphi\right) \quad (1$$

Substituting the numerical values of a, b, E_i^* we derive two ratios:

$$\gamma_1 \approx \frac{0.23}{\epsilon_F}(0.75E_i - 2\varphi), \qquad (16a)$$

$$\gamma_2 \approx \frac{0.18}{\epsilon_F}(0.83E_i - 2\varphi) \qquad (16b)$$

These approximated expressions for $\gamma(\varphi, \epsilon_F, E_i)$ are very close and differ by about $\pm 10\%$ at $E_i > 3\varphi$, i.e. they are as correct as the ratios (12) and (13). Using the precise dependences $\gamma_{\varphi, E_i}(\epsilon_F)$ and $\gamma_{\epsilon_F, E_i}(\varphi)$ we would have derived two coinciding ratios instead of (16a) and (16b). Hence with an accuracy of $\pm 10\%$

$$\gamma(\varphi, \epsilon_F, E_i) \approx \frac{0.2}{\epsilon_F}(0.8E_i - 2\varphi) \qquad (17)$$

Greater deviations from this dependence should be observed within the range of the potential emission threshold. As shown by the detailed analysis of ratios (4) and (10) at $E_i < 3\varphi$

$$\gamma \approx \frac{0.093(E_i - 2\varphi)^{5/2}}{\mathscr{T}_1} \qquad (18)$$

Deviations from ratio (17) should also be observed at higher values of E_i, for instance for deep levels in multiple charged ions. At $E_i > 2\epsilon_0$ coefficient γ changes in a non-linear fashion and has a tendency to saturation:

$$\gamma \approx 0.5 - 0.16\sqrt{\frac{\epsilon_0}{E_i - \epsilon_0 - \varphi}}. \qquad (19)$$

Within the intermediary ionization energy range ratio (17) is applicable and the ranges of application of (17) (18) (19) overlap.

8) The ratios derived provide a satisfactory explanation for the experimental results.

a) In experiments[5] there has been observed a linear dependence of $\gamma(\varphi)$ during the bombardment of a number of metal targets by Ne^+ and Ar^+ ions. This is in agreement with ratio (17). Deviations of experimental points from the linear dependence in the above-mentioned work have been small. This is due to the fact that the Fermi energy ϵ_F changes but little during the transition from one metal to another. However it is possible that the dependence of the PIEE coefficient on ϵ_F is weaker than follows from the above-mentioned calculations.

b) The approximate linear dependence $\gamma(E_i)$ during the bombardment of one metal by various inert gas ions is confirmed by experiments.[1,3]

c) At small values of the difference $E_i - 2\varphi$ the experimental values of $\gamma(E_i)^3$ and $\gamma(\varphi)^5$ deviate from the linear dependence in qualitative agreement with the ratio (18). The calculated γ values fit the experimental ones[1,3,5] as good as the calculations in Ref. 4.

REFERENCES

1. U. A. Arifov, "Vzaimodeistviye atomnikh chastits s poverkhnostyu tverdogo tela", Nauka, M. (1968).
2. S. S. Shekhter, *JETP* **7**, 750 (1937).
3. H. D. Hagstrum, *Phys. Rev.* **96**, 325 (1954).
4. H. D. Hagstrum, *Phys. Rev.* **96**, 336 (1954).
5. U. A. Arifov and R. R. Rakhimov, *Izv. AN SSSR. Seria phys.* **34**, 657 (1960).
6. P. M. Propst, *Phys. Rev.* **129**, 7 (1963).
7. H. D. Hagstrum, Y. Takeishy and D. D. Pretzer. *Phys. Rev.* **139A**, 526 (1965).
8. D. Sternberg, Ph.D. Thesis. Dept. of Phys., Columbia Univ., (1957).
9. I. M. Bronstein and B. S. Freiman, "Vtorichnaya elektronaya emissia", Nauka, M. (1969).

DEPENDENCE OF INELASTIC ENERGY LOSS ON THE ATOMIC NUMBERS OF THE IONS

B. E. BAKLITSKY, E. S. PARILIS and V. K. FERLEGER

Institute for Electronics, Uzbek Academy of Sciences, Tashkent, USSR.

The present paper provides a general analytical expression for inelastic losses taking into account the shell effects and also tracing the Z-oscillation of kinetic ion-electron emission from the surface.

The Firsov theory[1] suggested a mechanism for electron "friction" and on the basis of the Thomas–Fermi model study calculated the inelastic energy losses which cause excitation and ionization of electron shells during the collision of heavy atoms within the velocity range of 10^7–10^8 cm/sec.

This method was also used for elaborating a theory of kinetic electron emission from metals[2,3] and luminescence of crystals[4,5] under the impact of ion bombardment.

The Firsov model was specified: the plane which divides the colliding atoms was drawn through a point in the minimum of electron density,[3] which made it possible to obtain a better dependence of energy losses and electron yield on the atomic number of the ion and to apply this theory to light ions.

However, the Thomas–Fermi model, at least its initial form, does not take into account the shell structure of the atom. (Recently an attempt was made[6] to describe shells in the Thomas–Fermi model.) The inelastic energy losses estimated on the basis of this model increase monotonically with the increasing atomic numbers of the colliding particles.

Meanwhile a number of experiments connected with the passage of fast ions through matter[7,8] indicate the existence of maxima and minima for certain atomic numbers.

In order to remove this contradiction it was suggested that either the Slater wave functions[9,10] or the Hartree–Fock–Slater wave functions are to be used within the framework of the Firsov theory. This made it possible to explain Z-oscillations in the ion energy losses which had been observed experimentally. Later this technique was also applied in Refs. 13 and 14 to calculate the stopping power of crystals.

The non-monotony caused by shell effects could be clearly traced in kinetic electron emission or crystal luminescence under the ion bombardment[15]–phenomena which are directly determined by the cross-section for inelastic (electron) ion energy losses in the surface atomic layers.

The present paper provides a general analytical expression for inelastic losses taking into account the shell effects and also tracing the Z-oscillation of the coefficient of kinetic ion-electron emission from the surface of a crystal and in passage.

In order to calculate the inelastically transmitted energy during atomic collisions, we use the Firsov formula:[1]

$$\epsilon = m \int_{R_0}^{\infty} \dot{R} \, dR \int_s \tfrac{1}{4} nv \, ds \qquad (1)$$

Here m—mass of the electron; R—distance between atoms; $\dot{R}$—relative movement velocity; R_0—distance of closest approach; n—density of electrons according to the Thomas–Fermi model; s—surface passing through the minimum of potential energy.

In Ref. 9 the shell structure of atoms was taken into account by substituting for n the electron density which was estimated on the basis of the Slater functions. In this approximation (1) takes the form:

$$\epsilon = \frac{m}{4} \int_{R_0}^{\infty} \dot{R} \, dR \int_s \left\{ \sum_i V_i |\psi_i|^2 + \sum_j V_j |\psi_j|^2 \right\} \, ds \qquad (2)$$

Summing involves all electrons of colliding atoms. Index i refers to electrons of the first atom and j—to electrons of the second atom.

$$V_i = \frac{\hbar}{m} \xi_i \left(1 + 2 \frac{l_i(l_i + 1) - n_i(n_i - 1)}{n_i(2n_i - 1)} \right)^{1/2}$$

mean electron velocity in orbit.

$$\psi_i = \left[\frac{(2\xi_i)^{2n_i+1}}{(2n_i)!} \right]^{1/2} \cdot z^{n_i-1} \cdot e^{-\xi_i \cdot z}$$

Slater radial wave function.

$$\xi_i = \frac{Z - S}{n_i a_0}$$

Z—atomic number; S—screening parameter; n, l—radial and orbital quantum numbers.

In Ref. 9 Eq. (2) was solved by an electronic computer for a number of cases of channeling. Good agreement was obtained with experiment.[7]

We shall assume that surface s divides the distance between atoms in the ratio of $\alpha : (1 - \alpha)$, where

$$\alpha = \left[1 + \left(\frac{Z_2}{Z_1} \right)^{1/6} \right]^{-1} \tag{3}$$

does not depend on R. Integration in (2) along plane s produces the following expression for $\epsilon(p)$

$$\epsilon(p) = \frac{mV_0}{8} \int_{R_0}^{\infty} \frac{1 - V(R)/E}{\sqrt{1 - V(R)/E - P^2/R^2}}$$

$$\left\{ \sum_i V_i \xi_i / n_i \sum_{K=0}^{2n_i-1} \frac{[2\alpha R \xi_i]^K}{K!} e^{-2\alpha R \xi_i} \right.$$

$$+ \sum_j V_j \xi_j / n_j \sum_{K=0}^{2n_j-1} \frac{(2R(1-\alpha)\xi_j)^K}{K!}$$

$$\times e^{-2(\alpha-1)R\xi_j} \Bigg\} \, dR \tag{4}$$

In the following integration on R, the elastic scattering on a screened potential is taken into account:

$$V = \frac{(Z_1 + Z_2) e}{R} \chi[1, 13(Z_1 + Z_2)^{1/3} R/a_0] \tag{5}$$

The Firsov potential V within the investigated region of atomic spacing

$$0.7 \frac{a_{TF}}{(\sqrt{Z_1} + \sqrt{Z_2})^{2/3}} < R < 7 \frac{a_{TF}}{(\sqrt{Z_1} + \sqrt{Z_2})^{2/3}} \tag{6}$$

is described with a 10% accuracy by the expression:

$$V(R) = \frac{0.45 Z_1 Z_2 \, e^2 a_{TF}}{(\sqrt{Z_1} + \sqrt{Z_2})^{2/3} R^2} = \frac{A(Z_1, Z_2)}{R^2} \tag{7}$$

e—electron charge; a_{TF}—Thomas–Fermi parameter.

After the introduction of the dimensionless parameter $z_i = 2\alpha R_0 \xi_i$ and $z_j = 2(1 - \alpha) R_0 \xi_j$ we derive:

$$\epsilon(Z, E) = \frac{mV_0}{8} \left\{ \frac{1}{2\alpha} \sum_i V_i \left[\varphi_1(n_i, z_i) - \frac{(2\alpha\xi_i)^2 A}{E} \right. \right.$$

$$\times \varphi_2(n_i, z_i) \left. \right] + \frac{1}{2(1-\alpha)} \sum_j V_j$$

$$\times \left. \left[\varphi_1(n_j, z_j) - \frac{(2(1-\alpha)\xi_j)^2 A}{E} \varphi_2(n_j, z_j) \right] \right\} \tag{8}$$

Functions

$$\varphi_1(n_i, z_i) = \frac{1}{n_i} \sum_{K=0}^{2n_i-1} (-1)^{K+1} \frac{Z_i^{K+1}}{K!} \frac{d^{K+1}}{dZ_i^{K+1}} K_0(z_i)$$

$$\varphi_2(n_i, z_i) = \frac{1}{n_i} \sum_{K=0}^{2n_i-1} (-1)^{K+1} \frac{Z_i^{K-1}}{K!} \frac{d^{K-1}}{dZ_i^{K-1}} K_0(z_i) \tag{9}$$

$K_0(Z_i)$—the MacDonald function.

$\varphi_1(n_i, z_i)$ which depend only on dimensionless parameters were tabulated and represented graphically on Figure 1. These normalized functions characterize the compression of electron shells corresponding to the given Z during their successive filling.

Formula (8) determines the inelastically transmitted energy during the collision with given impact parameter.

In order to obtain the average inelastically transmitted energy, it is necessary to integrate $\epsilon(p)$ along all the impact parameters within the elementary cell of the crystal:

$$\epsilon(E) = \frac{\pi m V_0}{2} \left\{ \frac{1}{2\alpha} \sum_i V_i \left[\frac{1}{(2\alpha\xi_i)^2} \Psi_1(n_i, a_i) \right. \right.$$

$$\left. - \frac{A}{E} \Psi_2(n_i, a_i) \right] + \frac{1}{2(1-\alpha)} \sum_j V_j$$

$$\times \left. \left[\frac{1}{(2(\alpha-1)\xi_j)^2} \Psi_1(n_j, a_j) - \frac{A}{E} \Psi_2(n_j, a_j) \right] \right\} \tag{10}$$

In this formula

$$\Psi_1(n_i, a_i) = \frac{1}{n_i} \sum_{K=0}^{2n_i-1} (-1)^{K+1} \int_{a_i}^{\infty} \frac{z_i^{K+2}}{K!}$$

$$\times \frac{d^{K+1}}{dz_i^{K+1}} K_0(z_i) \, dz_i$$

$$\Psi_2(n_j, a_j) = \frac{1}{n_j} \sum_{K=0}^{2n_j-1} (-1)^{K+1} \int_{a_j}^{\infty} \frac{z_i^K}{K!}$$

$$\times \frac{d^{K-1}}{dz_j^{K-1}} K_0(z_j) \, dz_j \qquad (11)$$

During integration along the parameters, the expansion of the integration region to infinity is justifiable by the negligible contribution to ϵ beyond the limits of real atomic spacing distances in the crystal.

The dimensionless parameters a_i, a_j are related to the distances of closest approach of atoms which are determined by their trajectories. In the case of channeling, a_i, a_j—distances to the walls of the channel. The function Ψ_1 is illustrated in Figure 2.

The inelastic losses depending on the parameter or as the average in the crystal cell are calculated very simply when the atomic numbers of the colliding atoms differ by no more than a factor 5. In this case, the dependence of ϵ on α can be neglected, and the formulas for inelastic energy losses are reduced to simply a polynom with constant coefficients determined by electron momentum in orbit.

Figure 3 (curve 1) illustrates the dependence $\epsilon(p)$ for collisions Si–Si. For the sake of comparison the same graph illustrates curve 2 calculated on the basis of formula (1) taking scattering into account.[4]

The region of distant collisions, which is characteristic of the passage of particles along the channel axis, is shown in Figure 3. The dotted line denotes boundary parameters which correspond to the movement of the ion along the channel axes [100] and [110] in a silicon single crystal. In the given region, the inelastic losses have been calculated on the basis of formula (8), and they are twice as large as those calculated by formula (1).

Of particular interest is the investigation of Z-oscillation in ion-electron emission. As is known the electron escapes only from a few of the surface atomic layers.[2] This makes it possible to measure the inelastic losses directly within the surface layer.

According to formula (10) we calculated energy losses for the first six layers during bombardment of a silicon single crystal by ions with Z = from 3 to 18 in the case of normal incidence on faces [100] and [110] for $V_0 = 2.10^7$ cm/sec. Figure 4 illustrates the stopping power of the first, second and sixth layers depending on the atomic number of the incident ion.

The non-monotonic dependence of the atomic number reflects the competing effect of two factors: $\epsilon(p)$ increasing with the increasing number of electrons and the reduction of $\epsilon(p)$ due to the compression of

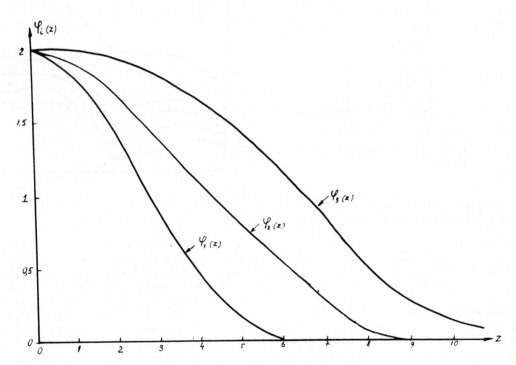

FIGURE 1 Z-dependence of the function $\varphi_i(Z)$, subscript i indicates the number of the shell.

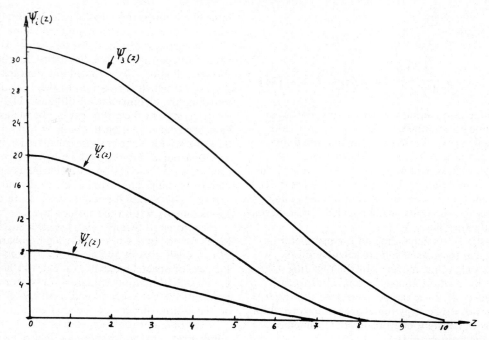

FIGURE 2 Z-dependence of the function $\Psi_i(Z)$, subscript i indicates the number of the shell.

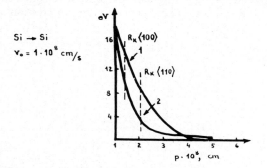

FIGURE 3   Curve 1: $\epsilon(p)$ for Si → Si. Curve 2: Dependence obtained from formula (1).

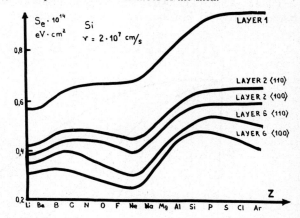

FIGURE 4   Stopping power of the first, second and sixth layer of Si.

the shell during filling. In calculating the curves on Figure 4 a_i was derived from the formula

$$a_i = R_{T,i} = 1,5 \sqrt[3]{\frac{\pi \, dR_0{}^2}{\mu_i + 1}} \qquad (12)$$

in accordance with Ref. 16 where the shape of the shadow in the sub-surface layers of the single crystal was calculated. Taking into account the exponential drop in the number of electrons along depth which are emitted from the crystal (for silicon ≈ 10 atomic

layers[17]), the contribution of layers 1–6 was summed up (Figure 5). The resulting curve has a weaker periodical dependence than the curve corresponding, for instance to the sixth layer. This periodical dependence is clearly evident only at R_T = the radius of the outer shell. Under bombardment of the non-shaded first layer, a considerable lapping of shells occurs, due to which the Z-oscillations weaken.

A more characteristic non-monotonic dependence of electron emission "from the opposite side" can be

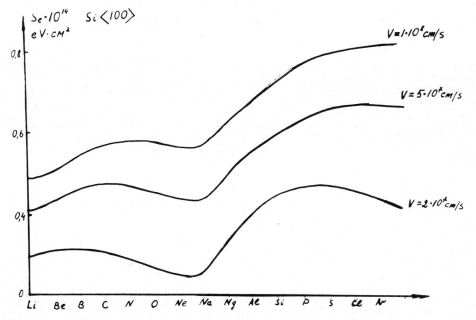

FIGURE 5   *Z*-dependence of the sum of the contributions of layers 1 to 6.

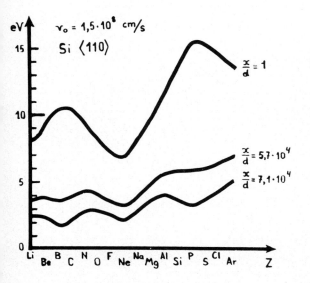

FIGURE 6   Energy loss due to the last layer of Si for different thickness of the crystal.

axis and participating only in distant collisions. When the thickness of the crystal is increased, however, different values of the kinetic energy retained by the ions emitted from the crystal will be obtained. Under channelling, this energy is determined by the value of the inelastic energy losses in each layer. That is the reason why the non-monotonic dependence $\epsilon(Z)$ is balanced with increasing thickness of the crystal and then takes the paradoxal form. Figure 6 illustrates the losses in the last layer of a Si single crystal with ions passing through channel [110] at Z from 3 to 18 and $V_0 = 1.5 \cdot 10^8$ cm/sec at various Z. x—the thickness of the crystal; d—constant of the lattice.

It would be interesting to study experimentally the dependence $\epsilon(Z)$ both directly and from the inverse side of a thin single crystal which would make it possible to follow the Z-oscillation in inelastic energy losses.

observed during the passage along the channel of the single crystal.

Emission from the opposite side of a single crystal is indeed caused by ions concentrated close to the channel

REFERENCES

1. O. B. Firsov, *JETP* **36**, 1517 (1959).
2. E. S. Parilis and L. M. Kishinevsky, *PTT* **3**, 1219 (1961).
3. L. M. Kishinevsky and E. S. Parilis, *Izv. AN SSSR. Ser. Phys.* **26**, 1409 (1961).
4. B. E. Baklitsky and E. S. Parilis, *Optika i spektroskopia* **27**, 169 (1969).

5. B. E. Baklitsky and E. S. Parilis, *Izv. AN UzSSR. Ser. phys. mat. nauk.* **2**, 89 (1970).

6. D. A. Kirzhnits and G. V. Shpatkovskaya, *JETP* **62**, 2082 (1972).

7. F. H. Eisen, *Can. J. Phys.* **46**, 561 (1968).

8. P. Hvelplund and B. Fastrup, *Phys. Rev.* **165**, 908 (1968).

9. I. M. Cheshire, G. Dearneley and J. M. Poate, *Phys. Let.* **27A**, 304 (1968).

10. I. M. Cheshire, G. Dearneley and J. M. Poate, *Proc. Royal Soc.* **A311**, 47 (1969).

11. C. P. Bhalla and J. N. Bradford, *Phys. Let.* **27A**, 318 (1968).

12. G. Reese, C. P. Bhalla and J. N. Bradford, *Bull. Amer. Phys. Soc.* **14**, 184 (1969).

13. V. S. Kesselman, *JTP* **41**, 1708 (1971).

14. N. A. Kumakhov. Trudi III Vsesoyuznogo soveschaniya po phizike vzaimodeistviya zaryazhennikh chastits s mono-kristalami.

15. B. E. Baklitsky and E. S. Parilis, *Radiation Effects* **12**, 137 (1972).

16. Y. V. Martinenko, *PTT* **8**, 637 (1966).

17. I. M. Bronstein and B. S. Freiman, *Vtorichnaya elektronnay emissiya* (Nauka, M., 1969).

ENERGY SPECTRUM OF ELECTRONS EMITTED BY ALKALI HALIDE CRYSTALS UNDER IMPACT OF ION AND ATOM BOMBARDMENT

U. A. ARIFOV, S. GAIPOV and R. R. RAKHIMOV

Institute for Electronics, Uzbek Academy of Sciences, Tashkent, USSR.

The energy distribution of electrons emitted by crystals of KBr, NaCl and LiF under bombardment by Kr^+ and $Kr^°$, Ar^+ and $Ar^°$, He^+ and $He^°$ in the energy range of 100–6000 eV was studied by the method of analysis in the field of a 127° cylindrical condenser. The separation of electron and ion components from the general spectrum of secondary particles was carried out by a weak magnetic field directed at right angle to the electrical field of the condenser.

A comparison of the electron spectrum for ions and atoms illustrates the behaviour of electron spectra of potential and kinetic emission. The energy spectra of electrons of kinetic electron emission are explained on the basis of the mechanism of the Auger phenomenon.

In Refs. 1 and 2 a spherical system of electrodes was used to study the electron energy spectrum emitted by alkali–halide crystals under bombardment by inert gas ions. This method does not take into consideration the energy spectrum of negative ions which constitute a substantial share of emission under the bombardment of these crystals by ions.

This paper studies the energy spectrum of potential and kinetic electron emission taking into account the contribution by the spectrum of negative ions in the case of bombardment of KBr, NaCl and LiF crystals by ions and atoms He^+ and $He^°$, Ar^+ and $Ar^°$, Kr^+ and $Kr^°$ within the energy range of 100–6000 eV.

EXPERIMENTAL TECHNIQUE

Experiments were conducted with a vacuum apparatus demonstrated in Figure 1. The basic units of the apparatus are an ion gun, a recharging chamber for producing a flux of neutral atoms, a target-collector system and an electron energy analyser. A 127° cylindrical condenser with a 57 mm deflection radius was used as the electron energy analyser. The entrance and exit slits of the analyser were 0.8 mm and this ensured a resolving power of 2%. The beam of primary particles was incident on the target at 45° and it was the secondary negative particles at a normal direction to the target surface which were subjected to energy analysis. After passing through the condenser the electrons entered the beam catcher which was linked up with the input of the VI-2 type electrometer amplifier. The amplified signal is fed to the electron automatic potentiometer of the EPP-09 type whose recorder chart registered the distribution curves.

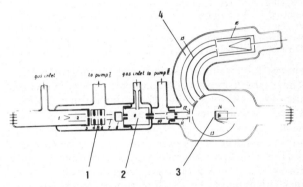

FIGURE 1   Scheme of the vacuum apparatus. 1–tungsten filament-cathode; 2–anode; 3–vacuum diaphragm; 4, 5, 6–electrostatic lens; 7, 8–flat condenser for correction of the trajectory of the ion beam; 9–recharging chamber; 10–flat condenser for separation of atoms from the ions; 11, 12–diaphragm; 13–collector of secondary particles; 14–target; 15–electron analyser; 16–Faraday cylinder.

The current density of the primary beam on the target did not exceed 10^{-10} a/cm^2. During measurement the target was at a temperature of 350–400°C. The magnetic field of the earth within the target zone was compensated by a Helmholtz coil.

The energy spectrum was divided into the electron and negative ion spectra in the following way: first we recorded the summary curve of energy distribution for negative ions and electrons (curve 1 in Figure 2). A weak magnetic field was then applied normal to the electrical field of the cylindrical condenser and the electrons were deflected to one of the condenser linings. This produced a curve for the energy distribution of negative ions (curve 2). The difference in spectra represented by curves 1 and 2 produces a curve for the

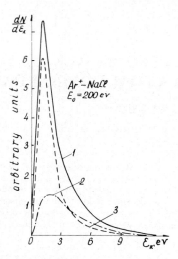

FIGURE 2 Energy spectra of secondary particles emitted from crystal NaCL during bombardment by Ar^+ ions at 200 eV energy. Curve 1–energy spectrum for total secondary negative emission; Curve 2–energy spectrum of negative ions; Curve 3–electron energy spectrum.

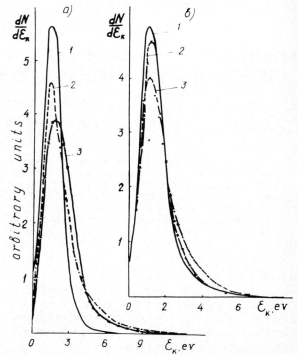

FIGURE 3 Energy distribution curves under the bombardment of LiF (a) and KBr (b) crystals by Kr^+ (1), Ar^+ (2) and He^+ (3) at 200 eV.

energy distribution of secondary electrons (curve 3). The magnetic intensity was adjusted such as to prevent the fastest electrons from penetrating through the exit slit of the condenser and at the same time to avoid changing the trajectory of the slow negative ions to any substantial degree. This was confirmed on the basis of studying the behaviour of energy distributions curves for negative ions with two opposing directions of the magnetic intensity vector. Experiments show that a change in the direction of the magnetic field does not result in any noticeable alterations in the form of the energy spectrum for negative ions. It appears that the negative ions display a relatively rigid energy spectrum as against that of the electrons. That is why in the case of heavy bombarding particles, where negative ions account for a substantial share of negative emission, they distort the spectra for electron energy considerably.

As the energy of primary particles grew the distortions in the energy spectrum of electrons introduced by negative ions diminished.

EXPERIMENTAL RESULTS AND DISCUSSION

Figure 3-a and 3-b illustrate the energy distribution of electrons emitted from KBr and LiF under bombardment by He^+, Ar^+ and Kr^+ at 200 eV. For the sake of convenient comparison the curves are normalized to one area. As is known[3] atoms of the same type at this energy (with the exception of atoms of $He°$) produce a relatively small electron emission from the above-mentioned crystals. Thus the given data mainly

characterizes the energy spectrum of electrons which had been excited by the inner energy of the incident ion.

It is apparent that the position of maxima in the distribution curves does not depend much on the type of the bombarding ion. In the case of He^+ ions on LiF it is slightly shifted in the direction of greater energies as compared with the curve maxima for Ar^+ and Kr^+. However the high-energy section of the electron energy spectrum undergoes certain changes, depending on the type of ion, which are due to the influence of the ionization potential (V_i) of the incident particles. An increase in the semi-widths of the distribution curves is also observed resulting from the increase of V_i particles. The only exception is the case when a KBr crystal is bombarded by Ar^+ and Kr^+ ions which have similar half-widths.

The fraction of relatively fast electrons in the energy spectrum grows along with the increased ionization potential of primary particles. In the case of bombardment of LiF the fraction of electrons whose energy lies within the interval from 5 to 10 eV is greater for Ar^+ than for He^+ ions. It has also been observed that the maximum electron energy (ϵ_{max}) grows with the increase of V_i of the ion.

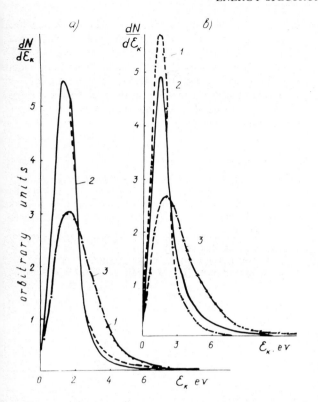

FIGURE 4   Energy distribution curves for electrons during the bombardment of KBr (1), NaCl (2) and LiF (3) crystals by Kr° (a) and Ar° (b) atoms at 6 keV.

Figure 4a and 4b illustrate the dependence of energy spectrum of electrons for Kr° and Ar° atoms on the type of crystal. It is seen that in the case of Kr° atoms the energy spectra of electrons emitted from KBr and NaCl are close while in the case of Ar° the energy spectrum for NaCl is somewhat enriched with fast electrons as compared with the spectrum for KBr. Electrons emitted from LiF have a more rigid spectrum and this difference is all the more pronounced in the case of Ar° atoms. In the case of Kr° the half-width of distribution curves increases from 1.48 (for KBr) to 2.55 eV for LiF, whereas in the case of Ar° these values are 1.62 and 3.68 eV accordingly. There are also certain differences in the values of ϵ_{max} for Ar° (8.6 and 9.6 eV) in the case of crystals KBr and LiF respectively and for Kr° (8 and 14.4 eV in the case of KBr and LiF). Examination of the data for a given crystal shows that the electron energy spectra change depending on the type of incident particle. Electrons emitted under the bombardment of the given crystal by Kr° atoms have the softest energy spectrum.

As the mass of the incident atoms decrease the fraction of fast electrons in the energy spectrum shows an increase. While the position of the maxima in the distribution curves remain almost uninfluenced by the type of atom, the half-widths of the curves and the values of maximum electron energy increase as the mass of the incident particles decreases. However in the case of LiF there occurs a certain "anomaly", i.e. in the energy spectrum the fraction of faster electrons is greater for Ar° than for He° although the values of ϵ_{max} for both atoms are close (−15 eV). In the case of KBr the values of ϵ_{max} are close for Ar° and Kr° atoms.

A comparison of energy spectra of electrons emitted by KBr, NaCl and LiF crystals under the bombardment by ions and atoms shows that at low energies the maxima peaks of the distribution curves for atoms are greater than for ions. Along with this the energy spectrum of electrons for ions is slightly enriched by faster electrons than the electron spectrum for atoms. As the energy of primary particles grows the difference in the energy spectra of electrons for ions and for atoms is reduced. Within the kV range of energies the energy spectra of electrons for ions and atoms in most cases show good agreement. In some cases (for instance Kr⁺ on LiF) the energy spectrum is enriched by faster electrons under atom bombardment.

By analogy with data for metals it would be expected that if ions result in potential and kinetic electron emission, then the atoms should produce only kinetic electron emission. But in fact the bombardment of the investigated crystals by ions of relatively low energy displayed a certain correlation between the nature of the spectrum on the one hand and the ratio between the ionization energy and the width of the prohibited zone in the crystal on the other.

These peculiarities within the energy spectra of kinetic emission electrons can be explained on the basis of the concepts of the mechanism of ionization during the collision of heavy atoms and ions. In alkali-halide crystals the anions F⁻, Cl⁻ and Br⁻ have a structure of the electron shell similar to that of atoms of noble gases Ne, Ar and Kr while the cations Li⁺, Na⁺ and K⁺ bear similarity with the atoms of He, Ne and Ar. Hence the ionization process during the collision of ions (and also atoms) of He⁺, Ar⁺ and Kr⁺ with crystals LiF, NaCl and KBr should be similar to that which occurs during the collision of ions with the atoms of the above-mentioned gases.

As is known[4] ionization under atomic collisions within the examined energy range occurs in two stages:

1) Formation of multiple excited "quasimolecules" and

2) Relaxation by single or cascading Auger-transitions during which electrons are emitted.

It is believed that in the first stage about twice as many electrons are excited as emitted in the second stage.[5] (During multiple ionization two electrons are excited). When applied to our case it means that two electrons are emitted from the valence band into the conduction band. At the next stage there is an Auger transition during which one of these electrons is transferred into the valency band while the second electron receiving the energy thus released is emitted into vacuum. The extreme energy of the Auger electron will be equal to $\Delta E + \delta + \chi$ where ΔE-width of the band gap δ- the width of the valency band and χ—the affinity of the crystal to the electron. Hence with the growth of the crystal band gap the number of relatively fast electrons should increase (Figure 4). However it must be borne in mind that at the moment of collision of the fast ion with the atoms of the target there occurs a local violation of the bands and the energy of the electron may not necessarily fit in the above-mentioned limits. Besides electrons may also be released from the incident ion whose level is slightly outside of the zone limit. Experimental data indicate that here we have the same regularity—the deeper the level the more energy is emitted by the electrons.

REFERENCES

1. A. I. Kondrashev and N. N. Petrov, *Trudi Leningradskogo politekhnicheskogo institute im. Kalinina.* **311**, 110 (1970).
2. U. A. Arifov, R. R. Rakhimov, S. Gaipov and I. Abdulkhako *10th Int. conf. on Phen. in Ionized Gases* (Oxford, England, 1971) p. 70.
3. U. A. Arifov, R. R. Rakhimov and S. Gaipov, *Izv. AN SSSR. Ser. Phys.* **35**, 562 (1971).
4. E. S. Parilis, *Auger Effet.* Tashkent (1969).
5. L. M. Kishinevsky and E. S. Parilis, *JTP* **38**, 760 (1968).

APPLICATION OF CHARACTERISTIC SECONDARY ION MASS SPECTRA TO A DEPTH ANALYSIS OF COPPER OXIDE ON COPPER

H. W. WERNER, H. A. M. DE GREFTE and J. VAN DEN BERG

Philips Research Laboratories, Eindhoven, The Netherlands

The study of molecular and cluster ions in secondary ion mass spectrometry has led to the concept of characteristic (fingerprint) spectra for this mode of ionization.

Thin films of copper oxide with decreasing oxygen concentration as a function of depth have been investigated. The dependence of all mass peaks as a function of depth can be calculated from the measured values of just one peak by means of a linear superposition of the characteristic mass spectra of copper oxide and copper.

Finally it is shown that the concentration of copper and copper oxide as a function of depth can be calculated from any mass peak when using the appropriate fingerprint spectra.

1 INTRODUCTION

The analysis of gas mixtures by means of electron-impact mass spectrometry is a well established method.[1-3] This method is based on the phenomenon that every component of the mixture gives rise to a specific cracking pattern in the mass spectrometer, usually referred to as characteristic mass spectrum or fingerprint spectrum. The mass spectrum obtained from the gas mixture is a linear superposition of the contributions of the individual components. Knowing the fingerprint spectra of the individual components, their concentration can therefore be calculated from the measured mass spectrum of the mixture.

In the analysis of solids by means of low resolution secondary ion mass spectrometry,[4-18] the situation is similar when the sample consists of a mixture of different compounds of two elements. The cracking patterns of the different compounds overlap more or less across the whole mass range.

An analysis of the concentration as a function of depth can be carried out in that case, however, if the fingerprint spectra, characteristic of the different compounds, do exist and further, if the measured mass spectrum is produced by a linear superposition of the different fingerprint spectra. In the present paper it will be shown from measurements on a copper oxide/copper sample that the measured spectrum is a linear superposition of the fingerprint spectra of the individual components. Finally the concept of fingerprint spectra will be used to calculate the copper oxide and copper concentration as a function of depth.

2 DESCRIPTION OF THE APPARATUS

All samples were analysed in a secondary ion mass spectrometer (Cameca S.A., Type IMS 300) by means of Ar^+ ion bombardment (angle of incidence 45°, energy of the argon ions: 5.5 keV when measuring positive secondary ions, 14.5 keV when measuring negative secondary ions). The primary ion beam of $i_p = 0.5 \mu A$ was focused on a spot of 300 μm diameter on the sample. The maximum power dissipated to the target i.e. when measuring negative secondary ions, thus amounts to about 7.5 mW. A temperature rise of the target as a whole is therefore not to be expected. The residual gas pressure in front of the target during bombardment amounted to 10^{-7} torr (nitrogen equivalent) $\approx 10^{-5}$ Pa (1 Pa = 1 Pascal = 1 N/m^2).

The ion mirror[18] was set at 20 V, i.e. only ions with initial energies between 0 and 20 V were detected. No surface charging up was observed during the experiments

3 DEPTH ANALYSIS OF A COPPER OXIDE/COPPER SAMPLE

An OFHC Cu sample was oxidized by heating in a gas flame. A depth analysis was then carried out, i.e. the sample was continuously bombarded in the secondary ion mass spectrometer with a constant beam of Ar^+ ions. Mass spectra from different depths were registered as they appeared during the bombardment. From these mass spectra the intensities of the different positive ion species as a function of time were determined. In

Figure 1 best fits through these measured values are given as solid lines.

From Figure 1 we see the following: the outermost layer ($0 < t < 30$ min) is characterized by the occurrence of Cu_2O_2H cluster ions. At a bombardment time $t = 30$ min these cluster ions have decreased to negligibly low values. We shall restrict our considerations to layers further inside the sample, i.e. $t > 30$ min, corresponding approximately to a depth of $z > 0.5$ μm.

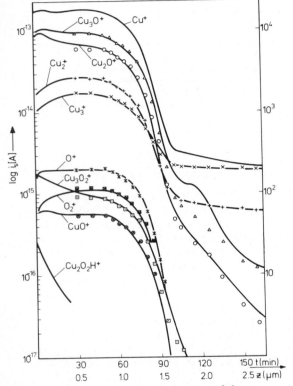

FIGURE 1   Positive secondary ion current of the copper oxide/copper sample log i_s^+ plotted as a function of bombardment time t. The depth scale z, which throughout this layer of changing composition need not always be proportional to the bombardment time is given to indicate approximately the thickness of the oxide layer. The points indicated by symbols ($\circ$, $\times$, $+$) etc were computer-calculated. The solid lines are bestfits to experimental data.

It is obvious that the relation between bombardment time and depth is not linear throughout the whole layer, since one is sputtering first an oxide, than an intermediate region consisting of various proportions of two phases and finally the metal. Satkiewicz[19] has given the following relation for converting bombardment time into depth of penetration:

$$R_c = R_I c_I + R_{II} c_{II}, \quad \text{with } c_I + c_{II} = 1,$$

where R_c is the sputtering rate at any depth, R_I and R_{II} are the sputtering rates of phases I and II, and c_I and c_{II} are the surface fractions covered with phase I and II.

At a time $t = 30$ min we have a fingerprint spectrum with high values of the isogenous cluster ions Cu^+, Cu_2^+, Cu_3^+ and also of the molecular ions CuO^+, Cu_2O^+, Cu_3O^+ and $Cu_3O_2^+$ (see also Table Ia) in good agreement with earlier experiments.[12]

This fingerprint spectrum agrees within the reproducibility of the experimental data (factor 2.5) with the fingerprint spectrum of a CuO reference sample (Table Ib). This reference sample was obtained by pressing a powder of CuO (5N-quality, Koch Light Ltd.) into a pellet. We may therefore conclude that the layer at $t = 30$ min, further to be indicated as phase I, consists of CuO. From a decrease of these ions with depth we may conclude that this phase decreases in concentration with depth. Regarding the poor reproducibility of the experimental data (factor 2.5) we are studying at the moment the influence of the residual gas atmosphere on the peak intensities and in particular on the copper hydroxile peak.

At a time $t = 165$ min, corresponding to a depth where no oxidation should have taken place, the above mentioned peaks have decreased or even disappeared (see Table IIa). This can be explained by the low degree of oxidation[12] at this depth. From a comparison of this fingerprint spectrum with one obtained from a non-oxidized OFHC-Cu sample (Table IIb) we can conclude that at this depth we have mainly the original OFHC-bulk, further indicated by phase II and a small amount (ca. 0.1%, see Figure 2) of phase I-copper oxide, the latter giving contributions to the Cu_2O^+ and Cu_3O^+ peaks as shown in Figure 1 and Table IIa.

From a comparison of the fingerprint spectra from the CuO reference sample (Table Ib) and the oxygen free Cu sample (Table IIb) one can see, that the CuO is characterized by oxygen bearing cluster ions, which are not found in the oxygen free Cu sample.

In the intermediate zones ($30 < t < 165$) we assume a mixture of copper oxide (phase I) and copper (phase II). Assuming that there are only two phases, the relation

$$c_I(t) + c_{II}(t) = 1 \tag{1}$$

holds for the whole range $30 < t < 165$. As we have seen that at $t = 30$ min only phase I exists and that at $t = 165$ min only phase II exists we have moreover $c_I(30) = 1$, and $c_{II}(165) = 1$.

In order to give evidence of the validity of the superposition theorem for this special case we must

TABLE Ia
Fingerprint spectrum from the oxidized OFHC–Cu sample. Positive secondary ion currents at $t = 30$ min (Phase I)

Ion species	Cu	CuO	Cu_2	Cu_2O	Cu_2O_2	Cu_3	Cu_3O	Cu_3O_2
m/e	63	79	126	142	158	191	207	233
Intensities $\times 10^{17}$ [A]	16500	55	2600	7000	<10	1650	10000	110
Normalized intensities	2380	7.9	37.2	1000	<1.4	238	1430	15.7

TABLE Ib
Fingerprint spectrum from a CuO reference sample. Positive secondary ion currents

Ion species	Cu	CuO	Cu_2	Cu_2O	Cu_2O_2	Cu_3	Cu_3O	Cu_3O_2
m/e	63	79	126	142	158	191	207	223
Intensities $\times 10^{17}$ [A]	2150	≈10	200	700	<1	80	440	≈5
Normalized intensities	3080	≈14	28.6	1000	<1.4	114	630	≈7

TABLE IIa
Fingerprint spectrum from the oxidized OFHC-Cu sample. Positive secondary ion currents at $t = 165$ min (Phase II)

Ion species	Cu	CuO	Cu_2	Cu_2O	Cu_2O_2	Cu_3	Cu_3O	Cu_3O_2
m/e	63	79	126	142	158	191	207	223
Intensities $\times 10^{17}$ [A]	200		60	2.5		190	12	
Normalized intensities	3320		1000	41.5		3150	200	

TABLE IIb
Fingerprint spectrum from an oxygen free Cu sample (reference). Positive secondary ion currents

Ion species	Cu	CuO	Cu_2	Cu_2O	Cu_2O_2	Cu_3	Cu_3O	Cu_3O_2
m/e	63	79	126	142	158	191	207	223
Intensities $\times 10^{17}$ [A]	365		115			240		
Normalized intensities	3170		1000			2080		

consider the following: let us first denote the time dependence in the spectra of the various ion species e.g. $Cu(t)$, $Cu_2(t)$, $Cu_2O(t)$ etc. by $i_1(t), i_2(t), i_3(t) \ldots$ etc. If these lines are produced by a linear superposition of the contributions of phase I and II we can write:

$$i_1(t) = c_I(t)i_{1,I} + c_{II}(t)i_{1,II}$$
$$i_2(t) = c_I(t)i_{2,I} + c_{II}(t)i_{2,II}$$
$$\vdots$$
$$\vdots$$
$$i_n(t) = c_I(t)i_{n,I} + c_{II}(t)i_{n,II}$$

(2a)

This can be written in matrix form as:

$$\begin{pmatrix} i_1(t) \\ i_2(t) \\ \cdot \\ \cdot \\ \cdot \\ i_n(t) \end{pmatrix} = \begin{pmatrix} i_{1,I} & i_{1,II} \\ i_{2,I} & i_{2,II} \\ & \\ & \\ i_{n,I} & i_{n,II} \end{pmatrix} \times \begin{pmatrix} c_I(t) \\ c_{II}(t) \end{pmatrix} \quad (2b)$$

or:

$$\vec{I}_n = \vec{\vec{K}}_{n2} \times \vec{C}_2$$

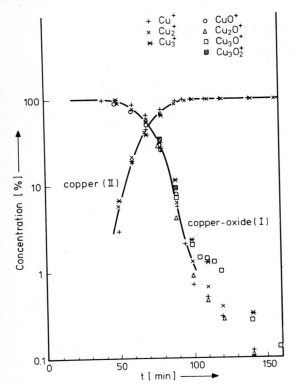

FIGURE 2   Concentration $c(t)$ of copper and copper oxide as a function of the bombardment time, calculated from different mass peaks.

The vector $\vec{I}_n$ represents the measured ion currents as a function of time, the first column of the matrix $\vec{K}_{n2}$ is the fingerprint spectrum of phase I, the second column that of phase II, and the vector $\vec{C}_2$ represents the concentrations as a function of time. One can see from Eq. (2a) that e.g. the fingerprint spectrum of phase I is found from the spectrum measured at $t = 30$ min, as we have: $c_I(30) = 1$ and $c_{II}(30) = 0$, giving $i_k(30) = i_{k,I}$.

If the superposition theorem is valid we can, for instance, calculate $c_I(t) = 1 - c_{II}(t)$ from any line in Eq. (2a) from the measured values $i_k(t)$. If we arbitrarily choose the Cu-line ($i_1(t)$) we obtain:

$$c_I(t) = \frac{i_{Cu}(t) - i_{Cu,II}}{i_{Cu,I} - i_{Cu,II}} \qquad (3)$$

The contributions $i_{Cu,I}$ and $i_{Cu,II}$ of phase I and II to the Cu peak we find from Table Ia and IIa as 1.6×10^{-13} A and 2×10^{-15} A respectively. For $t = 90$ min, e.g. we find from Figure 1: $i_{Cu}(90) = 1.2 \times 10^{-14}$ A and hence from Eq. (3): $c_I(90) = 0.063$. From these calculated values of $c_I(t)$ and from the known fingerprint spectra of copper (phase II) and copper oxide

(phase I) we can calculate *all* other ion currents $i_k(t)$ as a function of time, where $k = 2, 3, \ldots, n$. In Figure 1 these calculated ion currents $i_k(t)$ are indicated by symbols; the lines are best fits to experimental data.

From the good agreement between calculated and measured values one may consider the validity of the superposition theorem as proved, and also that two independent phases *viz.* copper oxide and copper coexist in the sample.

For the calculation of $c_I(t)$ and $c_{II}(t)$ we have (see Eq. (2a)) in principle *n*-equations from which we had arbitrarily chosen the first one (Cu). According to Eq. (2a) the concentrations $c_I(t)$ and $c_{II}(t)$ calculated from any line should be the same. In order to check this we have calculated the concentration from several lines of Eq. (2a). Note that e.g. the CuO$^+$ peak as a function of time (Figure 1) is proportional to the $c_I(t)$-dependence (Figure 2) as there is no contribution of phase II (Copper) to this mass peak: $i_{CuO} = c_I \cdot i_{CuO,I} + c_{II} \cdot O$. The results are given in Figure 2, showing that within the accuracy of the measurement (about 1%), there is good agreement between all calculated $c(t)$ values from any one line, whereas the actual values of the measured ion currents differ up to a factor of 100.

In conclusion it can be said that the method described, appears to work for two component systems consisting of reactive metals and their oxides. Moreover it can be seen that fingerprint spectra may be used to identify, out of a given number of compounds, those which are present in a given sample.

ACKNOWLEDGEMENTS

The authors wish to thank Dr. Venema for his interest and many helpful discussions. Our thanks are also due to Dr. van de Stolpe and Mr. A. W. Witmer for critically reading the manuscript and to Dr. Kwestroo for providing the CuO samples.

REFERENCES

1. H. Ewald and H. Hintenberger, *Methoden und Anwendungen der Massenspektroskopie* (Verlag Chemie, Weinheim/Bergstrasse, 1953), pp. 35–42.
2. G. P. Barnard, *Modern Mass Spectrometry* (The Institute of Physics, London, 1953), pp. 192–230.
3. H. Kienitz, *Hrsg, Massenspektrometrie* (Verlag Chemie, Weinheim/Bergstrasse, 1968), pp. 233–286.
4. R. F. K. Herzog and F. Viehböck, *Phys. Rev.* **76**, 855 (1949).
5. A. J. Smith, D. J. Marshall, L. A. Cambey and J. Michael, *Vacuum* **14**, 263 (1964).
6. A. Benninghoven, *Ann. Physik* **15**, 113 (1965).
7. H. Beske, Dissertation, Univ., Mainz, 1966.
8. H. W. Werner, *Philips Techn. Review* **27**, 344 (1966).
9. R. F. K. Herzog, W. P. Poschenrieder, F. G. Rüdenauer and F. G. Satkiewicz, *15th Annual Conf. Mass Spectr. and Allied Topics*, (ASTM E-14) (Denver, Col., 1967), p. 301.

10. A. Benninghoven, *Zs. f. Phys.* **199,** 141 (1967).
11. Ch. A. Evans, Jr., *Advances in Mass Spectrometry*, Vol. 5, edited by A. Quayle (The Inst. of Petroleum, London, 1971), p. 436.
12. H. W. Werner, in *Dev. Appl. Spectrosc.*, Vol. 7A, edited by E. L. Grove and A. J. Perkins (Plenum Press, New York, 1969), pp. 243–247.
13. H. E. Beske, *Zs. f. Natf.* **22a,** 459 (1967).
14. A. Benninghoven, *Zs. f. Natf.* **22a,** 841 (1967).
15. H. J. Liebl, *Advances in Mass Spectrometry*, Vol. 5, edited by A. Quayle (The Institute of Petroleum, London, 1971), p. 433.
16. C. A. Anderson, *Int. J. Mass Spectrom. Ion Phys.* **2,** 61 (1969).
17. C. A. Anderson, *Int. J. Mass Spectrom. Ion Phys.* **3,** 493 (1970).
18. R. Castaing and J. F. Hennequin, *Advances in Mass Spectrometry,* Vol. 5, edited by A. Quayle (The Institute of Petroleum, London, 1971), p. 419.
19. F. G. Satkiewicz, private communication, Nov. 1972 referring to Techn. Report AFAL-TR69-332, 1970 (Air Force Avionics Lab., WPFAB, Ohio).

List of Participants

Akaishi, K.
Institute of Plasma Physics
Nagoya University
Nagoya, JAPAN

Almén, O.
Chalmers University of Technology, Fack
S-40220 Göteborg 5, SWEDEN

Andersen, H. H.
Institute of Physics
University of Aarhus
DK-8000 Aarhus, DENMARK

Art, A.
Université de Bruxelles
Faculté des Sciences
Physique des Surfaces
Avenue F. Roosevelt 50
Bruxelles 1050, BELGIUM

Bach, H.
Jenaer Glaswerk
Schott & Gen.
D-65 Mainz, GERMANY

Baumgärtner, F.
Kernforschungszentrum Karlsruhe
D-75 Karlsruhe
Postfach 3640, GERMANY

Bay, H. L.
Institute of Physics
University of Aarhus
DK-8000 Aarhus, DENMARK

Bayly, A. R.
Sussex University
Department of Physics
Brighton BN1 9QH, U.K.

Behrisch, R.
Max-Planck-Institut für Plasmaphysik
D-8046 Garching, GERMANY

Bernheim, M.
Bat. 510, Faculté des Sciences
F-91 Orsay, FRANCE

Beske, H.
Kernforschungsanlage Jülich GmbH.
Zentralinstitut f. Analytische Chemie
D-517 Jülich, Postfach 365, GERMANY

Betz, G.
II. Institut für Experimentalphysik
Technische Hochschule
A-1040 Wien IV
Karlsplatz 13, AUSTRIA

Biersack, J. P.
Hahn-Meitner-Institut
f. Kernforschung Berlin GmbH.
D-1 Berlin 39
Glienicker Str. 100, Abt. C-r, GERMANY

Blaise, G.
Université de Paris
Faculté des Sciences d'Orsay
Physique des Solides, Bat. 510
F-91, Orsay, FRANCE

Blewer, R. S.
Sandia Laboratories
P.O. Box 5800
Albuquerque, New Mexico, U.S.A.

Bodansky,
EURATOM
Ispra, ITALY

Bøgh, E.
Institute of Physics
University of Aarhus
DK-8000 Aarhus, DENMARK

Boiziau, C.
Centre d'Etudes Nucléaires de Saclay
BP No. 2
F-91190 Gif-sur-Yvette, FRANCE

Brandt, W.
New York University
New York 10003
4 Washington Place, RM 912, U.S.A.

Van Breugel, P.
CAMECA
103 Boulevard Saint-Denis
F-92 Courbevoie, FRANCE

Browning, R.
University of Sussex
MAPS University of Sussex,
Brighton, U.K.

Carter, G.
Department of Electrical Engineering
University of Salford
Salford M5 4WT, Lancashire, U.K.

de Chateau Thierry, A.
Commissariat a l'Energie Atomique
Inst. National des Sciences et Techn. Nucléaires
BP No. 6, F-91190 Gif-sur-Yvette, FRANCE

Christensen, J. J.
Institute of Physics
University of Aarhus
DK-8000 Aarhus C, DENMARK

Clipstone, C. F.
Gillette Research and Development Lab.
454 Basingstoke Road
Reading, Berks, U.K.

Colligon, J. S.
Electrical Engineering Department
University of Salford
Salford M5 4WT, U.K.

Colombié, Nicole
Lab. de Physique des Solides
Université Paul Sabatier
118 route de Narbonne
F-31 Toulouse, FRANCE

Dehon, Marianne
Ecole Polytechnique
50 Av. F. D. Roosevelt
B-1050 Bruxelles, BELGIUM

Dittmann, J.
August Thyssen Hütte AG Chem. Lab.
D-41 Duisburg-Hamborn,
Postfach 67, GERMANY

Dittrich, I.
Ges. f. Strahlenforschung Univ.
München
D-8042 Neuherberg, Abt. PTA, GERMANY

Eckstein, W.
Max-Planck-Institut f. Plasmaphysik
D-8046 Garching, GERMANY

Evans, Jr. C. A.
Materials Res. Lab. of the
University of Illinois
Illinois, Urbana Ill, 61801, U.S.A.

Ewald, H.
II. Physikalisches Institut
Justus-Liebig-Universität
D-63 Gießen
Arndstr. 2, GERMANY

Fairbanks, J. W.
(Code 6146)
US Navy Ship Engineering Center
Prince Georges Center
Hyattsville, Maryland, U.S.A.

Farmery, B. W.
University of Sussex
MAPS University of Sussex
Falmer, Brighton BN1 9QH, U.K.

Feijen, H. H. W.
FOM, Techn. Mat. Lab.
Universiteitscomplex
Paddepool, Groningen, NETHERLANDS

Fert, C.
Lab. de Physique des Solides
Université Paul Sabatier
118 route de Narbonne
F-31 Toulouse, FRANCE

Fiedler, G.
II. Physikalisches Institut
Justus Liebig-Universität
D-63 Gießen, Arndtstr. 2, GERMANY

Finfgeld, C. R.
Dept. of Physics
Roanoke College
Salem, Va. 24153, U.S.A.

Fluit, J. M.
Stichting FOM
Fysisch Lab. der R.U.
Bijlhowerstraat 6
Utrecht, NETHERLANDS

Fontell, A.
Inst. of Physics
University of Helsinki
Helsinki, Sitavuorenpenger 20E, FINLAND

Frank, P. J.
General Electric Company
Pelikanstr. 37
CH-8001 Zürich, SWITZERLAND

Frank, R.
Commissariat á L'Energie Atomique
CEN Grenoble
SIG, Avenue des Martyrs
F-38 Grenoble, FRANCE

Gauthier, A.
CEA, CEN Grenoble
SIG, Avenue des Martyrs
F-38 Grenoble, FRANCE

Ginot, P.
EURATOM
CEA sur la Fusion
Cenfar BP No. 6
F-92 Fontenay-aux-Roses, FRANCE

Golovchenko, J. A.
Aarhus University
DK-8000 Aarhus C, DENMARK

Grundner, M.
Max-Planck-Institut f. Plasmaphysik
D-8046 Garching, GERMANY

Guckenberger, R.
Max-Planck-Institut f. Plasmaphysik
D-8046 Garching, GERMANY

Güttner, K.
II. Physikalisches Institut
Universität Gießen
D-63 Gießen, Arndtstr. 2, GERMANY

Haas, G.
Max-Planck-Institut f. Plasmaphysik
D-8046 Garching, GERMANY

Harrison, Jr. D. E.
US Naval Postgraduate School
Monterey, Ca. 93921, U.S.A.

Haymann
Université de Rouen
Blanc Hameau No. 3
F-76 Montigny, FRANCE

Heil, H.
Max-Planck-Institut für Plasmaphysik
D-8046 Garching, GERMANY

Heiland, W.
Max-Planck-Institut f. Plasmaphysik
D-8046 Garching, GERMANY

Herklotz
Heraeus GmbH.
D-645 Hanau, GERMANY

Hermanne, Nelly
Physique des Surfaces
Universite Libre de Bruxelles
Physique des Surfaces,
Fac. Sc. ULB
B-1050 Bruxelles, BELGIUM

Hernandez, R.
ONERA
29, Avenue de la Division Leclerc
F-92 Chatillon/Bagneux, FRANCE

Hertel, B.
Max-Planck-Institut f. Metallforschung
D-7 Stuttgart, GERMANY

Herzog, R. F. K.
GCA Technology Division
Bedford, Mass., U.S.A.

Hiesinger, L.
Dr. Johannes Heidenheim
D-8225 Traunreut, Postfach 1254, GERMANY

Heuschkel, J.
Kernforschungsanlage Jülich GmbH.
Inst. für Radiochemie
D-517 Jülich, Postfach 365, GERMANY

Hofer, W.
Max-Planck-Institut f. Plasmaphysik
D-8046 Garching, GERMANY

Holmen, G.
Chalmers University of Technology, Fack
S-40220 Göteborg 5, SWEDEN

Honig, R. E.
RCA Laboratories
David Sarnoff Research Center
Princeton, New Jersey 08540, U.S.A.

Hou, M.
Université Libre de Bruxelles
Faculté des Sciences
Physique des Surfaces
Avenue Roosevelt 50
B-1050 Bruxelles, BELGIUM

Hucks, P.
Kernforschungsanlage Jülich GmbH
Inst. für Radiochemie
D-517 Jülich, Postfach 365, GERMANY

Huber, W. K.
Balzers AG
FL-9496 Balzers, LIECHTENSTEIN

Jackson, D. P.
Atomic Research of Canada Ltd.
Chalk River, Ontario, CANADA

Johansen, A.
Physical Lab. II
HC Ørsted Institute
Universitetsparken 5
DK-2100 Copenhagen, DENMARK

Joyes, P.
Université Paris
Centre d'Orsay, Lab. de Phys.
des Solides, Bat. 510
F-91 Orsay, FRANCE

Jurela, Z.
Boris Kidric Institute
Beograd, P.O. Box 522, YUGOSLAVIA

Kaminsky, M. S.
Argonne National Lab.
9700 S. Cass Avenue
Argonne, Illin. 60439, U.S.A.

Kausche, H.
Siemens AG
D-8 München 80, Balanstr. 73, GERMANY

Kay, E.
IBM Research
San Jose, Cal.
Monterey and Cottle Rds.
Buildg. 028, U.S.A.

Kelly, J. C.
University of New South Wales
Kensington 2033, AUSTRALIA

Kelly, R.
Inst. for Materials Research
McMaster University
Hamilton, Ontario, CANADA

Kerkdyk, C. B. W.
FOM-Institut voor Atoom-en
Molecuulfysica
Kruislaan 407,
Amsterdam, NETHERLANDS

Kirschner, J.
Max-Planck-Institut f. Plasmaphysik
D-8046 Garching, GERMANY

Knauer, W.
Hughes Res. Lab.
3011 Malibu Canyon Rd.
Malibu, Ca. 90265, U.S.A.

Krimmel, E. F.
Siemens AG, Forschungslab. FL HLS
D-8 München, Balanstr. 73, GERMANY

Krüger, W.
Universität Gießen
D-63 Gießen, GERMANY

Kupschus
KFA Jülich
D-517 Jülich, Postfach 365, GERMANY

Labois, E.
Centre d'Etudes Nucléaires de Saclay
BP No. 2
91190 Gif-sur-Yvette, FRANCE

Laegreid, N.
Batelle-Northwest Labs.
231-Z Bld., 200-W Area
Richland, WA 99352, U.S.A.

Lanusse, P.
ONERA
29 Ave. de la Division Leclerc
92320 Chatillon, FRANCE

Laubert, R.
New York University, Phys. Dept.
New York, N.Y. 10003
4 Washington Place, RM 912, U.S.A.

Laurent, R.
Faculté des Sciences
Lab. de Physique des Solides
Bat. 510
F-91 Orsay, FRANCE

Lehmann, C.
Kernforschungsanlage Jülich
Inst. für Festkörperphysik
D-517 Jülich, Postfach 365, GERMANY

Leleyter, Mireille
Lab. de Physique des Solides
Bat. 510, Fac. des Sciences
F-91 Orsay, FRANCE

Lenoir, J. P.
CAMECA
103 Boulevard St. Denis
F-92 Courbevoie, FRANCE

Lettau, H.
Hahn-Meitner-Institut,
Sekt. Kernphysik
D-1 Berlin 39,
Glienickerstr. 100, GERMANY

Liebl, H.
Max-Planck-Institut f. Plasmaphysik
D-8046 Garching, GERMANY

Mader, L.
Siemens AG Abt. FL TEC 122
D-8 München 70, Hofmannstr. 51, GERMANY

Martel, J. G.
Centre de l'Energie,
INRS, Universite du Quebec
CP 1020, Varennes, Quebec, CANADA

Maul, J.
Ges. f. Strahlen- und Umweltforschung PTA
D-8042 Neuherberg, GERMANY

MacDonald, R. J.
The Australian National University,
Dept. of Physics, SGS
P.O. Box 4, Canberra ACT 2600, AUSTRALIA

McClanahan, E. D.
Battelle-Northwest Labs.,
231-Z Bldg., 200-W Area
Richland, WA 99352, U.S.A.

McCracken, G. M.
UKAEA Culham Laboratory
Abingdon, Berkshire, U.K.

Meier, St.
Varian MAT GmbH, Bremen
D-2800 Bremen, Postfach 4062, GERMANY

Meischner, P.
Max-Planck-Institut f. Plasmaphysik
D-8046 Garching, GERMANY

Menzel, D.
TU München
Inst. für Phys. Chemie
D-8046 Garching, GERMANY

Merkle, K. L.
Materials Science Div.
Argonne National Lab.
Argonne, Illinois 60439, U.S.A.

Mertens, H.
Atomica
D-8 München, GERMANY

Mertens, P.
Hahn-Meitner-Institut f. Kernforschung
D-1 Berlin 39, Glienickerstr. 100, GERMANY

Meyer, O.
Inst. für Angewandte Kernphysik
Kernforschungszentrum
D-75 Karlsruhe, GERMANY

Mischler, J.
Lab. de Physique des Solides
118, Route de Narbonne
F-31 Toulouse, FRANCE

Molchanov, V. A.
Institute of Nuclear Physics
Moscow State University
Moscow 117234, U.S.S.R.

Müller, N.
Max-Planck-Institut für Plasmaphysik
D-8046 Garching, GERMANY

Naguib, H. M.
Dept. of Electrical Engineering
University of Salford
Salford M5 4WT, Lancs., U.K.

Navinsek, B.
J. Stefan Institute
Jamova 39
Ljubljana, YUGOSLAVIA

Nelson, R. S.
AERE, Harwell
B. 393, Harwell, Didcot, Berks., U.K.

Niehus, H.
Phys. Inst. der TU Clausthal
D-3392 Clausthal-Zellerfeld 11
Leibnitzstr. 4, GERMANY

Oechsner, H.
Phys. Inst. der TU Clausthal
D-3392 Clausthal-Zellerfeld
Leibnitzstr. 4, GERMANY

Oetjen, G.-H.
Max-Planck-Institut f. Plasmaphysik
D-8046 Garching, GERMANY

Perkins, Harolyn, K.
Princeton University, Chem. Engineering
Engineering Quadrangle
Princeton, New Jersey 08540, U.S.A.

Pichlmayer, F.
Österr. Studien ges. für Atomenergie
A-1082 Wien, Lenaugasse 10, AUSTRIA

Poschenrieder, W.
Max-Planck-Institut für Plasmaphysik
D-8046 Garching, GERMANY

Prevot, F.
EURATOM-CEA sur la Fusion
Service de Confinement des
Plasmas, BP No 6,
F-92 Fontenay-aux-Roses, FRANCE

Rieger, Eva
FOM Instituut voor Atoom and Molecuulfysica
Kruislaan 407
Amsterdam, NETHERLANDS

Robinson, C. F.
Applied Res. Lab.
Hasler Research Center
95 La Patera Lane
Goleta, California 93017, U.S.A.

Robinson, M. T.
Oak Ridge National Lab.
P.O. Box X
Oak Ridge, Tenn. 37830, U.S.A.

Rödelsberger, K.
Universität Gießen
D-63 Gießen, GERMANY

Roosendall, H. E.
FOM-Institut voor Atoom & Molecuulfysica
Kruislaan 407
Amsterdam, NETHERLANDS

Roth, J.
Max-Planck-Institut für Plasmaphysik
D-8046 Garching, GERMANY

Roth, S.
Hahn-Meitner-Institut
D-1 Berlin 39,
Glienickerstr. 100, GERMANY

Rouberol, J. M.
CAMECA
103 Boulevard St. Denis
F-92 Courbevoie, FRANCE

Rüdenauer, F.
SGAE
A-Wien 10, Lenaugasse 10, AUSTRIA

Runge, H.
Siemens AG, Forschungslab. FL HLS
D-8 München 80, Balanstr. 73, GERMANY

Schäffler, H.
Max-Planck-Institut für Plasmaphysik
D-8046 Garching, GERMANY

Scharmann, A.
I. Phys. Inst. Universität Gießen
D-63 Gießen, GERMANY

Schauer, A.
Siemens AG
D-8 München 80, Balanstr. 73, GERMANY

Scherzer, B.M.U.
Max-Planck-Institut für Plasmaphysik
D-8046 Garching, GERMANY

Schillalies, H.
  c/o Leybold-Heraeus GmbH. & Co. KG
  D-5 Köln 51, Bonnerstr. 505, GERMANY

v.d. Schootbrugge, G. A.
  Stichting FOM
  Fysisch Lab. der RU
  Bijlhouwerstraat 6
  Utrecht, NETHERLANDS

Schram, A.
  Centre d'Etudes Nucleaires de Saclay,
  BP No. 2
  F-91 Gif-sur-Yvette, FRANCE

Schreiber, H. U.
  Inst. f. Halbleitertechnik,
  TH Aachen
  D-51 Aachen
  Lochnerstr. 4–20, GERMANY

Schulz, F.
  Ges. f. Strahlen- und Umweltforschung mbH
  D-8042 Neuherberg, GERMANY

Schurink, F.
  KEMA
  Utrechtseweg 310
  Arrtem, NETHERLANDS

Sennewald, G.
  Riber. S.A.
  F-92 Rue il Malmaison, BP 65, FRANCE

Sigmund, P.
  Phys. Lab. II
  HC Ørsted Institute
  Universitetsparken 5
  DK-2100 Copenhagen, DENMARK

Sizmann, R.
  Sektion Physik der Universität München
  D-8 München 13
  Amalienstr. 54/III, GERMANY

Sorensen, H.
  Danish AEC
  Risφ
  DK-4000 Roskilde, DENMARK

Staib, P.
  Max-Planck-Institut für Plasmaphysik
  D-8046 Garching, GERMANY

Staudenmaier, G.
  Sektion Physik, Lehrst. Sizmann
  D-8 München 13,
  Amalienstr. 54/III, GERMANY

Summers, A. J.
  Atomic Weapons Res. Est.
  Buildg. A81,
  Aldermaston, Berks. U.K.

Thompson, M. W.
  University of Sussex
  Brighton, U.K.

Tolk, N. H.
  Bell Telephone Lab.
  Murray Hill, New Jersey 07974, U.S.A.

Treitz, N.
  I. Physikalisches Institut, D-5 Köln, GERMANY

Turkenburg, W. C.
  FOM Institut voor Atoom en Molecuulfysica
  Kruislaan 407
  Amsterdam, NETHERLANDS

van Veen, A.
  Stichting FOM
  Fysisch Laboratorium der RU
  Bylhouwerstraat 6
  Utrecht, NETHERLANDS

Venker, H.
  ISPRA (Varese), ITALY

Verbeek, H.
  Max-Planck-Institut für Plasmaphysik
  D-8046 Garching, GERMANY

Vernickel, H.
  Max-Planck-Institut für Plasmaphysik
  D-8046 Garching, GERMANY

Verhey, R. K.
  FOM, Techn. Fys. Lab.
  Univ. Groningen
  Groningen, NETHERLANDS

Vetter, A.
  II. Phys. Inst. der Univ. Gießen
  D-63 Gießen, Arndtstr. 2, GERMANY

Viaris de Lesegno, P.
  CSP-Saint-Denis
  Place du 8 Mai 1945
  F-93 Saint-Denis, FRANCE

Viehböck, F.
  II. Inst. für Experimentalphysik TH Wien
  A-1040 Wien, Karlsplatz 13, AUSTRIA

Viel, L.
  Lab. de Phys. des Solides
  Université Paul Sabatier
  118 route de Narbonne
  F-31 Toulouse, FRANCE

Vogel, J.
  Balzers AG
  FL-9496 Balzers, LIECHTENSTEIN

Vogelbruch, K.
  KFA Jülich
  Inst. für Radiochemie
  D-517 Jülich, Postfach 365, GERMANY

Wagner, H.
KFA Jülich
Inst. für Technische Physik
D-517 Jülich, Postfach 365, GERMANY

van der Weg, W. F.
Philips Res. Labs. Dept. Amsterdam
Oosterringdyk 18
Amsterdam, NETHERLANDS

Wehner, G. K.
University of Minnesota
Minneapolis, U.S.A.

Weiser, J.
Siemens AG, DZL Lab. 112
D-8 München, Hoffmannstr. 51, GERMANY

Weißmann, R.
Max-Planck-Institut für Plasmaphysik
D-8046 Garching, GERMANY

Werner, H. W.
N. V. Philips Glueilampenfabricken
Philips Res. Labs.
Eindhoven, NETHERLANDS

Wilson, I. H.
University of Surrey
Dept. Electrical Engineering
Guildford, Surrey, U.K.

Winters, H. F.
IBM Research Laboratory
San Jose, Calif., U.S.A.

Wittmaack, K.
Ges. f. Strahlen- und Umweltforschung, PTA
D-8042 Neuherberg, GERMANY

Author Index